ERROR AND INFERENCE

Although both philosophers and scientists are interested in ways to obtain reliable knowledge in the face of error, a gap between their perspectives has been an obstacle to progress. Through a series of exchanges between the editors and leaders in philosophy of science, statistics, and economics, this volume offers a cumulative introduction that connects problems of traditional philosophy of science to problems of inference in statistical and empirical modeling practice. Philosophers of science and scientific practitioners are challenged to reevaluate the assumptions of their own theories – philosophical or methodological. Practitioners may better appreciate the foundational issues around which their questions revolve, thereby becoming better "applied philosophers." Conversely, new avenues emerge for finally solving recalcitrant philosophical problems of induction, explanation, and theory testing.

Deborah G. Mayo is a professor in the Department of Philosophy at Virginia Tech and holds a visiting appointment at the Center for the Philosophy of Natural and Social Science of the London School of Economics. She is the author of *Error and the Growth of Experimental Knowledge*, which won the 1998 Lakatos Prize, awarded to the most outstanding contribution to the philosophy of science during the previous six years. Professor Mayo coedited the volume *Acceptable Evidence: Science and Values in Risk Management* (1991, with R. Hollander) and has published numerous articles on the philosophy and history of science and on the foundations of statistics and experimental inference and interdisciplinary works on evidence relevant for regulation and policy.

Aris Spanos is the Wilson Schmidt Professor of Economics at Virginia Tech. He has also taught at Birkbeck College London, the University of Cambridge, the University of California, and the University of Cyprus. Professor Spanos is the author of *Probability Theory and Statistical Inference* (1999) and *Statistical Foundations of Econometric Modeling* (1986), both published by Cambridge University Press. Professor Spanos's research has appeared in journals such as the *Journal of Econometrics, Econometric Theory, Econometric Reviews,* and *Philosophy of Science*. His research interests include the philosophy and methodology of statistical inference and modeling, foundational problems in statistics, statistical adequacy, misspecification testing and respecification, resampling and simulation techniques, and modeling speculative prices.

Error and Inference

Recent Exchanges on Experimental Reasoning, Reliability, and the Objectivity and Rationality of Science

Edited by

DEBORAH G. MAYO

Virginia Tech

ARIS SPANOS

Virginia Tech

CAMBRIDGE UNIVERSITY PRESS
Cambridge, New York, Melbourne, Madrid, Cape Town,
Singapore, São Paulo, Delhi, Tokyo, Mexico City

Cambridge University Press
The Edinburgh Building, Cambridge CB2 8RU, UK

Published in the United States of America by Cambridge University Press, New York

www.cambridge.org
Information on this title: www.cambridge.org/9780521880084

© Cambridge University Press 2010

First published 2010

A catalogue record for this publication is available from the British Library

Library of Congress Cataloguing in Publication Data

Error and inference : recent exchanges on experimental reasoning,
reliability, and the objectivity and rationality of science / edited by
Deborah G. Mayo, Aris Spanos.
p. cm.
Includes bibliographical references and index.
ISBN 978-0-521-88008-4 (hardback)
1. Inference.   2. Science - Philosophy.   3. Science - Methodology.
I. Mayo, Deborah G.   II. Spanos, Aris, 1952-   III. Title.
Q175.32.I54E77   2009
501-dc22   2009012825

ISBN 978-0-521-88008-4 Hardback

To George W. Chatfield
For his magnificent support of reseacrch on E.R.R.O.R. in science

Contributors

Peter Achinstein (Philosophy, Yeshiva University, Johns Hopkins University) is the author of the books *Concepts of Science* (1968); *Law and Explanation* (1983); *Particles and Waves* (1991), which received the Lakatos Award in Philosophy of Science; and *The Book of Evidence* (2001). A collection of his essays, *Evidence, Explanation, and Realism*, will be published in 2010. He is Jay and Jeanie Schottenstein Professor of Philosophy, Yeshiva University, and has held Guggenheim, NEH, and NSF fellowships.

Alan Chalmers (Philosophy, Flinders University of South Australia) is the author of *What Is This Thing Called Science?* (first published in 1976 and now in its third edition), *Science and Its Fabrication* (1990), and *The Scientist's Atom and the Philosopher's Stone* (2009).

Sir David Cox (Statistics, Nuffield, Oxford). His books include *Planning of Experiments* (1958); *Statistical Analysis of Series of Events*, with P. A. W. Lewis (1966); *Theoretical Statistics*, with D. V. Hinkley (1974); *Applied Statistics*, with E. J. Snell (1981); *Inference and Asymptotics*, with O. E. Barndorff-Nielsen (1994); *Theory of the Design of Experiments*, with N. Reid (2000); and *Principles of Statistical Inference* (2006). He was editor of the journal *Biometrika* from 1966 to 1991.

Clark Glymour (Philosophy, Carnegie Mellon University). His books include *Theory and Evidence* (1980); *Foundations of Space-Time Theories*, with J. Earman (1986); *Discovering Causal Structure*, with R. Scheines, P. Spirtes, and K. Kelly (1987); *Causation, Prediction, and Search*, with P. Spirtes and R. Scheines (1993, 2001); and *Bayes Nets and Graphical Causal Models in Psychology* (2001).

Larry Laudan (Philosophy, National Autonomous University of Mexico; Law and Philosophy, University of Texas). His books include *Progress and Its*

Problems (1977), *Science and Hypothesis* (1981), *Science and Values* (1984), *Science and Relativism* (1991), *Beyond Positivism and Relativism* (1996), and *Truth, Error and Criminal Law* (2006).

Deborah G. Mayo (Philosophy, Virginia Tech) is the author of *Error and the Growth of Experimental Knowledge*, which received the 1998 Lakatos Prize award. She coedited, with R. Hollander, the book *Acceptable Evidence: Science and Values in Risk Management* (1991). Mayo is currently involved in supporting and organizing work on E.R.R.O.R. in Science, Statistics, and Modeling.

Alan Musgrave (Philosophy, University of Otago) coedited, with Imre Lakatos, the celebrated collection *Criticism and the Growth of Knowledge* (1970) and authored the books *Common Sense, Science, and Scepticism* (1993) and *Essays on Realism and Rationalism* (1999).

Aris Spanos (Economics, Virginia Tech) is the author of *Statistical Foundations of Econometric Modelling* (1986) and *Probability Theory and Statistical Inference: Econometric Modeling with Observational Data* (1999).

John Worrall (Philosophy, Logic and Scientific Method, London School of Economics) served as editor of the *British Journal for Philosophy of Science* from 1974 to 1983. He is author of numerous articles and of the forthcoming *Reason in "Revolution": A Study of Theory Change in Science.*

Preface

A central question of interest to both scientists and philosophers of science is, *How can we obtain reliable knowledge about the world in the face of error, uncertainty, and limited data?* The philosopher tackling this question considers a host of general problems: *What makes an inquiry scientific? When are we warranted in generalizing from data? Are there uniform patterns of reasoning for inductive inference or explanation? What is the role of probability in uncertain inference?* Scientific practitioners, by and large, just get on with the job, with a handful of favored methods and well-honed rules of proceeding. They may seek general principles, but largely they take for granted that their methods "work" and have little patience for unresolved questions of "whether the sun will rise tomorrow" or "whether the possibility of an evil demon giving us sensations of the real world should make skeptics of us all." Still, in their own problems of method, and clearly in the cluster of courses under various headings related to "scientific research methods," practitioners are confronted with basic questions of scientific inquiry that are analogous to those of the philosopher.

Nevertheless, there are several reasons for a gap between work in philosophy of science and foundational problems in methodological practice. First, philosophers of science tend to look retrospectively at full-blown theories from the historical record, whereas work on research methods asks how to set sail on inquiries and pose local questions. Philosophers might ask questions such as "What made it rational to replace the Newtonian theory of gravity with Einstein's General Theory of Relativity (and when)?" But the practitioner asks more localized questions: "Is HIV dormant or active during initial infection? Is the mechanism of Mad Cow Disease and CJD similar to other neurological conditions such as Alzheimer's disease?" Second, philosophers focus on characterizing abstract conceptions of

"*H* is confirmed by evidence *e*" for a given statement of evidence *e* but rarely engage methods actually used to obtain evidence *e* in the first place. Where philosophers tend to draw skeptical lessons from the fact that error is always possible, practitioners focus on specific threats that can sully the validity of their evidence and inferences. Third, philosophers of science themselves (at least in the past decade or so) often confess that they have given up on solving traditional philosophical riddles of induction and evidence. But unsolvable riddles, however interesting in their own right, scarcely seem relevant to the practical task of improving method.

Although we grant that, on the one hand, (1) current philosophy of science does not offer solutions to the problems of evidence and inference in scientific practice, at the same time we hold that (2) the resources of philosophy of science offer valuable tools for understanding and advancing solutions to the problems of evidence and inference in practice. These two assertions, however much they may seem to be in tension, we claim, are both true. The first, readily acknowledged in both the philosophy and the science camps, is generally taken as a basis for skepticism about assertion 2. Nevertheless, our experiences in debates about evidence and inference, and about method and statistics in practice, convince us of the truth of assertion 2 – even if the solutions do not have the form originally envisaged. What comes up short in methodological discussions in practice is a genuine comprehension of where the difficulties in solving the traditional philosophical problems lie – the reasons behind assertion 1. In dismissing traditional philosophical problems as esoteric, old-fashioned, or irrelevant, contemporary discussants of methodology may be unable to discern how the very same philosophical issue is, at least implicitly, raising its head in a contemporary methodological debate they care about. Making progress demands a meeting ground wherein the insights of the philosophical and scientific camps can be used to shed light on each other.

To get beyond the current impasse, we proposed to take a significant group of current representatives of philosophical schools and bring the philosophies to life, as it were. Initial dialogues grew into the *Experimental Reasoning, Reliability, Objectivity & Rationality: Induction, Statistics, Modelling* [ERROR 06] conference at Virginia Tech, June 1–5, 2006. Rather than regarding these diverse philosophies as closed "museum pieces," we needed to open them up to peer into them, see where they stand in 2006–2009, and see what perspectives they bring to bear on contemporary problems of modeling and inference.

Doing so, we recognized, required the continuation of the dialogue that we had initiated both before and during the conference. This volume reflects the results of these exchanges. Its contributions are directed to anyone interested in methodological issues that arise in philosophy of science, statistical sciences, or the social and natural sciences.

Origins of the Contributions to This Volume

The contributions in this volume, in one way or another, touch on issues of error and inference. Mayo's (1996) *Error and the Growth of Experimental Knowledge* (EGEK) provides a launching point or foil for addressing different problems of inference and evidence in the face of uncertainty and errors. The chapters reflect exchanges between the editors and the contributors that began at least two years before the ERROR 06 conference and continued almost that long afterward. Our goal is to highlight developments in the decade following EGEK and to point to open problems for future work. Our strategy required getting contributors, ourselves included, beyond our usual comfort zones. Whereas the Introduction extracts from EGEK by way of background, Mayo's Chapter 1 represents an attempt to move beyond its focus on local experimental tests to take up challenges regarding higher-level theories. Chapter 1 was a fixed target that remained unchanged throughout; however, the editors' reflections in the exchanges shifted and grew, as did the contributions, before arriving at their final form. The contributors were subjected, at times painfully, to our persistent attempts to elucidate positions, decipher disagreements, get beyond misunderstandings, and encourage moves, however small, from initial standpoints. Although open problems clearly remain, we think it is time to stop and take stock of where this dialectic has taken us, indicating where we have inched toward progress, and how we might get around remaining obstacles.

By presenting recent and ongoing exchanges between several representatives of contemporary movements in philosophy of science, statistics, and modeling, we hope to offer the reader glimpses into

- the struggles, arguments, issues, changes, and historical developments behind a cluster of deep and long-standing issues about scientific knowledge and inference, and
- the directions in which future philosophy and methodology of science might move.

Some of the untidiness that remains is instructive of where we have been and where one might go next. The exchanges do not report each stage of the dialogues with contributors, but rather they identify a set of general questions and responses that emerged from the numerous back-and-forth conversations. By highlighting the multitude of different ways these questions are answered, interpreted, and interrelated, we intend for this volume to offer a cumulative instruction for anyone interested in the philosophical and methodological issues of scientific inquiry.

Acknowledgments

We are indebted to many people, not the least to the contributors to this volume: Peter Achinstein, Alan Chalmers, David Cox, Clark Glymour, Larry Laudan, Alan Musgrave, and John Worrall. Without their willingness to engage openly in the extended dialectic out of which this volume grew, we would scarcely have gotten beyond restating previous positions and disagreements. In addition to the contributors to this volume, we wish to express our gratitude to the overall intellectual exchange provided by the contributed speakers at ERROR 06, the work of Kent Staley and Jean Miller in editing the corresponding special issue of *Synthese* (August 2008, Vol. 163, No. 3), and the presenters at a rather unique poster session on errors across a vast landscape of fields.

We wish to acknowledge the many contributions of H. Kyburg Jr. (1927–2007) to philosophy of science and foundations of probability; he was a specially invited speaker to ERROR 06, and we regret that illness prevented him from attending.

We would like to thank Scott Parris of Cambridge University Press for endorsing our project. We are deeply grateful to Eleanor Umali for providing extremely valuable help during the copyediting and proofreading stages of the volume and for maintaining calm in the midst of chaos.

We gratefully acknowledge the support of departments and colleges at Virginia Tech: Dean Niles of the College of Liberal Arts and Humanities, the College of Science, and the departments of economics and of philosophy.

Some of the research was conducted under a National Science Foundation Scholars Award (054 9886); Deborah Mayo gratefully acknowledges that support.

For valuable feedback and error-corrections on this manuscript, we thank several philosophy graduate students at Virginia Tech, notably, Emrah Aktunc, Tanya Hall, and Jean Miller. We are extremely grateful for the mammoth editorial assistance of Mary Cato and Jean Miller.

Finally, we thank George W. Chatfield for his enormous support and his generous award for the ongoing study of experimental reasoning, reliability, and the objectivity and rationality of science, statistics, and modeling.

References

Staley, K. (2008), "Introduction" *Synthese (Error and Methodology in Practice: Selected Papers from ERROR 2006)*, Vol. 163(3): 299–304.

Staley K., Miller J., and Mayo, D. (eds.) (2008), *Synthese (Error and Methodology in Practice: Selected Papers from ERROR 2006)*, Vol. 163(3).

Introduction and Background

Deborah G. Mayo and Aris Spanos

I Central Goals, Themes, and Questions

1 Philosophy of Science: Problems and Prospects

Methodological discussions in science have become increasingly common since the 1990s, particularly in fields such as economics, ecology, psychology, epidemiology, and several interdisciplinary domains – indeed in areas most faced with limited data, error, and noise. Contributors to collections on research methods, at least at some point, try to ponder, grapple with, or reflect on general issues of knowledge, inductive inference, or method. To varying degrees, such work may allude to philosophies of theory testing and theory change and philosophies of confirmation and testing (e.g., Popper, Carnap, Kuhn, Lakatos, Mill, Peirce, Fisher, Neyman-Pearson, and Bayesian statistics). However, the different philosophical "schools" tend to be regarded as static systems whose connections to the day-to-day questions about how to obtain reliable knowledge are largely metaphorical. Scientists might "sign up for" some thesis of Popper or Mill or Lakatos or others, but none of these classic philosophical approaches – at least as they are typically presented – provides an appropriate framework to address the numerous questions about the legitimacy of an approach or method.

Methodological discussions in science have also become increasingly sophisticated; and the more sophisticated they have become, the more they have encountered the problems of and challenges to traditional philosophical positions. The unintended consequence is that the influence of philosophy of science on methodological practice has been largely negative. If the philosophy of science – and the History and Philosophy of Science (HPS) – have failed to provide solutions to basic problems of evidence and inference, many practitioners reason, then how can they help scientists to look to philosophy of science to gain perspective? In this spirit, a growing tendency is to question whether anything can be said about what makes an enterprise scientific, or what distinguishes science from politics, art or other endeavors. Some works on methodology by practitioners look instead to sociology of science, perhaps to a variety of post-modernisms, relativisms, rhetoric and the like.

However, for the most part, scientists wish to resist relativistic, fuzzy, or postmodern turns; should they find themselves needing to reflect in a general way on how to distinguish science from pseudoscience, genuine tests from ad hoc methods, or objective from subjective standards in inquiry, they are likely to look to some of the classical philosophical representatives (and never mind if they are members of the list of philosophers at odds with the latest vogue in methodology). Notably, the Popperian requirement that our theories and hypotheses be testable and falsifiable is widely regarded to contain important insights about responsible science and objectivity; indeed, discussions of genuine versus ad hoc methods seem invariably to come back to Popper's requirement, even if his full philosophy is rejected. However, limiting scientific inference to deductive falsification without any positive account for warranting the reliability of data and hypotheses is too distant from day-to-day progress in science. Moreover, if we are to accept the prevalent skepticism about the existence of reliable methods for pinpointing the source of anomalies, then it is hard to see how to warrant falsifications in the first place.

The goal of this volume is to connect the methodological questions scientists raise to philosophical discussions on *Experimental Reasoning, Reliability, Objectivity, and Rationality* (E.R.R.O.R) of science. The aim of the "exchanges" that follow is to show that the real key to progress requires a careful unpacking of the central reasons that philosophy of science has failed to solve problems about evidence and inference. We have not gone far enough, we think, in trying to understand these obstacles to progress.

Achinstein (2001) reasons that, "scientists do not and should not take . . . philosophical accounts of evidence seriously" (p. 9) because they are based on a priori computations; whereas scientists evaluate evidence empirically. We ask: Why should philosophical accounts be a priori rather than empirical? Chalmers, in his popular book *What is This Thing Called Science?* denies that philosophers can say anything general about the character of scientific inquiry, save perhaps "trivial platitudes" such as "take evidence seriously" (Chalmers, 1999, p. 171). We ask: Why not attempt to answer the question of what it means to "take evidence seriously"? Clearly, one is not taking evidence seriously in appraising hypothesis H if it is predetermined that a way would be found to either obtain or interpret data as supporting H. If a procedure had little or no ability to find flaws in H, then finding none scarcely counts in *H*'s favor. One need not go back to the discredited caricature of the objective scientist as "disinterested" to extract an uncontroversial minimal requirement along the following lines:

Minimal Scientific Principle for Evidence. Data x_0 provide poor evidence for *H* if they result from a method or procedure that has little or no ability of finding flaws in *H*, even if *H* is false.

As weak as this is, it is stronger than a mere falsificationist requirement: it may be logically possible to falsify a hypothesis, whereas the procedure may make it virtually impossible for such falsifying evidence to be obtained.

It seems fairly clear that this principle, or something very much like it, undergirds our intuition to disparage ad hoc rescues of hypotheses from falsification and to require hypotheses to be accepted only after subjecting them to criticism. *Why then has it seemed so difficult to erect an account of evidence that embodies this precept without running aground on philosophical conundrums?* By answering this question, we hope to set the stage for new avenues for progress in philosophy and methodology. Let us review some contemporary movements to understand better where we are today.

2 Current Trends and Impasses

Since breaking from the grip of the logical empiricist orthodoxy in the 1980s, the philosophy of science has been marked by attempts to engage dynamically with scientific practice:

1. Rather than a "white glove" analysis of the logical relations between statements of evidence *e* and hypothesis *H*, philosophers of science would explore the complex linkages among data, experiment, and theoretical hypotheses.
2. Rather than hand down pronouncements on ideally rational methodology, philosophers would examine methodologies of science empirically and naturalistically.

Two broad trends may be labeled the "new experimentalism" and the "new modeling." Moving away from an emphasis on high-level theory, the new experimentalists tell us to look to the manifold local tasks of distinguishing real effects from artifacts, checking instruments, and subtracting the effects of background factors (e.g., Chang, Galison, Hacking). Decrying the straightjacket of universal accounts, the new modelers champion the disunified and pluralistic strategies by which models mediate among data, hypotheses, and the world (Cartwright, Morgan, and Morrison). The historical record itself is an important source for attaining relevance to practice in the HPS movement.

Amid these trends is the broad move to tackle the philosophy of methodology empirically by looking to psychology, sociology, biology, cognitive science, or to the scientific record itself. As interesting, invigorating, and right-headed as the new moves have been, the problems of evidence and inference remain unresolved. *By and large, current philosophical work and the conceptions of science it embodies are built on the presupposition that we cannot truly solve the classic conundrums about induction and inference.* To give up on these problems, however, does not make them go away; moreover, the success of naturalistic projects demands addressing them. Appealing to "best-tested" theories of biology or cognitive science calls for critical evaluation of the methodology of appraisal on which these theories rest.

The position of the editors of this volume takes elements from each of these approaches (new experimentalism, empirical modeling, and naturalism). We think the classic philosophical problems about evidence and inference are highly relevant to methodological practice and, furthermore, *that they are solvable.* To be clear, we do not pin this position on any of our contributors! However, the exchanges with our contributors elucidate this stance. Taking naturalism seriously, we think we should appeal to the conglomeration of research methods for collecting, modeling, and learning from data in the face of limitations and threats of error – including modeling strategies and probabilistic and computer methods – all of which we may house under the very general rubric of the methodology of inductive-statistical modeling and inference. For us, statistical science will always have this broad sense covering experimental design, data generation and modeling, statistical inference methods, and their links to scientific questions and models. We also regard these statistical tools as lending themselves to informal analogues in tackling general philosophical problems of evidence and inference. Looking to statistical science would seem a natural, yet still largely untapped, resource for a naturalistic and normative approach to philosophical problems of evidence. Methods of experimentation, simulation, model validation, and data collection have become increasingly subtle and sophisticated, and we propose that philosophers of science revisit traditional problems with these tools in mind. In some contexts, even where literal experimental control is lacking, inquirers have learned how to determine "what it would be like" if we were able to intervene and control – at least with high probability. Indeed "the challenge, the fun, of outwitting and outsmarting drives us to find ways to learn what it would be like to control, manipulate, and change, in situations where we cannot" (Mayo, 1996, p. 458). This perspective lets us broaden the umbrella of what we regard as an "experimental" context. When we need to restore the more usual distinction between experimental and observational research, we may dub the former "manipulative experiment" and the latter "observational experiment."

The tools of statistical science are plagued with their own conceptual and epistemological problems – some new, many very old. It is important to our goals to interrelate themes from philosophy of science and philosophy of statistics.

- The first half of the volume considers issues of error and inference in philosophical problems of induction and theory testing.
- The second half illuminates issues of errors and inference in practice: in formal statistics, econometrics, causal modeling, and legal epistemology.

These twin halves reflect our conception of philosophy and methodology of science as a "two-way street": on the one hand there is an appeal to methods and strategies of local experimental testing to grapple with philosophical problems of evidence and inference; on the other there is an appeal to philosophical analysis to address foundational problems of the methods and models used in practice; see Mayo and Spanos (2004).

3 Relevance for the Methodologist in Practice

An important goal of this work is to lay some groundwork for the methodologist in practice, although it must be admitted that our strategy at first appears circuitous. We do not claim that practitioners' general questions about evidence and method are directly answered once they are linked to what professional philosophers have said under these umbrellas. Rather, we claim that it is by means of such linkages that practitioners may better understand the foundational issues around which their questions revolve. In effect, practitioners themselves may become better "applied philosophers," which seems to be what is needed in light of the current predicament in philosophy of science. Some explanation is necessary.

In the current predicament, methodologists may ask, if each of the philosophies of science have unsolved and perhaps insoluble problems about evidence and inference, then how can they be useful for evidential problems in practice? "If philosophers and others within science theory can't agree about the constitution of the scientific method ... doesn't it seem a little dubious for economists to continue blithely taking things off the [philosopher's] shelf?" (Hands, 2001, p. 6). Deciding that it does, many methodologists in the social sciences tend to discount the relevance of the principles of scientific legitimacy couched within traditional philosophy of science. The philosophies of science are either kept on their shelves, or perhaps dusted off for cherry-picking from time to time. Nevertheless, practitioners still (implicitly or explicitly) wade into general questions about evidence or principles of inference and by elucidating the philosophical dimensions of such problems we hope to empower practitioners to appreciate and perhaps solve them. In a recent lead article in the journal *Statistical Science*, we read that "professional agreement on statistical philosophy is not on the immediate horizon, but this should not stop us from agreeing on methodology" (Berger, 2003, p. 2). But we think "what is correct methodologically" depends on "what is correct philosophically" (Mayo, 2003). Otherwise, choosing between competing methods and models may be viewed largely as a matter of pragmatics without posing deep philosophical problems or inconsistencies of principle. For the "professional agreement" to have weight, it cannot be merely an agreement to use methods with similar numbers when the meaning and import of such numbers remain up in the air (see Chapter 7). We cannot wave a wand and bring into existence the kind of philosophical literature that we think is needed. What we can do is put the practitioner in a better position to support, or alternatively, question the basis for professional agreement or disagreement.

Another situation wherein practitioners may find themselves wishing to articulate general principles or goals is when faced with the need to modify existing methods and to make a case for the adoption of new tools. Here, practitioners may serve the dual role of both inventing new methods and providing them with a principled justification – possibly by striving to find, or adapt features from, one or another philosophy of science or philosophy of statistics. Existing philosophy of science may not provide off-the-shelf methods for answering methodological problems in practice, but, coupled with the right road map, it may enable understanding, or even better, solving those problems.

An illustration in economics is given by Aris Spanos (Chapter 6). Faced with the lack of literal experimental controls, some economic practitioners attempt to navigate between two extreme positions. One position is the prevailing theory-dominated empirical modeling, largely limited to quantifying theories presupposed to be true. At the other extreme is data-driven modeling, largely limited to describing the data and guided solely by goodness-of-fit criteria. The former stays too close to the particular theory chosen at the start; the second stays too close to the particular data. Those practitioners seeking a "third way" are implicitly thrust into the role of striving to locate a suitable epistemological foundation for a methodology seemingly at odds with the traditional philosophical image of the roles of theory and data in empirical inquiry. In other words, the prescriptions on method in practice have trickled down from (sometimes competing) images of good science in traditional philosophy. We need to ask the question: *What are the threats to reliability and objectivity that lay behind the assumed prescriptions to begin with?* If data-dependent methods are thought to require the assumption of an overarching theory, or else permit too much latitude in constructing theories to fit data, then much of social science appears to be guilty of violating a scientific canon. But in practice, some econometricians work to develop methods whereby the data may be used to provide independent constraints on theory testing by means of intermediate-level statistical models with a "life of their own," as it were. This is the key to evading threats to reliability posed by theory-dominated modeling. By grasping the philosophical issues and principles, such applied work receives a stronger and far less tenuous epistemological foundation.

This brings us to a rather untraditional connection to traditional philosophy of science. In several of the philosophical contributions in this volume, we come across the very conceptions of testing that practitioners may find are in need of tweaking or alteration in order to adequately warrant methods they wish to employ. By extricating the legitimate threats to reliability and objectivity that lie behind the traditional stipulations, practitioners may ascertain where and when violations of established norms are justifiable. The exchange essays relating to the philosophical contributions deliberately try to pry us loose from rigid adherence to some of the standard prescriptions and prohibitions.

In this indirect manner, the methodologists' real-life problems are connected to what might have seemed at first an arcane philosophical debate. Insofar as these connections have not been made, practitioners are dubious that philosophers' debates about evidence and inference have anything to do with, much less help solve, their methodological problems. We think the situation is otherwise – that getting to the underlying philosophical issues not only increases the intellectual depth of methodological discussions but also paves the way for solving problems.

We find this strategy empowers students of methodology to evaluate critically, and perhaps improve on, methodologies in practice. Rather than approach alternative methodologies in practice as merely a menu of positions from which to choose, they may be grasped as attempted solutions to problems with deep philosophical roots. Conversely, progress in methodology may challenge philosophers of science to reevaluate the assumptions of their own philosophical theories. That is, after all, what a genuinely naturalistic philosophy of method would require. A philosophical problem, once linked to methodology in practice, enjoys solutions from the practical realm. For example, philosophers tend to assume that there are an infinite number of models that fit finite data equally well, and so data underdetermine hypotheses. Replacing "fit" with more rigorous measures of adequacy can show that such underdetermination vanishes (Spanos, 2007). This brings us to the last broad topic we consider throughout the volume.

We place it under the heading of *metaphilosophical themes.* Just as we know that evidence in science may be "theory-laden" – interpreted from the perspective of a background theory or set of assumptions – our philosophical theories (about evidence, inference, science) often color our philosophical arguments and conclusions (Rosenberg, 1992). The contributions in this volume reveal a good deal about these "philosophy-laden" aspects of philosophies of science. These revelations, moreover, are directly relevant to what is needed to construct a sound foundation for methodology in practice. The payoff is that understanding the obstacles to solving philosophical problems (the focus of Chapters 1–5) offers a clear comprehension of how to relate traditional philosophy of science to contemporary methodological and foundational problems of practice (the focus of Chapters 6–9).

4 Exchanges on E.R.R.O.R.

We organize the key themes of the entire volume under two interrelated categories:

(1) *experimental reasoning (empirical inference) and reliability,* and
(2) *objectivity and rationality of science.*

Although we leave these terms ambiguous in this introduction, they will be elucidated as we proceed. Interrelationships between these two categories immediately emerge. Scientific rationality and objectivity, after all, are generally identified by means of scientific methods: one's conception of objectivity and rationality in science leads to a conception of the requirements for an adequate account of empirical inference and reasoning. The perceived ability or inability to arrive at an account satisfying those requirements will in turn direct one's assessment of the possibility of objectivity and rationality in science. Recognizing the intimate relationships between categories 1 and 2 propels us toward both understanding and making progress on recalcitrant foundational problems about scientific inference. If, for example, empirical inference is thought to demand reliable rules of inductive inference, and if it is decided that such rules are unobtainable, then one may either question the rationality of science or instead devise a different notion of rationality for which empirical methods exist. On the other hand, if we are able to show that some methods are more robust than typically assumed, we may be entitled to uphold a more robust conception of science. Under category 1, we consider the nature and justification of experimental reasoning and the relationship of experimental inference to appraising large-scale theories in science.

4.1 Theory Testing and Explanation

Several contributors endorse the view that scientific progress is based on accepting large-scale theories (e.g., Chalmers, Musgrave) as contrasted to a view of progress based on the growth of more localized experimental knowledge (Mayo). Can one operate with a single overarching view of what is required for data to warrant an inference to H? Mayo says yes, but most of the other contributors argue for multiple distinct notions of evidence and inference. They do so for very different reasons. Some argue for a distinction between large-scale theory testing and local experimental inference. When it comes to large-scale theory testing, some claim that the most one can argue is that a theory is, comparatively, the best tested so far (Musgrave), or that a theory is justified by an "argument from coincidence" (Chalmers). Others argue that a distinct kind of inference is possible when the data are "not used" in constructing hypotheses or theories ("use-novel" data), as opposed to data-dependent cases where an inference is, at best, conditional on a theory (Worrall). Distinct concepts of evidence might be identified according to different background knowledge (Achinstein). Finally, different standards of evidence may be thought to emerge from the necessity of considering different costs (Laudan). The relations between testing and explanation often hover in the background of the discussion, or they may arise explicitly (Chalmers, Glymour, Musgrave). What are the explanatory virtues? And how do they relate to those of testing? Is there a tension between explanation and testing?

4.2 What Are the Roles of Probability in Uncertain Inference in Science?

These core questions are addressed both in philosophy of science, as well as in statistics and modeling practice. Does probability arise to assign degrees of epistemic support or belief to hypotheses, or to characterize the reliability of rules? A loose analogy exists between Popperian philosophers and frequentist statisticians, on the one hand, and Carnapian philosophers and Bayesian statisticians on the other. The latter hold that probability needs to supply some degree of belief, support, or epistemic assignment to hypotheses (Achinstein), a position that Popperians, or critical rationalists, dub 'justificationism' (Musgrave). Denying that such degrees may be usefully supplied, Popperians, much like frequentists, advocate focusing on the rationality of rules for inferring, accepting, or believing hypotheses. But what properties must these rules have?

In formal statistical realms, the rules for inference are reliable by dint of controlling error probabilities (Spanos, Cox and Mayo, Glymour). Can analogous virtues be applied to informal realms of inductive inference? This is the subject of lively debate in Chapters 1 to 5 in this volume. However, statistical methods and models are subject to their own long-standing foundational problems. Chapters 6 and 7 offer a contemporary update of these problems from the frequentist philosophy perspective. Which methods can be shown to ensure reliability or low long-run error probabilities? Even if we can show they have good long-run properties, how is this relevant for a particular inductive inference in science? These chapters represent exchanges and shared efforts of the authors over the past four years to tackle these problems as they arise in current statistical methodology. Interwoven throughout this volume we consider the relevance of these answers to analogous questions as they arise in philosophy of science.

4.3 Objectivity and Rationality of Science, Statistics, and Modeling

Despite the multiplicity of perspectives that the contributors bring to the table, they all find themselves confronting a cluster of threats to objectivity in observation and inference. Seeing how analogous questions arise in philosophy and methodological practice sets the stage for the meeting ground that creates new synergy.

- Does the fact that observational claims themselves have assumptions introduce circularity into the experimental process?
- Can one objectively test assumptions linking actual data to statistical models, and statistical inferences to substantive questions?

On the one hand, the philosophers' demand to extricate assumptions raises challenges that the practitioner tends to overlook; on the other hand, progress in methodology may point to a more subtle logic that gets around the limits that give rise to philosophical skepticism.

What happens if methodological practice seems in conflict with philosophical principles of objectivity? Some methodologists reason that if it is common, if not necessary, to violate traditional prescriptions of scientific objectivity in practice, then we should renounce objectivity (and perhaps make our subjectivity explicit). That judgment is too quick. If intuitively good scientific practice seems to violate what are thought to be requirements of good science, we need to consider whether in such cases scientists guard against the errors that their violation may permit.

To illustrate, consider one of the most pervasive questions that arises in trying to distinguish genuine tests from ad hoc methods:

Is it legitimate to use the same data in both constructing and testing hypotheses?

This question arises in practice in terms of the legitimacy of data-mining, double counting, data-snooping, and hunting for statistical significance. In philosophy of science, it arises in terms of novelty requirements. Musgrave (1974) was seminal in tackling the problems of how to define, and provide a rationale for, preferring novel predictions in the Popper-Lakatos traditions. However, these issues have never been fully resolved, and they continue to be a source of debate. The question of the rationale for requiring novelty arises explicitly in Chapter 4 (Worrall) and the associated exchange.

Lurking in the background of all of the contributions in this volume is the intuition that good tests should avoid double-uses of data, that would result in violating what we called the *minimal scientific principle for evidence*. Using the same data to construct as well as test a hypothesis, it is feared, makes it too easy to find accordance between the data and the hypothesis even if the hypothesis is false. By uncovering how reliable learning may be retained despite double-uses of data, we may be able to distinguish legitimate from illegitimate double counting.

The relevance of this debate for practice is immediately apparent in the second part of the volume where several examples of data-dependent modeling and non-novel evidence arise: in accounting for selection effects, in testing assumptions of statistical models, in empirical modeling in economics, in algorithms for causal model discovery, and in obtaining legal evidence.

This leads to our third cluster of issues that do not readily fit under either category (1) or (2) – the host of "meta-level" issues regarding philosophical assumptions (theory-laden philosophy) and the requirements of a successful two-way street between philosophy of science and methodological practice.

The questions listed in Section 6 identify the central themes to be taken up in this volume. The essays following the contributions are called "exchanges" because they are the result of a back-and-forth discussion over a period of several years. Each exchange begins by listing a small subset of these questions that is especially pertinent for reflecting on the particular contribution.

5 Using This Volume for Teaching

Our own experiences in teaching courses that blend philosophy of science and methodology have influenced the way we arrange the material in this volume. We have found it useful, for the first half of a course, to begin with a core methodological paper in the given field, followed by selections from the philosophical themes of Chapters 1–5, supplemented with 1–2 philosophical articles from the references (e.g., from Lakatos, Kuhn, Popper). Then, one might turn to selections from Chapters 6–9, supplemented with discipline-specific collections of papers.* The set of questions listed in the next section serves as a basis around which one might organize both halves of the course. Because the exchange that follows each chapter elucidates some of the key points of that contribution, readers may find it useful to read or glance at the exchange first and then read the corresponding chapter.

6 Philosophical and Methodological Questions Addressed in This Volume

6.1 Experimental Reasoning and Reliability

Theory Testing and Explanation

- Does theory appraisal demand a kind of reasoning distinct from local experimental inferences?
- Can generalizations and theoretical claims ever be warranted with severity?
- Are there reliable observational methods for discovering or inferring causes?
- How can the gap between statistical and structural (e.g., causal) models be bridged?

---

* A variety of modules for teaching may be found at the website: http://www.econ.vt .edu/faculty/facultybios/spanos_error_inference.htm.

- Must local experimental tests always be done within an overarching theory or paradigm? If so, in what sense must the theory be assumed or accepted?
- When does *H*'s successful explanation of an effect warrant inferring the truth or correctness of *H*?
- How do logical accounts of explanation link with logics of confirmation and testing?

How to Characterize and Warrant Methods of Experimental Inference

- Can inductive or "ampliative" inference be warranted?
- Do experimental data so underdetermine general claims that warranted inferences are limited to the specific confines in which the data have been collected?
- Can we get beyond inductive skepticism by showing the existence of reliable test rules?
- Can experimental virtues (e.g., reliability) be attained in nonexperimental contexts?
- How should probability enter into experimental inference and testing: by assigning degrees of belief or by characterizing the reliability of test procedures?
- Do distinct uses of data in science require distinct criteria for warranted inferences?
- How can methods for controlling long-run error probabilities be relevant for inductive inference in science?

6.2 Objectivity and Rationality of Science

- Should scientific progress and rationality be framed in terms of large-scale theory change?
- Does a piecemeal account of explanation entail a piecemeal account of testing?
- Does an account of progress framed in terms of local experimental inferences entail a nonrealist role for theories?
- Is it unscientific (ad hoc, degenerating) to use data in both constructing and testing hypotheses?
- Is double counting problematic only when it leads to unreliable methods?
- How can we assign degrees of objective warrant or rational belief to scientific hypotheses?

- How can we assess the probabilities with which tests lead to erroneous inferences (error probabilities)?
- Can an objective account of statistical inference be based on frequentist methods? On Bayesian methods?
- Can assumptions of statistical models and methods be tested objectively?
- Can assumptions linking statistical inferences to substantive questions be tested objectively?
- What role should probabilistic/statistical accounts play in scrutinizing methodological desiderata (e.g., explanatory virtues) and rules (e.g., avoiding irrelevant conjunction, varying evidence)?
- Do explanatory virtues promote truth, or do they conflict with well-testedness?
- Does the latitude in specifying tests and criteria for accepting and rejecting hypotheses preclude objectivity?
- Are the criteria for warranted evidence and inference relative to the varying goals in using evidence?

6.3 Metaphilosophical Themes

Philosophy-Laden Philosophy of Science

- How do assumptions about the nature and justification of evidence and inference influence philosophy of science? In the use of historical episodes?
- How should we evaluate philosophical tools of logical analysis and counterexamples?
- How should probabilistic/statistical accounts enter into solving philosophical problems?

Responsibilities of the "Two-Way Street" between Philosophy and Practice

- What roles can or should philosophers play in methodological problems in practice? (Should they be in the business of improving practice as well as clarifying, reconstructing, or justifying practice?)
- How does studying evidence and methods in practice challenge assumptions that may go unattended in philosophy of science?

II The Error-Statistical Philosophy

The Preface of *Error and the Growth of Experimental Knowledge* (EGEK) opens as follows:

Despite the challenges to and changes in traditional philosophy of science, one of its primary tasks continues to be to explain if not also to justify, scientific methodologies for learning about the world. To logical empiricist philosophers (Carnap, Reichenbach) the task was to show that science proceeds by objective rules for appraising hypotheses. To that end many attempted to set out formal rules termed inductive logics and confirmation theories. Alongside these stood Popper's methodology of appraisal based on falsification: evidence was to be used to falsify claims deductively rather than to build up inductive support. Both inductivist and falsificationist approaches were plagued with numerous, often identical, philosophical problems and paradoxes. Moreover, the entire view that science follows impartial algorithms or logics was challenged by Kuhn (1962) and others. What methodological rules there are often conflict and are sufficiently vague as to "justify" rival hypotheses. Actual scientific debates often last for several decades and appear to require, for their adjudication, a variety of other factors left out of philosophers' accounts. The challenge, if one is not to abandon the view that science is characterized by rational methods of hypothesis appraisal, is either to develop more adequate models of inductive inference or else to find some new account of scientific rationality. (Mayo, 1996, p. ix)

Work in EGEK sought a more adequate account of induction based on a cluster of tools from statistical science, and this volume continues that program, which we call the error-statistical account.

Contributions to this volume reflect some of the "challenges and changes" in philosophy of science in the dozen years since EGEK, and the ensuing dialogues may be seen to move us "Toward an Error-Statistical Philosophy of Science" – as sketchily proposed in EGEK's last chapter. Here we collect for the reader some of its key features and future prospects.

15

7 What Is Error Statistics?

Error statistics, as we use the term, has a dual dimension involving philosophy and methodology. It refers to a standpoint regarding both (1) a general philosophy of science and the roles probability plays in inductive inference, and (2) a cluster of statistical tools, their interpretation, and their justification. It is unified by a general attitude toward a fundamental pair of questions of interest to philosophers of science and scientists in general:

- *How do we obtain reliable knowledge about the world despite error?*
- *What is the role of probability in making reliable inferences?*

Here we sketch the error-statistical methodology, the statistical philosophy associated with the methods ("error-statistical philosophy"), and a philosophy of science corresponding to the error-statistical philosophy.

7.1 Error-Statistical Philosophy

Under the umbrella of error-statistical methods, one may include all standard methods using error probabilities based on the relative frequencies of errors in repeated sampling – often called *sampling theory*. In contrast to traditional confirmation theories, probability arises not to measure degrees of confirmation or belief in hypotheses but to quantify how frequently methods are capable of discriminating between alternative hypotheses and how reliably they facilitate the detection of error. These probabilistic properties of inference procedures are *error frequencies* or *error probabilities*. The statistical methods of significance tests and confidence-interval estimation are examples of formal error-statistical methods. Questions or problems are addressed by means of hypotheses framed within statistical models.

A statistical model (or family of models) gives the probability distribution (or density) of the sample $\mathbf{X} = (X_1, \ldots, X_n)$, $f_X(\mathbf{x}; \boldsymbol{\theta})$, which provides an approximate or idealized representation of the underlying data-generating process. Statistical hypotheses are typically couched in terms of an unknown parameter, $\boldsymbol{\theta}$, which governs the probability distribution (or density) of $\mathbf{X}$. Such hypotheses are claims about the data-generating process. In error statistics, statistical inference procedures link special functions of the data, $d(\mathbf{X})$, known as *statistics*, to hypotheses of interest. All error probabilities stem from the distribution of $d(\mathbf{X})$ evaluated under different hypothetical values of parameter θ.

Consider for example the case of a random sample $\mathbf{X}$ of size n from a Normal distribution ($N(\mu,1)$) where we want to test the hypotheses:

$$H_0: \mu = \mu_0 \text{ vs. } H_1: \mu > \mu_0.$$

The test statistic is $d(\mathbf{X}) = (\overline{X} - \mu_0)/\sigma_x$, where $\overline{X} = (1/n)\sum_{i=1}^{n} X_i$ and $\sigma_x = (\sigma/\sqrt{n})$. Suppose the test rule T construes data $\mathbf{x}$ as evidence for a discrepancy from μ_0 whenever $d(\mathbf{x}) > 1.96$. The probability that the test would indicate such evidence when in fact μ_0 is true is $P(d(\mathbf{X}) > 1.96; H_0) = .025$. This gives us what is called the *statistical significance level*. Objectivity stems from controlling the relevant error probabilities associated with the particular inference procedure. In particular, the claimed error probabilities approximate the actual (long-run) relative frequencies of error. (See Chapters 6 and 7.)

Behavioristic and Evidential Construal. By a "statistical philosophy" we understand a general concept of the aims and epistemological foundations of a statistical methodology. To begin with, two different interpretations of these methods may be given, along with diverging justifications. The first, and most well known, is the *behavioristic construal*. In this case, tests are interpreted as tools for deciding "how to behave" in relation to the phenomena under test and are justified in terms of their ability to ensure low long-run errors. A nonbehavioristic or *evidential construal* must interpret error-statistical tests (and other methods) as tools for achieving inferential and learning goals. How to provide a satisfactory evidential construal has been the locus of the most philosophically interesting controversies and remains the major lacuna in using these methods for philosophy of science. This is what the severity account is intended to supply. However, there are contexts wherein the more behavioristic construal is entirely appropriate, and it is retained within the "error-statistical" umbrella.

Objectivity in Error Statistics. The inferential interpretation forms a central part of what we refer to as *error-statistical philosophy*. Underlying this philosophy is the concept of scientific objectivity: although knowledge gaps leave plenty of room for biases, arbitrariness, and wishful thinking, in fact we regularly come up against experiences that thwart our expectations and disagree with the predictions and theories we try to foist upon the world – this affords objective constraints on which our critical capacity is built. Getting it (at least approximately) right, and not merely ensuring internal consistency or agreed-upon convention, is at the heart of objectively orienting ourselves toward the world. Our ability to recognize when data fail to match anticipations is what affords us the opportunity to systematically improve our orientation in direct response to such disharmony. Failing to falsify hypotheses, while rarely allowing their acceptance as true, warrants the exclusion of various discrepancies, errors, or rivals, provided the test had a high probability of uncovering such flaws, if they were present. In those cases, we may infer that the discrepancies, rivals, or errors are ruled out with *severity*.

We are not stymied by the fact that inferential tools have assumptions but rather seek ways to ensure that the validity of inferences is not much threatened by what is currently unknown. This condition may be secured either because tools are robust against flawed assumptions or that subsequent checks will detect (and often correct) them with high probability. Attributes that go unattended in philosophies of confirmation occupy important places in an account capable of satisfying error-statistical goals. For example, explicit attention needs to be paid to communicating results to set the stage for others to check, debate, and extend the inferences reached. In this view, it must be part of any adequate statistical methodology to provide the means to address critical questions and to give information about which conclusions are likely to stand up to further probing and where weak spots remain.

Error-Statistical Framework of "Active" Inquiry. The error-statistical philosophy conceives of statistics (or statistical science) very broadly to include the conglomeration of systematic tools for collecting, modeling, and drawing inferences from data, including purely "data-analytic" methods that are normally not deemed "inferential." For formal error-statistical tools to link data, or *data models*, to *primary scientific hypotheses*, several different statistical hypotheses may be called upon, each permitting an aspect of the primary problem to be expressed and probed. An auxiliary or "secondary" set of hypotheses is called upon to check the assumptions of other models in the complex network.

The error statistician is concerned with the critical control of scientific inferences by means of stringent probes of conjectured flaws and sources of unreliability. Standard statistical hypotheses, while seeming oversimplified in and of themselves, are highly flexible and effective for the piecemeal probes our error statistician seeks. Statistical hypotheses offer ways to couch canonical flaws in inference. We list six overlapping errors:

1. Mistaking spurious for genuine correlations,
2. Mistaken directions of effects,
3. Mistaken values of parameters,
4. Mistakes about causal factors,
5. Mistaken assumptions of statistical models,
6. Mistakes in linking statistical inferences to substantive scientific hypotheses.

The qualities we look for to express and test hypotheses about such inference errors are generally quite distinct from those traditionally sought in appraising substantive scientific claims and theories. Although the overarching goal is to find out what is (truly) the case about aspects of phenomena, the hypotheses erected in the actual processes of finding things out are generally approximations and may even be deliberately false. Although we cannot fully formalize, we can systematize the manifold steps and interrelated checks that, taken together, constitute a full-bodied experimental inquiry. Background knowledge enters the processes of designing, interpreting, and combining statistical inferences in informal or semiformal ways – not, for example, by prior probability distri-butions.

The picture corresponding to error statistics is one of an activist learner in the midst of an inquiry with the goal of finding something out. We want hypotheses that will allow for stringent testing so that if they pass we have evidence of a genuine experimental effect. The goal of attaining such well-probed hypotheses differs crucially from seeking highly probable ones (however probability is interpreted). This recognition is the key to getting a handle on long-standing Bayesian–frequentist debates.

The error statistical philosophy serves to guide the use and interpretation of frequentist statistical tools so that we can distinguish the genuine foundational differences from a host of familiar fallacies and caricatures that have dominated 75 years of "statistics wars." The time is ripe to get beyond them.

7.2 Error Statistics and Philosophy of Science

The *error-statistical philosophy* alludes to the general methodological principles and foundations associated with frequentist error-statistical methods;

it is the sort of thing that would be possessed by a statistician, when thinking foundationally, or by a philosopher of statistics. By an *error-statistical philosophy of science*, on the other hand, we have in mind the use of those tools, appropriately adapted, to problems of philosophy of science: to model scientific inference (actual or rational), to scrutinize principles of inference (e.g., preferring novel results, varying data), and to frame and tackle philosophical problems about evidence and inference (how to warrant data, pinpoint blame for anomalies, and test models and theories). Nevertheless, each of the features of the error-statistical philosophy has direct consequences for the philosophy of science dimension.

To obtain a philosophical account of inference from the error-statistical perspective, one would require forward-looking tools for finding things out, not for reconstructing inferences as "rational" (in accordance with one or another view of rationality). An adequate philosophy of evidence would have to engage statistical methods for obtaining, debating, rejecting, and affirming data. From this perspective, an account of scientific method that begins its work only once well-defined evidence claims are available forfeits the ability to be relevant to understanding the actual processes behind the success of science. Because the contexts in which statistical methods are most needed are ones that compel us to be most aware of the strategies scientists use to cope with threats to reliability, the study of the nature of statistical method in the collection, modeling, and analysis of data is an effective way to articulate and warrant principles of evidence. In addition to paving the way for richer and more realistic philosophies of science, we think, examining error-statistical methods sets the stage for solving or making progress on long-standing philosophical problems about evidence and inductive inference.

Where the recognition that data are always fallible presents a challenge to traditional empiricist foundations, the cornerstone of statistical induction is the ability to move from less accurate to more accurate data.

Where the best often thought "feasible" means getting it right in some asymptotic long run, error-statistical methods enable specific precision to be ensured in finite samples and supply ways to calculate how large the sample size n needs to be for a given level of accuracy.

Where pinpointing blame for anomalies is thought to present insoluble "Duhemian problems" and "underdetermination," a central feature of error-statistical tests is their capacity to evaluate error probabilities that hold regardless of unknown background or "nuisance" parameters.

We now consider a principle that links (1) the error-statistical philosophy and (2) an error-statistical philosophy of science.

7.3 The Severity Principle

A method's error probabilities refer to their performance characteristics in a hypothetical sequence of repetitions. How are we to use error probabilities of tools in warranting particular inferences? This leads to the general question:

When do data x_0 provide good evidence for or a good test of hypothesis H?

Our standpoint begins with the intuition described in the first part of this chapter. We intuitively deny that data x_0 are evidence for H if the inferential procedure had very little chance of providing evidence against H, even if H is false. We can call this the "weak" severity principle:

Severity Principle (Weak): Data x_0 do not provide good evidence for hypothesis H if x_0 result from a test procedure with a very low probability or capacity of having uncovered the falsity of H (even if H is incorrect).

Such a test, we would say, is insufficiently stringent or severe. The onus is on the person claiming to have evidence for H to show that the claim is not guilty of at least so egregious a lack of severity. Formal error-statistical tools provide systematic ways to foster this goal and to determine how well it has been met in any specific case. Although one might stop with this negative conception (as perhaps Popperians do), we continue on to the further, positive conception, which will comprise the full severity principle:

Severity Principle (Full): Data x_0 provide a good indication of or evidence for hypothesis H (just) to the extent that test T has severely passed H with x_0.

The severity principle provides the rationale for error-statistical methods. We distinguish the severity rationale from a more prevalent idea for how procedures with low error probabilities become relevant to a particular application; namely, since the procedure is rarely wrong, the probability it is wrong in this case is low. In that view, we are justified in inferring H because it was the output of a method that rarely errs. It is as if the long-run error probability "rubs off" on each application. However, this approach still does not quite get at the reasoning for the particular case at hand, at least in nonbehavioristic contexts. The reliability of the rule used to infer H is at most a necessary and not a sufficient condition to warrant inferring H. All of these ideas will be fleshed out throughout the volume.

Passing a severe test can be encapsulated as follows:

A hypothesis H passes a severe test T with data x_0 *if*

   (S-1) x_0 *agrees with H, (for a suitable notion of "agreement") and*
   (S-2) *with very high probability, test T would have produced a result that accords less well with H than does* x_0, *if H were false or incorrect.*

Severity, in our conception, somewhat in contrast to how it is often used, is not a characteristic of a test in and of itself, but rather of the test T, a specific test result x_0, and a specific inference being entertained, H. Thereby, the severity function has three arguments. We use SEV(T, x_0, H) to abbreviate "the severity with which H passes test T with data x_0" (Mayo and Spanos, 2006).

   The existing formal statistical testing apparatus does not include severity assessments, but there are ways to *use* the error-statistical properties of tests, together with the outcome x_0, to evaluate a test's severity. This is the key for our inferential interpretation of error-statistical tests. The severity principle underwrites this inferential interpretation and addresses chronic fallacies and well-rehearsed criticisms associated with frequentist testing. Among the most familiar of the often repeated criticisms of the use of significance tests is that with large enough sample size, a small significance level can be very probable, even if the underlying discrepancy γ from null hypothesis $\mu = \mu_0$ is substantively trivial. Why suppose that practitioners are incapable of mounting an interpretation of tests that reflects the test's sensitivity? The severity assessment associated with the observed significance level [p-value] directly accomplishes this.

   Let us return to the example of test T for the hypotheses: H_0: $\mu = 0$ vs. H_1: $\mu > 0$. We see right away that the same value of $d(x_0)$ (and thus the same p-value) gives different severity assessments for a given inference when n changes.

   In particular, suppose one is interested in the discrepancy $\gamma = .2$, so we wish to evaluate the inference $\mu > .2$. Suppose the same $d(x_0) = 3$ resulted from two different sample sizes, $n = 25$ and $n = 400$. For $n = 25$, the severity associated with $\mu > .2$ is .97, but for $n = 400$ the severity associated with $\mu > .2$ is .16. So the same $d(x_0)$ gives a strong warrant for $\mu > .2$ when $n = 25$, but provides very poor evidence for $\mu > .2$ when $n = 400$.

   More generally, an α-significant difference with n_1 passes $\mu > \mu_1$ less severely than with n_2 where $n_1 > n_2$. With this simple interpretive tool, all of the variations on "large n criticisms" are immediately scotched (Cohen, 1994, Lindley, 1957, Howson and Urbach, 1993, inter alia). (See Mayo and Spanos, 2006, and in this volume, Chapter 7).

Getting around these criticisms and fallacies is essential to provide an adequate philosophy for error statistics as well as to employ these ideas in philosophy of science.

The place to begin, we think, is with general philosophy of science, as we do in this volume.

8 Error-Statistical Philosophy of Science

Issues of statistical philosophy, as we use that term, concern methodological and epistemological issues surrounding statistical science; they are matters likely to engage philosophers of statistics and statistical practitioners interested in the foundations of their methods. Philosophers of science generally find those issues too specialized or too technical for the philosophical problems as they are usually framed. By and large, this leads philosophers of science to forfeit the insights that statistical science and statistical philosophy could offer for the general problems of evidence and inference they care about. To remedy this, we set out the distinct category of an error-statistical philosophy of science. An error-statistical philosophy of science alludes to the various interrelated ways in which error-statistical methods and their interpretation and rationale are relevant for three main projects in philosophy of science: to characterize scientific inference and inquiry, solve problems about evidence and inference, and appraise methodological rules.

The conception of inference and inquiry would be analogous to the piecemeal manner in which error statisticians relate raw data to data models, to statistical hypotheses, and to substantive claims and questions. Even where the collection, modeling, and analysis of data are not explicitly carried out using formal statistics, the limitations and noise of learning from limited data invariably introduce errors and variability, which suggests that formal statistical ideas are more useful than deductive logical accounts often appealed to by philosophers of science. Were we to move toward an error-statistical philosophy of science, statistical theory and its foundations would become a new formal apparatus for the philosophy of science, supplementing the more familiar tools of deductive logic and probability theory.

The indirect and piecemeal nature of this use of statistical methods is what enables it to serve as a forward-looking account of ampliative (or inductive) inference, not an after-the-fact reconstruction of past episodes and completed experiments. Although a single inquiry involves a network of models, an overall "logic" of experimental inference may be identified: *data* x_0 *indicate the correctness of hypothesis H to the extent that H passes a stringent or severe test with* x_0. Whether the criterion for warranted inference is put in terms of severity or reliability or degree of corroboration, problems of induction become experimental problems of how to control and assess the error probabilities needed to satisfy this requirement. Unlike the traditional "logical problem of induction," this experimental variant is solvable.

Methodological rules are regarded as claims about strategies for coping with and learning from errors in furthering the overarching goal of severe testing. Equally important is the ability to use *in*severity to learn what is *not* warranted and to pinpoint fruitful experiments to try next. From this perspective, one would revisit philosophical debates surrounding double counting and novelty, randomized studies, the value of varying the data, and replication. As we will see in the chapters that follow, rather than give all-or-nothing pronouncements on the value of methodological prescriptions, we obtain a more nuanced and context-dependent analysis of when and why they work.

8.1 Informal Severity and Arguing from Error

In the quasi-formal and informal settings of scientific inference, the severe test reasoning corresponds to the basic principle that *if a procedure had very low probability of detecting an error if it is present, then failing to signal the presence of the error is poor evidence for its absence.* Failing to signal an error (in some claim or inference *H*) corresponds to the data being in accord with (or "fitting") some hypothesis *H*. This is a variant of the minimal scientific requirement for evidence noted in part I of this chapter. Although one can get surprising mileage from this negative principle alone, we embrace the positive side of the full severity principle, which has the following informal counterpart:

Arguing from Error: An error or fault is absent when (and only to the extent that) a procedure of inquiry with a high probability of detecting the error if and only if it is present, nevertheless detects no error.

We argue that an error is absent if it fails to be signaled by a highly severe error probe.

The strongest severity arguments do not generally require formal statistics. We can retain the probabilistic definition of severity in the general context that arises in philosophical discussions, so long as we keep in mind that it serves as a brief capsule of the much more vivid context-specific arguments that flesh out the severity criterion when it is clearly satisfied or flagrantly violated.

We can inductively infer the absence of any error that has been well probed and ruled out with severity. It is important to emphasize that an "error" is understood as any mistaken claim or inference about the phenomenon being probed – theoretical or non-theoretical (see exchanges with Chalmers and Musgrave). Doubtless, this seems to be a nonstandard use of "error." We introduce this concept of error because it facilitates the assessment of severity appropriate to the particular local inference – it directs one to consider the particular inferential mistake that would have to be ruled out for the data to afford evidence for *H*. Although "*H* is false" refers to a specific error, it is meant to encompass erroneous claims about underlying causes and mistaken understandings of any testable aspect of a phenomenon of interest. Often the parameter in a statistical model directly parallels the theoretical quantity in a substantive theory or proto-theory.

Degrees of severity might be available, but in informal assessments it suffices to consider qualitative classifications (e.g., highly, reasonably well, or poorly probed). This threshold-type construal of severity is all that will be needed in many of the discussions that follow. In our philosophy of inference, if *H* is not reasonably well probed, then it should be regarded as poorly probed. Even where *H* is known to be true, a test that did a poor job in probing its flaws would fail to supply good evidence for *H*.

Note that we choose to couch all claims about evidence and inference in testing language, although one is free to deviate from this. Our idea is to emphasize the need to have done something to check errors before claiming to have evidence; but the reader must not suppose our idea of inference is limited to the familiar view of tests as starting out with hypotheses, nor that it is irrelevant for cases described as estimation. One may start with data and arrive at well-tested hypotheses, and any case of statistical estimation can be put into testing terms.

Combining Tests in an Inquiry. Although it is convenient to continue to speak of a severe test *T* in the realm of substantive scientific inference (as do several of the contributors), it should be emphasized that reference to "test *T*" may actually, and usually does, combine individual tests and inferences together; likewise, the data may combine results of several tests. To avoid confusion, it may be necessary to distinguish whether we have in mind several tests or a given test – a single data set or all information relevant to a given problem.

Severity, Corroboration, and Belief. Is the degree of severity accorded *H* with x_0 any different from a degree of confirmation or belief? While a hypothesis that passes with high severity may well warrant the belief that it is correct, the entire logic is importantly different from a "logic of belief" or confirmation. For one thing, I may be warranted in strongly believing H and yet deny that this particular test and data warrant inferring H. For another, the logic of probability does not hold. For example, that H is poorly tested does not mean "not H" is well tested. There is no objection to substituting "H passes severely with x_0 from test T" with the simpler form of "data x_0 from test T corroborate H" (as Popper suggested), so long as it is correctly understood. A logic of severity (or corroboration) could be developed – a futuristic project that would offer a rich agenda of tantalizing philosophical issues.

8.2 Local Tests and Theory Appraisal

We have sketched key features of the error statistical philosophy to set the stage for the exchanges to follow. It will be clear at once that our contributors take issue with some or all of its core elements. True to the error-statistical principle of learning from stringent probes and stress tests, the contributors to this volume serve directly or indirectly to raise points of challenge. Notably, while granting the emphasis on local experimental testing provides "a useful corrective to some of the excesses of the theory-dominated approach" (Chalmers 1999, p. 206), there is also a (healthy) skepticism as to whether the account can make good on some of its promises, at least without compromising on the demands of severe testing. The tendency toward "theory domination" in contemporary philosophy of science stems not just from a passion with high-level physics (we like physics too) but is interestingly linked to the felt shortcomings in philosophical attempts to solve problems of evidence and inference. If we have come up short in justifying inductive inferences in science, many conclude, we must recognize that such inferences depend on accepting or assuming various theories or generalizations and laws. It is only thanks to already accepting a background theory or paradigm T that inductive inferences can get off the ground. How then to warrant theory T? If the need for an empirical account to warrant T appears to take one full circle, T's acceptance may be based on appeals to explanatory, pragmatic, metaphysical, or other criteria. One popular view is that a theory is to be accepted if it is the "best explanation" among existing rivals, for a given account of explanation, of which there are many. The error-statistical account of local testing, some may claim, cannot escape the circle: it will invariably require a separate account of theory appraisal if it is to capture and explain the success of science. This takes us to the question asked in Chapter 1 of this volume: What would an adequate error-statistical account of large-scale theory testing be?

References

Achinstein, P. (2001), *The Book of Evidence*, Oxford University Press, Oxford.

Berger, J. O. (2003), "Could Fisher, Jeffreys and Neyman Have Agreed on Testing?" *Statistical Science*, 18: 1–12.

Cartwright, N. (1983), *How the Laws of Physics Lie*, Oxford University Press, Oxford.

Chalmers, A. (1999), *What is This Thing Called Science? 3rd edition*, University of Queensland Press.

Chang, H. (2004), *Inventing Temperature: Measurement and Scientific Progress*, Oxford University Press, Oxford.

Cohen, J. (1994), "The Earth Is Round (p < .05)," *American Psychologist*, 49: 997–1003.

Galison, P. L. (1987), *How Experiments End*, The University of Chicago Press, Chicago.

Hacking, I. (1983), *Representing and Intervening: Introductory Topics in the Philosophy of Natural Science*, Cambridge University Press, Cambridge.

Hands, W. D. (2001), *Reflection Without Rules: Economic Methodology and Contemporary Science Theory*, Cambridge University Press, Cambridge.

Howson, C. and Urbach, P. (1993), *Scientific Reasoning: A Bayesian Approach*, 2nd ed., Open Court, Chicago.

Kuhn, T. S. (1962), *The Structure of Scientific Revolutions*, Chicago University Press, Chicago.

Lindley, D. V. (1957), "A Statistical Paradox," *Biometrika*, 44:187–92.

Mayo, D. G. (1996), *Error and the Growth of Experimental Knowledge*, The University of Chicago Press, Chicago.

Mayo, D. G. (2003), "Could Fisher, Jeffreys and Neyman Have Agreed on Testing? Commentary on J. Berger's Fisher Address," *Statistical Science*, 18: 19–24.

Mayo, D. G. and Spanos, A. (2004), "Methodology in Practice: Statistical Misspecification Testing," *Philosophy of Science*, 71: 1007–25.

Mayo, D. G. and Spanos, A. (2006), "Severe testing as a basic concept in a Neyman–Pearson philosophy of induction," *British Journal for the Philosophy of Science*, 57: 323–57.

Morgan, M. S. and Morrison, M. (1999), *Models as Mediators: Perspectives on Natural and Social Science*, Cambridge University Press, Cambridge.

Morrison, M. (2000), *Unifying Scientific Theories: Physical Concepts and Mathematical Structures*, Cambridge University Press, Cambridge.

Musgrave, A. (1974), "Logical versus Historical Theories of Confirmation," *British Journal for the Philosophy of Science*, 25: 1–23.

Rosenberg, A. (1992), *Economics – Mathematical Politics or Science of Diminishing Returns?* (Science and Its Conceptual Foundations series) University of Chicago Press, Chicago.

Spanos, A. (2007), "Curve-Fitting, the Reliability of Inductive Inference and the Error-Statistical Approach," *Philosophy of Science*, 74: 1046–66.

ONE

Learning from Error, Severe Testing, and the Growth of Theoretical Knowledge

Deborah G. Mayo

I regard it as an outstanding and pressing problem in the philosophy of the natural sciences to augment the insights of the new experimentalists with a correspondingly updated account of the role or roles of theory in the experimental sciences, substantiated by detailed case studies. (Chalmers, 1999, p. 251)

1 Background to the Discussion

The goal of this chapter is to take up the aforementioned challenge as it is posed by Alan Chalmers (1999, 2002), John Earman (1992), Larry Laudan (1997), and other philosophers of science. It may be seen as a first step in taking up some unfinished business noted a decade ago: "How far experimental knowledge can take us in understanding theoretical entities and processes is not something that should be decided before exploring this approach much further" (Mayo, 1996, p. 13). We begin with a sketch of the resources and limitations of the "new experimentalist" philosophy.

Learning from evidence, in this experimentalist philosophy, depends not on appraising large-scale theories but on local experimental tasks of estimating backgrounds, modeling data, distinguishing experimental effects, and discriminating signals from noise. The growth of knowledge has not to do with replacing or confirming or probabilifying or "rationally accepting" large-scale theories, but with testing specific hypotheses in such a way that there is a good chance of learning something – whatever theory it winds up as part of. This learning, in the particular experimental account we favor, proceeds by testing experimental hypotheses and inferring those that pass probative or *severe* tests – tests that would have unearthed some error in, or discrepancy from, a hypothesis H, were H false. What enables this account of severity to work is that the immediate hypothesis H under test by means of data is designed to be a specific and local claim (e.g., about parameter values, causes, the reliability of an effect, or experimental assumptions). "*H* is false" is not a disjunction of all possible rival explanations of the data, at all levels of complexity; that is, it is not the so-called catchall hypothesis but refers instead to a specific error being probed.

1.1 What Is the Problem?

These features of piecemeal testing enable one to exhaust the possible answers to a specific question; the price of this localization is that one is not entitled to regard full or large-scale theories as having passed severe tests, so long as they contain hypotheses and predictions that have not been well probed. If scientific progress is thought to turn on appraising high-level theories, then this type of localized account of testing will be regarded as guilty of a serious omission, unless it is supplemented with an account of theory appraisal.

1.2 The Comparativist Rescue

A proposed remedy is to weaken the requirement so that a large-scale theory is allowed to pass severely so long as it is the "best-tested" theory so far, in some sense. Take Laudan:

[W]hen we ask whether [the General Theory of Relativity] GTR can be rationally accepted, we are not asking whether it has passed tests which it would almost certainly fail if it were false. As Mayo acknowledges, we can rarely if ever make such judgments about most of the general theories of the science. But we can ask "Has GTR passed tests which none of its known rivals have passed, while failing none which those rivals have passed." Answering such a question requires no herculean enumeration of all the possible hypotheses for explaining the events in a domain. (Laudan, 1997, p. 314)

We take up this kind of comparativist appraisal and argue that it is no remedy; rather, it conflicts with essential ingredients of the severity account – both with respect to the "life of experiment" and to the new arena, the "life of theory."

1.3 Is Severity Too Severe?

One of the main reasons some charge that we need an account showing acceptance of high-level theories is that scientists in fact seem to accept them; without such an account, it is said, we could hardly make sense of scientific practice. After all, these philosophers point out, scientists set about probing and testing theories in areas beyond those in which they have been well tested. While this is obviously true, we question why it is supposed that in doing so scientists are implicitly accepting all of the theory in question. On the contrary, we argue, this behavior of scientists seems to underscore the importance of distinguishing areas that are from those that are not (thus far) well tested; such a distinction would be blurred if a full theory is accepted when only portions have been well probed. Similarly, we can grant Earman's point that "in 1918 and 1919 physicists were in no position to be confident that the vast and then unexplored space of possible gravitational theories denoted by [not-GTR] does not contain alternatives to GTR that yield that same prediction for the bending of light as GTR" (Earman, 1992, p. 117), while asking why this shows our account of severity is too severe rather than being a point in its favor. It seems to us that being prohibited from regarding GTR as having passed severely, at that stage, is just what an account ought to do. At the same time, the existence of what Earman aptly dubs a "zoo of alternatives" to GTR did not prevent scientists from severely probing and passing claims about light-bending and, more generally, extending their knowledge of gravity. We shall return to consider GTR later.

1.4 The Challenge

We welcome the call to provide the "life of experiment" with a corresponding "life of theory": the challenge leads to extending the experimental testing account into that arena in ways that we, admittedly, had not been sufficiently clear about or had not even noticed. In particular, taking up the large-scale theory challenge leads to filling in some gaps regarding the issues of (1) how far a severity assessment can extend beyond the precise experimental domain tested and (2) what can be said regarding hypotheses and claims that *fail* to have passed severe tests. Regarding the first issue, we argue that we can inductively infer the absence of any error that has been well probed and ruled out with severity. Although "*H* is false" refers to a specific error, this may and should encompass erroneous claims about underlying causes and mistaken understandings of any testable aspect of a phenomenon of interest. Concerning the second issue, we wish to explore the value of understanding why evidence may prohibit inferring a full theory severely – how it helps in systematically setting out rivals and partitioning the ways we can be in error regarding the claims that have so far agreed with data.

Thus, we accept the challenge in the epigraph, but in addition wish to "raise the stakes" on what an adequate account of theory appraisal should provide. More than affording an after-the-fact reconstruction of past cases of theory appraisal, an adequate account should give forward-looking methods for making progress in both building and appraising theories. We begin in Section 2 by considering the severity account of evidence; then in Section 3, we consider some implications for high-level theory. In Section 4, we examine and reject the "comparativist rescue" and in Section 5, we take up the case of theory testing of GTR. Our issue – let me be clear at the outset – is not about whether to be a realist about theories; in fact the same criticisms are raised by philosophers on both sides of this divide. Thus, in what follows we try to keep to language used by realists and nonrealists alike.

2 Error-Statistical Account of Evidence

2.1 Severity Requirement

Let us begin with a very informal example. Suppose we are testing whether and how much weight has been gained between now and the time George left for Paris, and we do so by checking if any difference shows up on a series of well-calibrated and stable weighing methods, both before his leaving and upon his return. If no change on any of these scales is registered, even though, say, they easily detect a difference when he lifts a 0.1-pound potato, then this may be regarded as grounds for inferring that George's weight gain is negligible within limits set by the sensitivity of the scales. The hypothesis *H* here might be that George's weight gain is no greater than δ, where δ is an amount easily detected by these scales. *H*, we would say, has passed a severe test: were George to have gained δ pounds or more (i.e., were *H* false), then this method would almost certainly have detected this. Clearly *H* has been subjected to, and has passed, a more stringent test than if, say, *H* were inferred based solely on his still being able to button elastic-waist pants. The same reasoning abounds in science and statistics (p. 256).

Consider data on light-bending as tests of the deflection effect λ given in Einstein's GTR. It is clear that data based on very long baseline radio interferometry (VLBI) in the 1970s taught us much more about, and provided much better evidence for the Einsteinian-predicted light deflection (often set these days at 1) than did the passing result from the celebrated 1919 eclipse tests. The interferometry tests are far more capable of uncovering a variety of errors, and discriminating values of the deflection, λ, than were the crude eclipse tests. Thus, the results set more precise bounds on how far a gravitational theory can differ from the GTR value for λ. Likewise, currently-planned laser interferometry tests would probe discrepancies even more severely than any previous tests.

We set out a conception of evidence for a claim or hypothesis H:

Severity Principle (SP): Data **x** (produced by process G) provides a good indication or evidence for hypothesis H if and only if **x** results from a test procedure T which, taken as a whole, constitutes H having passed a severe test – that is, a procedure which would have, at least with very high probability, uncovered the falsity of, or discrepancies from H, and yet no such error is detected.

Instead, the test produces results that are in accord with (or fit) what would be expected under the supposition that H is correct, as regards the aspect probed.

While a full explication of severity is developed throughout this volume (e.g., introductory chapter), we try to say enough for current purposes. To begin with, except for formal statistical contexts, "probability" here may serve merely to pay obeisance to the fact that all empirical claims are strictly fallible. Take, for example, the weighing case: if the scales work reliably and to good precision when checked on test objects with known weight, we would ask, rightly, what sort of extraordinary circumstance could cause them to all go systematically astray just when we do not know the weight of the test object (George)? We would infer that his weight gain does not exceed such-and-such amount, without any explicit probability model.[1] Indeed, the most forceful severity arguments usually do not require explicit reference to probability or statistical models. We can retain the probabilistic definition of severity so long as it is kept in mind that it covers this more informal use of the term. Furthermore, the role of probability where it does arise, it is important to see, is not to assign degrees of confirmation or support or belief to hypotheses but to characterize how frequently methods are capable of detecting and discriminating errors, called error frequencies or *error probabilities*. Thus, an account of evidence broadly based on error probabilities may be called an *error-statistical account*, and a philosophy of science based on this account of evidence may be called an error-statistical philosophy of science (see Introduction and Background, Part II).

---

[1] Even in technical areas, such as in engineering, it is not uncommon to work without a well-specified probability model for catastrophic events. In one such variation, H is regarded as having passed a severe test if an erroneous inference concerning H could result only under extraordinary circumstances. (Ben-Haim, 2001, p. 214)

The severe test reasoning corresponds to a variation of an "argument from error" (p. 24):

Argument from Error: There is evidence that an error is absent when a procedure of inquiry with a high probability of detecting the error's presence nevertheless regularly yields results in accord with no error.

By "detecting" an error, we mean it "signals the presence of" an error; we generally do not know from the observed signal whether it has correctly done so. Since any inductive inference could be written as inferring the absence of an error of some type, the argument from error is entirely general. Formal error-statistical tests provide tools to ensure that errors will be correctly detected (i.e., signaled) with high probabilities.[2]

2.2 Some Further Qualifications

The simple idea underlying the severity principle (SP), once unpacked thoroughly, provides a very robust concept of evidence. We make some quick points of most relevance to theory testing: Since we will use T for theory, let E denote an experimental test.[3] First, although it is convenient to speak of a severe test E, it should be emphasized that E may actually, and usually does, combine individual tests and inferences together; likewise, data $\mathbf{x}$ may combine results of several tests. So long as one is explicit about the test E being referred to, no confusion results. Second, a severity assessment is a function of a particular set of data or evidence $\mathbf{x}$ and a particular hypothesis or claim. More precisely, it has three arguments: a test, an outcome or result, and an inference or a claim. "The severity with which H passes test E with outcome $\mathbf{x}$" may be abbreviated as SEV(Test E, outcome $\mathbf{x}$, claim H). When $\mathbf{x}$ and E are clear, we may write SEV(H). Defining severity in terms of three arguments is in contrast with a common tendency to speak of a "severe test" divorced from the specific inference at hand. This common tendency leads to fallacies we need to avoid. A test may be made so sensitive (or powerful) that discrepancies from a hypothesis H are inferred too readily. However, the severity associated with such an inference is *decreased* as test sensitivity

---

[2] Control of error rates, even if repetitions are hypothetical, allows the probativeness of *this* test to be assessed for reliably making *this* inference (see chapter 7). Nevertheless, low long-run error rates at individual stages of a complex inquiry (e.g., the error budgets in astronomic inferences) play an important role in the overall severity evaluation of a primary inference.

[3] Experiments, for us, do not require literal control; it suffices to be able to develop and critique arguments from error, which include the best practices in observational inquiries and model specification and validation. Nor need "thought experiments" be excluded.

increases (not the reverse). For example, we expect our interferometry test to yield some nonzero difference from the GTR prediction ($\lambda = 1$), the null hypothesis of the test, even if $\lambda = 1$. To interpret any observed difference, regardless of how small, as signaling a substantive discrepancy from the GTR prediction would be to infer a hypothesis with very *low* severity. That is because this test would very often purport to have evidence of a genuine discrepancy from $\lambda = 1$, even if the GTR prediction is correct (perhaps within a specified approximation).

The single notion of severity suffices to direct the interpretation and scrutiny of the two types of errors in statistics: erroneously rejecting a statistical (null) hypothesis h_0 – type I error – and erroneously failing to reject h_0 (sometimes abbreviated as "accepting" h_0) – type II error. The actual inference, H, will generally go beyond the stark formal statistical output. For example, from a statistical rejection of h_0, one might infer:

H: $\mathbf{x}$ is evidence of a discrepancy δ from h_0.

Then calculating SEV(H) directs one to consider the probability of a type I error.

If h_0 is not rejected, the hypothesis inferred might take the form:

H: $\mathbf{x}$ is evidence that any discrepancy from h_0 is less than δ.

Now the type II error probability (corresponding to δ) becomes relevant. Severity, as a criterion for evidence, avoids standard statistical fallacies due both to tests that are overly sensitive and to those insufficiently sensitive to particular errors and discrepancies (e.g., statistical vs. substantive differences; see Mayo, 1996; Mayo and Spanos, 2006).

Note that we always construe the question of evidence using testing language, even if it is described as an estimation procedure, because this is our general terminology for evidence, and any such question can be put in these terms. Also, the locution "severely tested" hypothesis H will always mean that H has *passed* the severe or stringent probe, not, for example, merely that H was subjected to one.

2.3 Models of Inquiry

An important ingredient of this account of testing is the insistence on avoiding oversimplifications of accounts that begin with statements of evidence and hypotheses overlooking the complex series of models required in inquiry, stretching from low-level theories of data and experiment to high-level hypotheses and theories. To discuss these different pieces, questions, or problems, we need a framework that lets us distinguish the steps involved in any realistic experimental inquiry and locate the necessary background information and the errors being probed – even more so when attempting to relate low-level tests to high-level theories. To organize these interconnected pieces, it helps to view any given inquiry as involving a *primary question* or *problem*, which is then embedded and addressed within one or more other models which we may call "experimental".[4] *Secondary questions* would include a variety of inferences involved in probing answers to the primary question (e.g., How well was the test run? Are its assumptions satisfied by the data in hand?). The primary question, couched in an appropriate experimental model, may be investigated by means of properly modeled data, not "raw" data. Only then can we adequately discuss the inferential move (or test) from the data (data model) to the primary claim H (through the experimental model E). Take the interferometric example. The primary question – determining the value of the GTR parameter, λ – is couched in terms of parameters of an astrometric model M which (combined with knowledge of systematic and nonsystematic errors and processes) may allow raw data, adequately modeled, to estimate parameters in M to provide information about λ (the deflection of light). We return to this in Section 5.

How to carve out these different models, each sometimes associated with a level in a hierarchy (e.g., Suppes, 1969) is not a cut-and-dried affair, but so long as we have an apparatus to make needed distinctions, this leeway poses no danger. Fortunately, philosophers of science have become increasingly aware of the roles of models in serving as "mediators," to use an apt phrase from Morrison and Morgan (1999), and we can turn to the central issue of this paper.[5]

3 Error-Statistical Account and Large-Scale Theory Testing

This localized, piecemeal testing does have something to say when it comes to probing large-scale theories, even if there is no intention to severely pass the entire theory. Even large-scale theories when we have them (in our account) are applied and probed only by a piecemeal testing of local

---

[4] This is akin to what Spanos calls the "estimable" model; see Chapter 6, this volume. See also note 3.
[5] Background knowledge, coming in whatever forms available – subject matter, instrumental, simulations, robustness arguments – enters to substantiate the severity argument. We think it is best to delineate such information within the relevant models rather than insert a great big "B" for "background" in the SEV relation, especially because these assumptions must be separately probed.

experimental hypotheses. Rival theories T_1 and T_2 of a given phenomenon or domain, even when corresponding to very different primary models (or rather, very different answers to primary questions), need to be applicable to the same data models, particularly if T_2 is to be a possible replacement for T_1. This constraint motivates the development of procedures for rendering rivals applicable to shared data models.

3.1 Implications of the Piecemeal Account for Large-Scale Testing

Several implications or groups of theses emerge fairly directly from our account, and we begin by listing them:

1. *Large-scale theories are not severely tested all at once.* To say that a given experiment E is a test of theory T is an equivocal way of saying that E probes what T says about a particular phenomenon or experimental effect (i.e., E attempts to discriminate the answers to a specific question, H). We abbreviate what theory T_i says about H as $T_i(H)$. This is consistent with the common scientific reports of "testing GTR" when in fact what is meant is that a particular aspect or parameter is going to be probed or delimited to a high precision. Likewise, the theory's passing (sometimes with "flying colors") strictly refers to the one piecemeal question or estimate that has passed severely (e.g., Will, 1993).

2. *A severity assessment is not threatened by alternatives at "higher levels."* If two rival theories, T_1 and T_2, say the same thing with respect to the effect or hypothesis H being tested by experimental test E (i.e., $T_1(H) = T_2(H)$), then T_1 and T_2 *are not rivals* with respect to experiment E. Thus, *a severity assessment can remain stable through changes in "higher level" theories*[6] or answers to different questions. For example, the severity with which a parameter is determined may remain despite changing interpretations about the cause of the effect measured (see Mayo, 1997b).

3. *Severity discriminates between theories that "fit" the data equally well.* T_1 is discriminated from T_2 (whether known, or a "beast lurking in the bush"[7]) by identifying and testing experimental hypotheses on which they disagree (i.e., where $T_1(H) \neq T_2(H)$). Even though *two rival hypotheses might "fit" the data equally well, they will not generally be equally severely tested by experimental test E.*

---

[6] Here we follow Suppes (1969) in placing the models in a vertical hierarchy from the closest to the farthest from data.

[7] We allude here to a phrase in Earman (1992).

The preceding points, as we will see, concern themselves with *reliability*, *stability*, and avoidance of serious *underdetermination*, respectively.

3.2 Contrast with a Bayesian Account of Appraisal

At this point, it is useful to briefly contrast these consequences with an approach, better known among philosophers to the inductive appraisal of hypotheses: the Bayesian approach. Data **x** may be regarded as strong evidence for, or as highly confirming of, theory *T* so long as the posterior probability of *T* given **x** is sufficiently high (or sufficiently higher than the prior probability in *T*),[8] where probability is generally understood as a measure of degree of belief, and P(*T*|**x**) is calculated by means of Bayes's theorem:

$$P(T|\mathbf{x}) = P(\mathbf{x}|T)P(T)/[P(\mathbf{x}|T)P(T) + P(\mathbf{x}|\text{not-}T)P(\text{not-}T)]$$

This calculation requires an exhaustive set of alternatives to *T* and prior degree-of-belief assignments to each, and an assessment of the term P(**x**|not-*T*), for "not-*T*," the *catchall hypothesis*. That scientists would disagree in their degree-of-belief probability assignments is something accepted and expected at least by subjectivist Bayesians.[9]

In one sense, it is simplicity itself for a (subjective) Bayesian to confirm a full theory *T*. For a familiar illustration, suppose that theory *T* accords with data **x** so that P(**x**|*T*) = 1, and assume equal prior degrees of belief for *T* and not-*T*. If the data are regarded as very improbable given that theory *T* is false – if a low degree of belief, say *e*, is accorded to what may be called the *Bayesian catchall factor*, P(**x**|not-*T*) – then we get a high posterior probability in theory *T*; that is, P(*T*|**x**) = 1/(1 + *e*). The central problem is this: What warrants taking data **x** as incredible under any theory other than *T*, when these would include all possible rivals, including those not even thought of? We are faced with the difficulty Earman raised (see 1.3), and it also raises well-known problems for Bayesians.

High Bayesian support does not suffice for well-testedness in the sense of the severity requirement. The severity requirement enjoins us to consider this Bayesian procedure: basically, it is to go from a low degree of belief in the Bayesian catchall factor to inferring *T* as confirmed. One clearly cannot vouch for the reliability of such a procedure – that it would rarely affirm theory *T* were *T* false – in contrast to point 1 above. Similar problems

---

[8] Several related measures of Bayesian confirmation may be given. See, for example, Good (1983).

[9] Some might try to assign priors by appealing to ideas about simplicity or information content, but these have their own problems (e.g., Cox, 2006; Kass and Wasserman, 1996). See Chapter 7, pp. 298–302.

confront the Bayesian dealing with data that are anomalous for a theory T (e.g., in confronting Duhemian problems). An anomaly x' warrants Bayesian disconfirmation of an auxiliary hypothesis A (used to derive prediction x), so long as the prior belief in T is sufficiently high and the Bayesian catchall factor is sufficiently low (see, e.g., Dorling, 1979). The correctness of hypothesis A need not have been probed in its own right. For example, strictly speaking, believing more strongly in Newton's than in Einstein's gravitational theory in 1919 permits the Bayesian to blame the eclipse anomaly on, say, a faulty telescope, even without evidence for attributing blame to the instrument (see Mayo, 1997a; Worrall, 1993; and Chapters 4 and 8, this volume).

Consider now the assurance about stability in point 2. Operating with a "single probability pie," as it were, the Bayesian has the difficulty of redistributing assignments if a new theory is introduced. Finally, consider the more subtle point 3. For the Bayesian, two theories that "fit" the data x equally well (i.e., have identical likelihoods) are differentially supported only if their prior probability assignments differ. This leads to difficulties in capturing methodological strictures that seem important in discriminating two equally well-fitting hypotheses (or even the same hypothesis) based on the manner in which each hypothesis was constructed or selected for testing. We return to this in Section 5. Further difficulties are well known (e.g., the "old evidence problem," Glymour, 1980; Kyburg, 1993) but will not be considered.

I leave it to Bayesians to mitigate these problems, if problems they be for the Bayesian. Of interest to us is that it is precisely to avoid these problems, most especially consideration of the dreaded catchall hypothesis and the associated prior probability assignments, that many are led to a version of a comparativist approach (e.g., in the style of Popper or Lakatos).

3.3 The Holist–Comparativist Rescue

One can see from my first point in Section 3.1 why philosophers who view progress in terms of large-scale theory change are led to advocate a comparative testing account. Because a large-scale theory may, at any given time, contain hypotheses and predictions that have not been probed at all, it would seem impossible to say that such a large-scale theory had severely passed a test as a whole.[10] A comparative testing account, however, would

---

[10] Note how this lets us avoid tacking paradoxes: Even if H has passed severely with data x, if x fails to probe hypothesis J, then x fails to severely pass H and J (see Chalmers, 1999). By contrast, Bayesians seem content to show that x confirms the irrelevant conjunction less strongly than the conjunct (see Chapter 8, this volume). For a recent discussion and references, see Fitelson (2002).

allow us to say that the theory is best tested so far, or, using Popperian terms, we should "prefer" it so far. Note that their idea is not merely that testing should be comparative – the severe testing account, after all, tests *H* against its denial within a given model or space – but rather that testing, at least testing large-scale theories, may and generally will be a comparison between *nonexhaustive* hypotheses or theories. The comparativist reasoning, in other words, is that since we will not be able to test a theory against its denial (regarded as the "catchall hypothesis"), we should settle for testing it against one or more *existing* rivals. Their position, further, is that one may regard a theory as having been well or severely tested as a whole, so long as it has passed more or better tests than its existing rival(s). To emphasize this we will allude to it as a *comparativist-holist* view:

The comparativist . . . insists on the point, which [Mayo] explicitly denies, that testing or confirming one "part" of a general theory provides, defeasibly, an evaluation of all of it. (Laudan, 1997, p. 315)

Alan Chalmers maintains, in an earlier exchange, that we must already be appealing to something akin to a Popperian comparativist account:

[Mayo's] argument for scientific laws and theories boils down to the claim that they have withstood severe tests better than any available competitor. The only difference between Mayo and the Popperians is that she has a superior version of what counts as a severe test. (Chalmers, 1999, p. 208)

Amalgamating Laudan and Chalmers's suggestions for "comparativist–holism" gives the following:

Comparativist (Holist) Testing: A theory has been well or *severely tested* provided that it has survived (local) severe tests that its known rivals have failed to pass (and not vice versa).

We argue that the comparativist–holist move is no rescue but rather conflicts with the main goals of the severity account, much as the Bayesian attempt does. We proceed by discussing a cluster of issues relating to the points delineated in Section 3.1.

4 Comparing Comparativists with Severe Testers

4.1 Point 1: Best Tested Does Not Entail Well Tested

One cannot say about the comparatively best-tested theory what severity requires – that the ways the theory or claim can be in error have been well-probed and found to be absent (to within the various error margins of the test). It seems disingenuous to say all of theory *T* is well tested (even to a degree) when it is known there are ways T can be wrong that have received no scrutiny or that there are regions of implication not checked at all. Being best tested is relative not only to existing theories but also to existing tests: they may all be poor tests for the inference to T as a whole. One is back to a problem that beset Popper's account – namely, being unable to say "What is so good about the theory that (by historical accident) happens to be the best tested so far?" (Mayo, 2006, p. 92).

Whereas we *can* give guarantees about the reliability of the piecemeal experimental test, we *cannot* give guarantees about the reliability of the procedure advocated by the comparativist-holist tester. Their procedure is basically to go from passing hypothesis H (perhaps severely in its own right) to passing all of T – but this is a highly *un*reliable method; anyway, it is unclear how one could assess its reliability. By contrast, we can apply the severity idea because the condition "given H is false" (even within a larger theory) always means "given H is false with respect to what T says about *this particular* effect or phenomenon" (i.e., $T(H)$).[11] If a hypothesis $T(H)$ passes a severe test we can infer something positive: that T gets it right about the specific claim H, or that given errors have been reliably ruled out. This also counts as evidence against any rival theory that conflicts with $T(H)$.

Granted, it may often be shown that ruling out a given error is connected to, and hence provides evidence for, ruling out others. The ability to do so is a very valuable and powerful way of cross-checking and building on results. Sometimes establishing these connections is achieved by using theoretical background knowledge; other times sufficient experimental knowledge will do. But whether these connections are warranted is an empirical issue that has to be looked into on a case-by-case basis – whereas the comparativist-holist would seem to be free of such an obligation, so long as theory T is the best tested so far. Impressive "arguments from coincidence" from a few successful hypotheses to the entire theory must be scrutinized for the case in hand. We return to this in Chapter 2.

Rational Acceptability. It is not that we are barred from finding a theory T "rationally acceptable," preferred, or worthy of pursuit – locutions often used by comparativists – upon reaching a point where T's key experimental predictions have been severely probed and found to pass. One could infer that T had solved a set of key experimental problems and take this as

---

[11] It is important to see that the severity computation is not a conditional probability, which would implicitly assume prior probability assignments to hypotheses which severity does not assume. Rather, severity should be understood as the probability of so good an agreement (between H and x) *calculated under the assumption that H is false.*

grounds for "deciding to pursue" it further. But these decisions are distinct from testing and would call for a supplement to what we are offering.[12]

As we see it, theories (i.e., theoretical models) serve a role analogous to experimental models in the tasks of learning from data. Just as experimental models serve to describe and analyze the relevance of any of the experimental data for the experimental phenomenon, theoretical models serve *to analyze the relevance of any of the experimental inferences (estimates and tests) for the theoretical phenomenon.* If a theory T_2 is a viable candidate to take the place of rival T_1, then it must be able to *describe and analyze the significance of the experimental outcomes that T_1 can.* We come back to this in considering GTR. We should be concerned, too, by the threat to the *stability* of severity assessments that the comparativist account would yield – the second point in Section 3.1.

4.2 Point 2: Stability

Suppose an experimental test E is probing answers to the question: What is the value of a given parameter λ? Then, if a particular answer or hypothesis severely passes, this assessment is not altered by the existence of a theory that gives the same answer to this question. More generally, our account lets us say that severely passing $T(H)$ (i.e., what T says about H) gives us experimental knowledge about this aspect of T, and this assessment remains even through improvements, revisions, and reinterpretations of that knowledge. By contrast, the entrance of a rival that passes all the tests T *does* would seem to force the comparativist to change the assessment of how well theory T had been tested.

On the severity account, if a rival theory T_2 agrees with T_1 with respect to the effect or prediction under test, then the two theories are not rivals *so far as this experimental test is concerned* – no matter how much they may differ from each other in their full theoretical frameworks or in prediction ranges not probed by the experimental test E. It is very important to qualify this claim. Our claim is not that two theories fail to be rivals just because the test is insufficiently sensitive to discriminate what they say about the phenomenon under test; our claim is that they fail to be rivals when the two say exactly the same thing with respect to the effect or hypothesis under test.[13] The severity assessment reflects this. If theory T_1 says exactly the

---

[12] Larry Laudan (1977) himself has always stressed that we should distinguish theory pursuit from other stances one might take toward theories.

[13] Of course, determining this might be highly equivocal, but that is a distinct matter.

same thing about H as T_2 – that is, $(T_1(H) = T_2(H))$ – then T_2 cannot alter the severity with which the test passes H.[14] Note, though, that this differs from saying $T_1(H)$ and $T_2(H)$ pass with equal severity. We consider this argument in Section 4.3.

4.3 Point 3: Underdetermination

Point 3 refers to a key principle of error statistics, which is also the basis for solving a number of philosophical problems. It is often argued that data underdetermine hypotheses because data may equally well warrant conflicting hypotheses according to one or another base measure of evidential relationship. However, we can distinguish, on grounds of severity, the well-testedness of two hypotheses and thereby get around underdetermination charges. We take this up elsewhere (e.g., Mayo, 1997b). Here our interest is in how the feature in point 3 bears on our question of moving from low-level experimental tests to higher level theories. In particular, two hypotheses may be nonrivals (relative to a primary question) and yet be tested differently by a given test procedure – indeed the same hypothesis may be better- or less-severely tested by means of (what is apparently) the "same" data because of aspects of either the data generation or the hypothesis construction procedure.

We can grant, for example, that a rival theory could always be erected to accommodate the data, but a key asset of the error-statistical account is its ability to distinguish the well-testedness of hypotheses and theories by the reliability or severity of the accommodation method. Not all fits are the same. Thus, we may be able to show, by building on individual hypotheses, that one theory *at some level* (in the series or models) or a close variant to this theory, severely passes. In so doing, we can show that no rival to this theory can also severely pass.

Admittedly, all of this demands an examination of the detailed features of the recorded data (the data models), not just the inferred experimental effect or phenomenon. It sounds plausible to say there can always be some rival, when that rival merely has to "fit" already-known experimental effects. The situation is very different if one takes seriously the constraints imposed

---

[14] Mistakes in regarding H as severely passed can obviously occur. A key set of challenges comes from those we group under "experimental assumptions." Violated assumptions may occur because the actual experimental data do not satisfy the assumptions of the experimental model or because the experimental test was not sufficiently accurate or precise to reliably inform about the primary hypothesis or question. Of course, "higher-lower" is just to distinguish primary questions; they could be arranged horizontally.

by the information in the detailed data coupled with the need to satisfy the severity requirement.

Finally, nothing precludes the possibility that so-called low-level hypotheses *could* warrant inferring a high-level theory with severity. Even GTR, everyone's favorite example, is thought to predict a unique type of gravitational radiation, such that affirming that particular "signature" with severity would rule out all but GTR (in its domain). With this tantalizing remark, let us look more specifically at the patterns of progress in experimental GTR.

5 Experimental Gravitation

This example is apt for two reasons. First, it is an example to which each of the philosophers we have mentioned allude in connection with the problem of using local experimental tests for large-scale theories. Second, the fact that robust or severe experiments on gravitational effects are so hard to come by led physicists to be especially deliberate about developing a theoretical framework in which to discuss and analyze rivals to GTR and to compare the variety of experiments that might enable their discrimination. To this end, they developed a kind of *theory of theories* for delineating and partitioning the space of alternative gravity theories, called the parameterized post-Newtonian (PPN) framework. The only philosopher of science to discuss the PPN framework in some detail, to my knowledge, is John Earman; although the program has been updated and extended since his 1992 discussion, the framework continues to serve in much the same manner. What is especially interesting about the PPN framework is its role in *inventing* new classes of rivals to GTR, beyond those that are known. It points to an activity that any adequate account of theories should be able to motivate, if it is to give forward-looking methods for making theoretical progress rather than merely after-the-fact reconstructions of episodes. Popperians point out that Popper had always advocated looking for rivals as part of his falsification mandate. Granted, but neither he nor the current-day critical rationalists supply guidance for developing the rivals or for warranting claims about where hypotheses are likely to fail if false – eschewing as they do all such inductivist claims about reliable methods (see Mayo, 2006).[15]

Experimental testing of GTR nowadays is divided into four periods: 1887–1919, 1920–1960, 1960–1980, and 1980 onward. Following Clifford

---

[15] Popper's purely deductive account is incapable, by his own admission, of showing the reliability of a method.

Will, the first is the period of *genesis*, which encompasses experiments on (1) the foundations of relativistic physics (Michelson-Morley and the Eötvös experiments) and the GTR tests on (2) the deflection of light and perihelion of Mercury (for excellent discussions, see Will, 1980, 1986, 1996, 2004). From the comparativist's perspective, 1920–1960 would plainly be an era in which GTR enjoyed the title of "best-tested" theory of gravity: it had passed the "classical" tests to which it had been put and no rival existed with a superior testing record to knock it off its pedestal. By contrast, from 1960 to 1980, a veritable "zoo" of rivals to GTR had been erected, all of which could be constrained to fit these classical tests. So in this later period, GTR, from the comparativist's perspective, would have fallen from its pedestal, and the period might be regarded as one of crisis, threatening progress or the like. But in fact, the earlier period is widely regarded (by experimental gravitation physicists) as the period of "stagnation," or at least "hibernation," due to the inadequate link up between the highly mathematical GTR and experiment. The later period, by contrast, although marked by the zoo of alternatives, is widely hailed as the "golden era" or "renaissance" of GTR.

The golden era came about thanks to events of 1959–1960 that set the stage for new confrontations between GTR's predictions and experiments. Nevertheless, the goals of this testing were not to decide if GTR was correct in all its implications, but rather, in the first place, to learn more about GTR (i.e., what does it really imply about experiments we can perform?) and, in the second place, to build models for phenomena that involve relativistic gravity (e.g., quasars, pulsars, gravity waves, and such). The goal was *to learn more about gravitational phenomena*.

Comparativist testing accounts, eager as they are to license the entire theory, ignore what for our severe tester is the central engine for making progress, for getting ideas for fruitful things to try next to learn more. This progress turned on distinguishing those portions of GTR that were and were not well tested. Far from arguing for GTR on the grounds that it had survived tests that existing alternatives could not, as our comparativist recommends, our severe tester would set about exploring just *why* we are *not* allowed to say that GTR is severely probed as a whole – in all the arenas in which gravitational effects may occur. Even without having full-blown alternative theories of gravity in hand we can ask (as they did in 1960): *How could it be a mistake to regard the existing evidence as good evidence for GTR?* Certainly we could be wrong with respect to predictions and domains that were not probed at all. But how could we be wrong even with respect to what GTR says about the probed regions, in particular, solar system tests? One must begin where one is.

Table 1.1. *The PPN parameters and their significance*

Parameter	What it measures relative to GTR	Values in GTR
λ	How much space-curvature produced by unit rest mass?	1
β	How much "nonlinearity" in the superposition law for gravity?	1
ξ	Preferred location effects?	0
α_1	Preferred frame effects?	0
α_2		0
α_3		0
α_3	Violation of conservation of total momentum?	0
ζ_1		0
ζ_2		0
ζ_3		0

Source: Adapted from Will (2005).

To this end, experimental relativists deliberately designed the PPN frame-work to prevent them from being biased toward accepting GTR prematurely (Will, 1993, p. 10), while allowing them to describe violations of GTR's hypotheses – discrepancies from what it said about specific gravitational phenomena in the solar system. The PPN framework set out a list of param-eters that allowed for a systematic way of describing violations of GTR's hypotheses. These alternatives, by the physicists' own admissions, were set up largely as straw men with which to set firmer constraints on these param-eters. The PPN formalism is used to get *relativistic* predictions rather than those from Newtonian theory – but in a way that is not biased toward GTR. It gets all the relativistic theories of gravity talking about the same things and to connect to the same data models (Mayo, 2002).

The PPN framework is limited to probing a portion or variant of GTR (see Table 1.1):

The PPN framework takes the slow motion, weak field, or post-Newtonian limit of metric theories of gravity, and characterizes that limit by a set of 10 real-valued parameters. Each metric theory of gravity has particular values for the PPN param-eters. (Will, 1993, p. 10)

The PPN framework permitted researchers to compare the relative merits of various experiments ahead of time in probing the solar system approxima-tion, or solar system variant, of GTR. Appropriately modeled astronomical data supply the "observed" (i.e., estimated) values of the PPN parameters, which could then be compared with the different values hypothesized by the diverse theories of gravity. This permitted the same PPN models of experiments to serve as intermediate links between the data and several alternative primary hypotheses based on GTR and its rival theories.

This mediation was a matter of measuring, or more correctly *inferring*, the values of PPN parameters by means of complex, statistical least-square fits to parameters in models of data. Although clearly much more would need to be said to explain how even one of the astrometric models is developed to design what are described as "high-precision null experiments," it is interesting to note that, even as the technology has advanced, the overarching reasoning shares much with the classic interferometry tests (e.g., those of Michelson and Morley). The GTR value for the PPN parameter under test serves as the null hypothesis from which discrepancies are sought. By identifying the null with the prediction from GTR, any discrepancies are given a very good chance to be detected; so, if no significant departure is found, this constitutes evidence for the GTR prediction with respect to the effect under test. Without warranting an assertion of zero discrepancy from the null GTR value (set at 1 or 0), the tests are regarded as ruling out GTR violations exceeding the bounds for which the test had very high probative ability. For example, λ, the deflection-of-light parameter, measures "spatial curvature;" by setting the GTR predicted value to 1, modern tests infer upper bounds to violations (i.e., $|1 - \lambda|$). (See "Substantive Nulls," this volume, p. 264.)

Some elements of the series of models for the case of λ are sketched in Table 1.2.

The PPN framework is more than a bunch of parameters; it provides a general way to interpret the significance of the piecemeal tests for primary gravitational questions, including deciding to which questions a given test discriminates answers. Notably, its analysis revealed that one of the classic tests of GTR (redshift) "was not a true test" of GTR but rather tested the *equivalence principle* – roughly the claim that bodies of different composition fall with the same accelerations in a gravitational field. This principle is inferred with severity by passing a series of null hypotheses (e.g., Eötvös experiments) that assert a zero difference in the accelerations of two differently composed bodies. The high precision with which these null hypotheses passed gave warrant to the inference that "gravity is a phenomenon of curved spacetime, that is, it must be described by a metric theory of gravity" (Will, 1993, p. 10).

For the comparativist, the corroboration of a part of GTR, such as the equivalence principle, is regarded as corroborating, defeasibly, GTR as a whole. In fact, however, corroborating the equivalence principle is recognized only as discriminating between so-called metric versus nonmetric gravitational theories, e.g., those gravity theories that do, versus those that

Table 1.2. *Elements of the series of models for the case of* λ

PRIMARY: Testing the post-Newtonian approximation of GTR
Parameterized post-Newtonian (PPN) formalism
Delineate and test predictions of the metric theories using the PPN parameters
Use estimates to set new limits on PPN parameters and on adjustable parameters in
  alternatives to GTR
Example: For λ, how much spatial curvature does mass produce?

EXPERIMENTAL MODELS: PPN parameters are modeled as statistical null
  hypotheses (relating to models of the experimental source)
Failing to reject the null hypothesis (identified with the GTR value) leads to setting
  upper and lower bounds, values beyond which are ruled out with high severity
Example: hypotheses about λ in optical and radio deflection experiments

---

DATA: Models of the experimental source (eclipses, quasar, moon, earth–moon system,
  pulsars, Cassini)
Least-squares fits of several parameters, one of which is a function of the observed
  statistic and the PPN parameter of interest (the function having known distribution)
Example: least-squares estimates of λ from "raw" data in eclipse and radio
  interferometry experiments.

DATA GENERATION AND ANALYSIS, EXPERIMENTAL DESIGN
Many details which a full account should include.

---

do not, satisfy this fundamental principle. This recognition only emerged once it was realized that all metric theories say the same thing with respect to the equivalence principle. Following point 2 above, they were not rivals with respect to this principle. More generally, an important task was to distinguish classes of experiments according to the specific aspects each probed and thus tested. An adequate account of the role and testing of theories must account for this, and the comparativist–holist view does not. The equivalence principle itself, more correctly called the Einstein equivalence principle, admitted of new partitions (e.g., into strong and weak, see later discussion), leading to further progress.[16]

---

[16] More carefully, we should identify the Einstein equivalence principle (EEP) as well as distinguish weak and strong forms; the EEP states that (1) the weak equivalence principle (WEP) is valid; (2) the outcome of any local nongravitational experiment is independent of the velocity of the freely falling reference frame in which it is performed (Lorentz invariance); and (3) the outcome of any local nongravitational experiment is independent of where and when in the universe it is performed (local position invariance). A subset of metric theories obeys a stronger principle, the strong equivalence principle (SEP). The SEP asserts that the stipulation of the equivalence principle also hold for self-gravitating bodies, such as the earth–moon system.

5.1 Criteria for a Viable Gravity Theory (during the "Golden Era")

From the outset, the PPN framework included not all logically possible gravity theories but those that passed the criteria for *viable* gravity theories.

   (i) *It must be complete*, i.e., it must be capable of analyzing from "first principles" the outcome of any experiment of interest. It is not enough for the theory to *postulate* that bodies made of different material fall with the same acceleration . . . [This does not preclude "arbitrary parameters" being required for gravitational theories to accord with experimental results.]
   (ii) *It must be self-consistent*, i.e., its prediction for the outcome of every experiment must be unique, so that when one calculates the predictions by two different, though equivalent methods, one always gets the same results . . .
   (iii) *It must be relativistic*, i.e., in the limit as gravity is 'turned off' . . . the nongravitational laws of physics must reduce to the laws of special relativity . . .
   (iv) *It must have the correct Newtonian limit*, i.e., in the limit of weak gravitational fields and slow motions, it must reproduce Newton's laws . . . (Will, 1993, pp. 18–21).

From our perspective, viable theories must (1) account for experimental results already severely passed and (2) show the significance of the experimental data for gravitational phenomena.[17] Viable theories would have to be able to analyze and explore experiments about as well as GTR; there is a comparison here but remember that what makes a view "comparativist" is that it regards the full theory as well tested by dint of being "best tested so far." In our view, viable theories are required to pass muster for the goals to which they are put at this stage of advancing the knowledge of gravitational effects. One may regard these criteria as intertwined with the "pursuit" goals – that a theory should be useful for testing and learning more.

The experimental knowledge gained permits us to infer that we have a correct parameter value – but in our view it does more. It also indicates we have a correct understanding of how gravity behaves in a given domain. Different values for the parameters correspond to different mechanisms,

---

[17] Under consistency, it is required that the phenomenon it predicts be detectable via different but equivalent procedures. Otherwise they would be idiosyncratic to a given procedure and would not give us genuine, repeatable phenomena.

however abstract, at least in viable theories. For example, in the Brans–Dicke theory, gravity couples both to a tensor metric and a scalar, and the latter is related to a distinct metaphysics (Mach's principle). Although theoretical background is clearly what provides the interpretation of the relevance of the experimental effects for gravity, no one particular theory needs to be accepted to employ the PPN framework – which is at the heart of its robustness. Even later when this framework was extended to include nonmetric theories (in the fourth period, labeled "the search for strong gravitational effects"), those effects that had been vouchsafed with severity remain (although they may well demand reinterpretations).

5.2 Severity Logic and Some Paradoxes regarding Adjustable Constants

Under the completeness requirement for viable theories there is an explicit caveat that this does not preclude "arbitrary parameters" from being necessary for gravitational theories to obtain correct predictions, even though these are deliberately set to fit the observed effects and are not the outgrowth of "first principles." For example, the addition of a scalar field in Brans–Dicke theory went hand-in-hand with an adjustable constant w: the smaller its value the larger the effect of the scalar field and thus the bigger the difference with GTR, but as w gets larger the two become indistinguishable. (An interesting difference would have been with evidence that w is small, such as 40; its latest lower bound is pushing 20,000!) What should we make of the general status of the GTR rivals, given that their agreement with the GTR predictions and experiment required adjustable constants? This leads us to the general and much debated question of when and why data-dependent adjustments of theories and hypotheses are permissible.

The debate about whether to require or at least prefer (and even how to define) "novel" evidence is a fascinating topic in its own right, both in philosophy of science and statistics (Mayo, 1991, 1996), and it comes up again in several places in this volume (e.g., Chapters 4, 6, and 7); here, we consider a specific puzzle that arises with respect to experimental GTR. In particular, we consider how the consequences of severity logic disentangle apparently conflicting attitudes toward such "data-dependent constructions." Since all rivals were deliberately assured of fitting the effects thanks to their adjustable parameters, whereas GTR required no such adjustments, intuitively we tend to think that GTR was better tested by dint of its agreement with the experimental effects (e.g., Worrall, 1989). This leads the comparativist to reject such parameter adjustments. How then to explain the permissive attitude toward the adjustments in experimental GTR? The comparativist cannot have it both ways.

By contrast, Bayesians seem to think they can. Those who wish to justify differential support look for it to show up in the prior probabilities, since all rivals fit the observed effects. Several Bayesians (e.g., Berger, Rosenkrantz) postulate that a theory that is free of adjustable parameters is "simpler" and therefore enjoys a higher prior probability; this would explain giving GTR higher marks for getting the predictions right than the Brans–Dicke theory or other rivals relying on adjustments (Jeffreys and Berger, 1992). But to explain why researchers countenance the parameter-fixing in GTR alternatives, other Bayesians maintain (as they must) that GTR should *not* be given a higher prior probability. Take Earman: "On the Bayesian analysis," this countenancing of parameter fixing "is not surprising, since it is not at all clear that GTR deserves a higher prior than the constrained Brans and Dicke theory" (Earman, 1992, p. 115). So Earman denies differential support is warranted in cases of parameter fixing ("why should the prior likelihood of the evidence depend upon whether it was used in constructing T?"; Earman, 1992, p. 116), putting him at odds with the Bayesian strategy for registering differential support (by assigning lower priors to theories with adjustable constants).

The Bayesian, like the comparativist, seems to lack a means to reflect, with respect to the *same* example, both (a) the intuition to give less credit to passing results that require adjustable parameters and (b) the accepted role, in practice, of deliberately constrained alternatives that are supported by the *same data* doing the constraining. Doubtless ways may be found, but would they avoid "ad hoc-ness" and capture what is actually going on?

To correctly diagnose the differential merit, the severe testing approach instructs us to consider the particular inference and the ways it can be in error in relation to the corresponding test procedure. There are two distinct analyses in the GTR case. First consider λ. The value for λ is fixed in GTR, and the data could be found to violate this fixed prediction by the procedure used for estimating λ (within its error margins). By contrast, in adjusting w, thereby constraining Brans–Dicke theory to fit the estimated λ, what is being learned regarding the Brans–Dicke theory is *how large would w need to be* to agree with the estimated λ? In this second case, inferences that pass with high severity are of the form "*w* must be at least 500." The questions, hence the possible errors, hence the severity differs.

But the data-dependent GTR alternatives play a second role, namely to show that GTR has not passed severely as a whole: They show that were a rival account of the mechanism of gravity correct, existing tests would not have detected this. In our view, this was the major contribution provided by the rivals articulated in the PPN framework (of viable rivals to GTR). Even without being fully articulated, they effectively block GTR from having passed with severity as a whole (while pinpointing why). Each GTR rival gives different underlying accounts of the behavior of gravity (whether one wishes to call them distinct "mechanisms" or to use some other term). This space of rival explanations may be pictured as located at a higher level than the space of values of this parameter (Table 1.2). Considering the λ effect, the constrained GTR rivals succeed in showing that the existing experimental tests did not rule out, with severity, alternative explanations for the λ effect given in the viable rivals.[18] But the fact that a rival, say Brans–Dicke theory, served to block a high-severity assignment to GTR, given an experiment E, is not to say that E accords the rival high severity; it does not.

5.3 Nordvedt Effect η

To push the distinctions further, the fact that the rival Brans–Dicke theory is not severely tested (with E) is not the same as evidence against it (the severity logic has all sorts of interesting consequences, which need to be drawn out elsewhere). Evidence against it came later. Most notably, a surprise discovery in the 1960s (by Nordvedt) showed that Brans–Dicke theory would conflict with GTR by predicting a violation of what came to be known as the strong equivalence principle (basically the weak equivalence principle for massive self-gravitating bodies, e.g., stars and planets; see Note 16). This recognition was welcomed (apparently, even by Dicke) as a new way to test GTR as well as to learn more about gravity experiments.

Correspondingly, a new parameter to describe this effect, the Nordvedt effect, was introduced into the PPN framework (i.e., η). The parameter η would be 0 for GTR, so the null hypothesis tested is that $\eta = 0$ as against $\eta \neq 0$ for rivals. Measurements of the round-trip travel times between the Earth and the Moon (between 1969 and 1975) enabled the existence of such an anomaly for GTR to be probed severely (the measurements continue today). Again, the "unbiased, theory-independent viewpoint" of the PPN framework (Will, 1993, p. 157) is credited with enabling the conflicting prediction to be identified. Because the tests were sufficiently sensitive, these measurements provided good evidence that the Nordvedt effect is absent, set upper bounds to the possible violations, and provided evidence

---

[18] Another way to see this is that the Brans–Dicke effect blocks high severity to the hypothesis about the specific nature of the gravitational cause of curvature – even without its own mechanism passing severely. For this task, they do not pay a penalty for accommodation; indeed, some view their role as estimating cosmological constants, thus estimating violations that would be expected in strong gravity domains.

for the correctness of what GTR says with respect to this effect – once again instantiating the familiar logic.[19]

5.4 Another Charge We Need to Tackle

According to Mayo, a test, even a severe test, of the light-bending hypothesis leaves us in the dark about the ability of GTR to stand up to tests of different ranges of its implications. For instance, should GTR's success in the light-bending experiments lend plausibility to GTR's claims about gravity waves or black holes? Mayo's strictures about the limited scope of severity seem to preclude a positive answer to that question. (Laudan, 1997, p. 313)

In our view, there will not be a single answer, positive or negative. Whether *T*'s success in one part or range indicates it is likely to succeed (and to what extent) in another is an empirical question that must be answered on a case-by-case basis. Moreover, because this question seems to us to be the motivation for a good part of what scientists do in exploring theories, a single context-free answer would not even be desirable. But consider GTR: although one splits off the piecemeal tests, we do not face a disconnected array of results; indeed the astrometric (experimental) models show that many of the parameters are functions of the others. For example, it was determined that the deflection effect parameter λ measures the same thing as the so-called time delay, and the Nordvedt parameter η gives estimates of several others. Because it is now recognized that highly precise estimates of λ constrain other parameters, λ is described as the fundamental parameter in some current discussions.

Putting together the interval estimates, it is possible to constrain the values of the PPN parameters and thus "squeeze" the space of theories into smaller and smaller volumes as depicted in Figure 1.1. In this way, entire chunks of theories can be ruled out at a time (i.e., all theories that predict the values of the parameter outside the interval estimates). By getting increasingly accurate estimates, more severe constraints are placed on how far theories can differ from GTR, in the respects probed. By 1980, it could be reported that "one can now regard solar system tests of post-Newtonian effects as measurements of the 'correct' values of these parameters" (Will, 1993).

---

[19] In the "secondary" task of scrutinizing the validity of the experiment, they asked, can other factors mask the η effect? Most, it was argued, can be separated cleanly from the η effect using the multiyear span of data; others are known with sufficient accuracy from previous measurements or from the lunar lasing experiment itself.

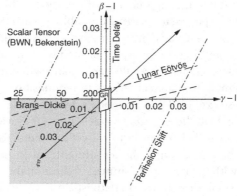

Figure 1.1.

5.5 Going beyond Solar System Tests

We can also motivate what happens next in this episode, although here I must be very brief. Progress is again made by recognizing the errors that are still not ruled out.

All tests of GTR within the solar system have this qualitative weakness: they say nothing about how the "correct" theory of gravity might behave when gravitational forces are very strong such as near a neutron star. (Will, 1996, p. 273)

The discovery (in 1974) of the binary pulsar 1913 + 16 opened up the possibility of probing new aspects of gravitational theory: the effects of gravitational radiation. Finding the decrease in the orbital period of this (Hulse-Taylor) binary pulsar at a rate in accordance with the GTR prediction of gravity wave energy loss is often regarded as the last event of the golden age. This example is fascinating in its own right, but we cannot take up a discussion here[20] (see Damour and Taylor, 1991; Lobo, 1996, pp. 212–15; Will, 1996).

There is clearly an interplay between theoretical and experimental considerations driving the program. For example, in the fourth and contemporary period, that of "strong gravity," a number of theoretical grounds indicate that GTR would require an extension or modification for strong gravitational fields – regions beyond the domains for which effects have been probed with severity. Although experimental claims (at a given level, as it

[20] For a brief discussion of how the hierarchy of models applies to the binary pulsar analysis, see Mayo (2000).

were) can remain stable through change of theory (at "higher" levels), it does not follow that experimental testing is unable to reach those theoretical levels. An error, as we see it, can concern any aspect of a model or hypothesis or mistaken understandings of an aspect of the phenomenon in question. For example, the severely tested results can remain while researchers consider alternative gravitational mechanisms in regimes not probed. Despite the latitude in these extended gravity models, by assuming only some general aspects on which all the extended models agree, they are able to design what are sometimes called "clean tests" of GTR; others, found sullied by uncertainties of the background physics, are entered in the logbooks for perhaps tackling with the next space shuttle![21] These analyses motivate new searches for very small deviations of relativistic gravity in the solar system that are currently present in the range of approximately 10^{-5}. Thus, probing new domains is designed to be played out in the solar system, with its stable and known results. This stability, however, does not go hand-in-hand with the kind of conservative attitude one tends to see in philosophies of theory testing: rather than hanker to adhere to well-tested theories, there seems to be a yen to find flaws potentially leading to new physics (perhaps a quantum theory of gravity).[22]

General relativity is now the "standard model" of gravity. But as in particle physics, there may be a world beyond the standard model. Quantum gravity, strings and branes may lead to testable effects beyond general relativity. Experimentalists will continue to search for such effects using laboratory experiments, particle accelerators, instruments in space and cosmological observations. At the centenary of relativity it could well be said that experimentalists have joined the theorists in relativistic paradise (Will, 2005, p. 27).

6 Concluding Remarks

Were one to pursue the error-statistical account of experiment at the level of large-scale theories, one would be interested to ask not "How can we severely pass high-level theories?" but rather, "How do scientists break

[21] Even "unclean" tests can rule out rivals that differ qualitatively from estimated effects. For example, Rosen's bimetric theory failed a "killing test" by predicting the reverse change in orbital period. "In fact we conjecture that for a wide class of metric theories of gravity, the binary pulsar provides the *ultimate* test of relativistic gravity" (Will, 1993, p. 287).

[22] According to Will, however, even achieving superunification would not overthrow the standard, macroscopic, or low-energy version of general relativity. Instead, any modifications are expected to occur at the Planck energy appropriate to the very early universe, or at singularities inside black holes.

down their questions about high-level theories into piecemeal questions that permit severe testing?" And how do the answers to these questions enable squeezing (if not exhausting) the space of predictions of a theory or of a restricted variant of a theory? We are not inductively eliminating one theory at a time, as in the typical "eliminative inductivism," but rather classes of theories, defined by giving a specified answer to a specific (experimental) question.

Note, too, that what is sought is not some way to talk about a measure of the degree of support or confirmation of one theory compared with another, but rather ways to measure how far off what a given theory says about a phenomenon can be from what a "correct" theory would need to say about it by setting *bounds on the possible violations*. Although we may not have a clue what the final correct theory of the domain in question would look like, the value of the experimental knowledge we can obtain now might be seen as giving us a glimpse of what a correct theory would say regarding the question of current interest, no matter how different the full theory might otherwise be.

References

Ben Haim, Y. (2001), *Information-Gap Decision Theory: Decisions Under Severe Uncertainty*, Academic Press, San Diego, CA.

Chalmers, A. (1999), *What Is This Thing Called Science?* 3rd ed., Open University Press, and University of Queensland Press.

Chalmers, A. (2002), "Experiment and the Growth of Experimental Knowledge," pp. 157–70 in *Proceedings of the International Congress for Logic, Methodology, and Philosophy of Science* (Vol. 1 of the 11th International Congress of Logic, Methodology, and Philosophy of Science, Cracow, August 1999), P. Gardenfors, J. Wolenski, and K. Kijania-Placek (eds.). Kluwer, Dordrecht, The Netherlands.

Cox, D.R. (2006), *Principles of Statistical Inference*, Cambridge University Press, Cambridge.

Damour, T., and Taylor, T.H. (1991), "On the Orbital Period Change of the Binary Pulsar PSR 1913 + 16," *Astrophysical Journal*, 366: 501–11.

Dorling, J. (1979), "Bayesian Personalism, the Methodology of Scientific Research Programmes, and Duhem's Problem," *Studies in History and Philosophy of Science*, 10: 177–87.

Earman, J. (1992), *Bayes or Bust: A Critical Examination of Bayesian Confirmation Theory*, MIT Press, Cambridge, MA.

Fitelson, B. (2002), "Putting the Irrelevance Back into the Problem of Irrelevant Conjunction," *Philosophy of Science*, 69: 611–22.

Glymour, C. (1980), *Theory and Evidence*, Princeton University Press, Princeton.

Good, I.J. (1983), *Good Thinking*, University of Minnesota Press, Minneapolis.

Jeffreys, W., and Berger, J. (1992), "Ockham's Razor and Bayesian Analysis," *American Scientist*, 80: 64–72.

Kass, R.E., and Wasserman, L. (1996), "Formal Rules of Selecting Prior Distributions: A Review and Annotated Bibliography," *Journal of the American Statistical Association*, 91: 1343–70.

Kyburg, H.E., Jr. (1993), "The Scope of Bayesian Reasoning," in D. Hull, M. Forbes, and K. Okruhlik (eds.), *PSA 1992*, Vol. II, East Lansing, MI.

Laudan, L. (1977), *Progress and Its Problems*, University of California Press, Berkeley.

Laudan, L. (1997), "How About Bust? Factoring Explanatory Power Back into Theory Evaluation," *Philosophy of Science*, 64:303–16.

Lobo, J. (1996), "Sources of Gravitational Waves," pp. 203–22 in G.S. Hall and J.R. Pulham (eds.), *General Relativity: Proceedings of the Forty-Sixth Scottish Universities Summer School in Physics*, SUSSP Publications, Edinburgh, and Institute of Physics, London.

Mayo, D.G. (1991), "Novel Evidence and Severe Tests." *Philosophy of Science*, 58(4): 523–52.

Mayo, D.G. (1996), *Error and the Growth of Experimental Knowledge*, University of Chicago Press, Chicago.

Mayo, D.G. (1997a), "Duhem's Problem, the Bayesian Way, and Error Statistics, or 'What's Belief Got to Do with It?'" and "Response to Howson and Laudan," *Philosophy of Science*, 64: 222–44, 323–33.

Mayo, D.G. (1997b), "Severe Tests, Arguing from Error, and Methodological Underdetermination," *Philosophical Studies*, 86: 243–66.

Mayo, D.G. (2000), "Experimental Practice and an Error Statistical Account of Evidence." *Philosophy of Science* 67, (Proceedings). Edited by D. Howard. Pages S193–S207.

Mayo, D.G. (2002), "Theory Testing, Statistical Methodology, and the Growth of Experimental Knowledge," pp. 171–90 in *Proceedings of the International Congress for Logic, Methodology, and Philosophy of Science* (Vol. 1 of the 11th International Congress of Logic, Methodology, and Philosophy of Science, Cracow, August 1999), P. Gardenfors, J. Wolenski, and K. Kijania-Placek (eds.). Kluwer, Dordrecht, The Netherlands.

Mayo, D.G. (2006), "Critical Rationalism and Its Failure to Withstand Critical Scrutiny," pp. 63–96 in C. Cheyne and J. Worrall (eds.), *Rationality and Reality: Conversations with Alan Musgrave*, Springer, Dordrecht.

Mayo, D.G., and Spanos, A. (2006), "Severe Testing as a Basic Concept in a Neyman–Pearson Philosophy of Induction," *British Journal of Philosophy of Science*, 57(2): 323–57.

Morrison, M., and Morgan, M. (eds.) (1999), *Models as Mediators: Perspectives on Natural and Social Science*, Cambridge University Press, Cambridge.

Suppes, P. (1969), "Models of Data," pp. 24–35 in *Studies in the Methodology and Foundations of Science*, D. Reidel, Dordrecht.

Will, C.M. (1980), "General Relativity," pp. 309–21 in J. Ehlers, J.J. Perry, and M. Walker (eds.), *Ninth Texas Symposium on Relativistic Astrophysics*, New York Academy of Sciences, New York.

Will, C.M. (1986), *Was Einstein Right?* Basic Books, New York (reprinted 1993).

Will, C.M. (1993), *Theory and Experiment in Gravitational Physics*, Cambridge University Press, Cambridge.

Will, C.M. (1996), "The Confrontation Between General Relativity and Experiment. A 1995 Update," pp. 239–81 in G.S. Hall and J.R. Pulham, *General Relativity: Proceedings of the Forty Sixth Scottish Universities Summer School in Physics*, SUSSP Publications, Edinburgh, and Institute of Physics, London.

Will, C.M. (2004), "The Confrontation Between General Relativity and Experiment," *Living Reviews in Relativity*, http://relativity.livingreviews.org/Articles/lrr-2001-4/title.html.

Will, C.M. (2005), "Relativity at the Centenary," *Physics World*, 18: 27.

Worrall, J. (1989), "Fresnel, Poisson and the White Spot: The Role of Successful Predictions in the Acceptance of Scientific Theories," pp. 135–57 in D. Gooding, T. Pinch, and S. Schaffer (eds.), *The Uses of Experiment: Studies in the Natural Sciences*, Cambridge University Press, Cambridge.

Worrall, J. (1993), "Falsification, Rationality and the Duhem Problem: Grünbaum vs Bayes," pp. 329–70 in J. Earman, A.I. Janis, G.J. Massey, and N. Rescher (eds.), *Philosophical Problems of the Internal and External Worlds*, University of Pittsburgh Press, Pittsburgh.

TWO

The Life of Theory in the New Experimentalism

Can Scientific Theories Be Warranted?

Alan Chalmers

1 Introduction

Deborah Mayo's "error-statistical" account of science and its mode of progress is an attempt to codify and capitalise on the "new experimentalist" approach to science that has made its appearance in the past few decades as an alternative to "theory-dominated" accounts of science. Rather than understanding scientific progress in terms of the replacement of one large-scale theory by another in the light of experiments designed to test them, new experimentalists view progress in terms of the accumulation of experimental discoveries and capabilities established independently of high-level theory. The new experimentalists aspire to an account of science and its mode of progress that evades problems associated with the theory dependence of observation, the underdetermination of theories by evidence, the Duhem problem and incommensurability that have beset the theory-dominant approach. Here is how Mayo (this volume, p. 28) herself characterizes the situation:

Learning from evidence, in this experimentalist philosophy, depends not on appraising large-scale theories but on local experimental tasks of estimating backgrounds, modeling data, distinguishing experimental effects, and discriminating signals from noise. The growth of knowledge has not to do with replacing or confirming or probabilifying or 'rationally accepting' large-scale theories, but with testing specific hypotheses in such a way that there is a good chance of learning something – whatever theory it winds up as part of.

Central to Mayo's version of the new experimentalism is her notion of a severe test. A hypothesis H is severely tested by evidence e produced by test T if and only if H fits e and T has a low probability of yielding e if H is false. Possible ways in which T might produce e were H false need to be identified and eliminated by appropriate error probes. Mayo gives a fairly

formal account of severe tests and error elimination in her error statistics –
a creative adaptation and development of Neyman–Pearson statistics. She
also extends the account in a less formal, but no less effective, way to some
classic examples of experimentation in physics. For example, her analysis of
the nitty-gritty details of Perrin's experiments on Brownian motion (Mayo,
1996, chapter 7) shows how Perrin was able to eliminate possible sources of
error from his highly consequential measurements of various characteristics
of the motion and also how he was able to do so without recourse to high-
level theory.

I have no doubt that there is much of value in Mayo's analysis of experimental science that should be accommodated by any adequate theory of
science, and it is certainly a useful and much-needed corrective to some
of the excesses of the theory-dominated approach. But Mayo has made it
quite clear that she is not happy with such an assessment of her error-
statistics. She aims for her position to constitute *the* account of science
and its progress, not a useful contribution or corrective to one. As she has
come to acknowledge, the biggest problem Mayo faces is finding an adequate home in science for theory. As illustrated in the preceding quotation,
Mayo's notion of a severe test comes into play in the arena of low-level
claims for which possible sources of error can be identified and eliminated.
"If scientific progress is thought to turn on appraising high-level theories,"
writes Mayo, "then this type of localized account of testing will be regarded
as guilty of serious omission, unless it is supplemented with an account
of theory appraisal" (this volume, p. 29). In the following, I argue that
high-level theory is indispensably implicated in science and its mode of
progress; there is no "if" about it. Some of these ways constitute a challenge
to the separation of low-level experimental claims from high-level theory
implicit in Mayo's presentation of the problem. I argue that Mayo's notion
of a severe test is too demanding for application to theories, and I show that
replacing it by a less demanding one does not have the dire consequences
for the reliability of science that she seems to fear.

2 Mayo on Reliability, Stability, and Severity

Mayo is concerned with giving an account of science that captures its special
status. She seeks to capture the sense in which scientific knowledge is confirmed in an especially demanding fashion that renders it reliable in a way
that lesser kinds of knowledge are not. She is justifiably scornful of Popper's
declaration that corroboration of scientific claims "says nothing whatever
about future performance, or about the 'reliability' of a theory" (1996,

p. 9). Claims that have passed Mayo's severe tests are reliable insofar as the ways in which they could be at fault have been probed and eliminated. They are reliable to the extent that they probably would not have passed the tests that they have in fact passed were they false. It is this focus on reliability that leads Mayo to reject comparativist accounts of confirmation according to which a theory is warranted if it has passed tests that its known rivals cannot pass. She is unhappy with this because being best-tested in this sense is no guarantee that a theory is free of error. Best tested does not imply well tested, as far as Mayo is concerned.

Mayo assumes that if scientific claims are to be warranted in a sense that makes them reliable, then the severity assessments that sanction assumptions about reliability must be stable. A claim that has been severely tested stays severely tested. She rejects the comparativist account because she sees it as violating this condition. A theory believed to be well tested because it has withstood tests its known rivals cannot withstand becomes no longer well tested once a previously unthought-of alternative that is able to match its ability to pass known tests is proposed. In Mayo's eyes, in the comparativist's account, the degree of testedness and, hence, reliability, of a scientific claim is dependent on which theories happen to have been thought of at that time. Her own account of severity is designed to avoid such a conclusion. Claims about the degree to which some experimental claim has survived a severe test are not designed to be revisable in the light of some newly proposed theory. "Whereas we can give guarantees about the reliability of the piecemeal experimental test, we cannot give guarantees about the reliability of the procedure advocated by the comparative tester" (2002, p. 189).

Mayo gives an account of severe testing which has the consequence that, although certain low-level experimental claims can survive such testing, high-level theories cannot. This account serves to highlight the issue that is the concern of this chapter (as well as Mayo's own chapter). What home for theory can be found in Mayo's account of science? How can scientific theories be warranted?

3 Theories Cannot Be Severely Tested

It must be stressed that theories, not just high-level theories but theories in general, cannot be severely tested in Mayo's sense. If scientific claims are warranted only by dint of surviving severe tests, then theories cannot be warranted.

Scientific theories are distinguished from low-level observational and experimental claims by virtue of their generality. It is precisely because of this feature that they cannot be severely tested in Mayo's sense. The content of theories goes beyond the empirical evidence in their support and, in doing so, opens up various ways in which theories could be wrong. Let us consider a theory that has passed some experimental test. In what ways could it have passed this test if it were false? We could answer, for example: if the theory got it right about the experiment in question but wrong in other contexts or if some alternative theory, which says the same about the experimental claim in question, is true. A tension exists between the demand for severity and the desire for generality. The many possible applications of a general theory constitute many possible ways in which that theory could conceivably break down.

No sensible doubt exists that Newtonian mechanics progressed success- fully and dramatically for two hundred years after its inception. If this fact does not constitute scientific progress, then nothing does. But Newton's theory had not passed severe tests in Mayo's sense. Each new application introduced a new area in which it could conceivably break down. When a feather and a stone were dropped on the moon and seen to fall together, no one was surprised. After all, Newtonian theory predicted it, and New- tonian theory was well confirmed. But it was not well confirmed in the sense of being severely tested. It is conceivable that the feather and stone may have fallen differently. In what circumstances could Newton's theory have survived two hundred years of experimental testing and yet be false? We could answer, for example: if the mechanical behaviour of heavy objects is different on the moon than it is on earth. We do not need to appeal to hypothetical examples to make the point that the success of Newtonian theory cannot be attributed to its having passed severe tests, because we now know of tests which it fails. Newton's theory gets it wrong about the path of fast-moving electrons, for instance.

It is not merely high-level theories such as Newton's that cannot be severely tested, Mayo style. Low-level experimental laws cannot be severely tested either. I illustrate with one of Mayo's own examples: the testing by Eddington's eclipse experiments of the light-bending that follows from Einstein's general relativity theory. Mayo rightly points out that, because a range of theories besides Einstein's predicts Einstein's law of gravity, testing the law of gravity by the eclipse experiments does not serve as a test of Einstein's theory as opposed to those alternatives. Eddington's experiments were a test of Einstein's law of gravity. Mayo (1996, p. 291) observes that before Eddington's results, which were in conformity with Einstein's law of gravity, could be taken as constituting a severe test of that law, various alternative explanations of the measured effect needed to be ruled out.

Once various alternatives had been ruled out, Einstein's law was retained "as the only satisfactory explanation." However, the fact that Einstein's law was shown to be preferable to available alternatives was not sufficient to show that it was preferable to all possible alternatives, as made clear by Mayo's own criticism of the comparativists. As it happens, within a few decades, a rival theory to Einstein's, the Brans-Dicke theory, predicted a law of gravity that differed from Einstein's but that could not be ruled out by the Eddington experiments because they are insufficiently sensitive.

The preceding considerations show that if we are to extract from science only those claims that have survived Mayo's version of a severe test, then we will be left with only some very low-level, and extremely qualified, statements about some experimental laws, such as, "Einstein's law of gravity is consistent with specified experimental tests to within [some specified degree of accuracy]."

In a sense, I have been unduly labouring the point that theories cannot be severely tested because it is one that Mayo herself accepts. It is precisely because this is so that she herself has seen it necessary to take up the question of the role of theory in science.

4 Mayo on the Role of Theory

Mayo rightly insists that the extent to which a theory is supported by some experimental evidence that fits it is an issue that needs to be carefully investigated on a case-by-case basis. A key claim of hers is a challenge to the idea that a theory is confirmed by successfully accounting for some experimentally confirmed effect. The fact that a theory gets it right about a particular phenomenon in its domain is no guarantee in itself that it gets it right about other phenomena in its domain. The fact that Einstein's theory was shown by Eddington to get it at least roughly right about light-bending is insufficient to establish that it gets it right about Mercury's orbit, the red shift, alteration of the effective speed of light by strong gravitational fields and so on. "It seems disingenuous to say all of theory T is well tested (even to a degree) when it is known there are ways in which T can be wrong that have received no scrutiny, or that there are regions of implication not checked at all" (this volume, pp. 39–40). Equally plausibly, Mayo insists that if two theories say the same thing about some experimental phenomenon, then neither of them can be severely tested against that phenomenon. Because a range of alternatives to Einstein's general theory of relativity exist that share its prediction of the law of gravity, tests of it by Eddington's experiment cannot serve as a test of Einstein's general theory. In addition, the mere fact that a theory fits a phenomenon is not in itself sufficient to indicate that it is severely tested by that phenomenon. The fit may be artificially contrived in a post hoc way by the choice of an otherwise arbitrary parameter, for example.

I fully endorse some version of all of these points. However, doing so does not help with the basic problem of accommodating theory into Mayo's account of science. Indeed, insofar as all the previously mentioned points serve to stress that (1) theories are not severely tested by experiment and (2) theoretical considerations are important, they serve to highlight the problem. Point 2 needs elaboration. The recognition that a theory has genuine alternatives, that some phenomena can be derived from some subset of a theory rather than the theory as a whole, or that some phenomena, believed to be independent, are intimately connected, requires work at the theoretical level. The fact that the equivalence principle follows from curved space-time alone, independently of more specific accounts of the degree and cause of the curvature, was not clear to Einstein and was clarified by subsequent theoretical work. The fact that a class of rivals to Einstein's theory exists compatible with the experimental evidence available in 1960, say, and the realisation that the equivalence principle and the red shift are not independent phenomena but are both a consequence of curved space-time were all discoveries made at the theoretical level. Appreciation of the significance of the results of experiments depends on appeal to theory.

Mayo's response to this situation is to make a distinction between severely tested experimental knowledge, on the one hand, and theory, on the other, and to attribute to the latter only some heuristic role. Theories are an important part of science insofar as they serve to guide experiment.

> It is not that we are barred from finding a theory T "rationally acceptable," preferred, or worthy of pursuit ... upon reaching a point where T's key experimental predictions have been severely probed and found to pass. One could infer that T had solved a set of key experimental problems and take this as grounds for "deciding to pursue" it further. But these decisions are distinct from testing and would call for a supplement to what we are offering. (this volume, pp. 40–1)

> From our perspective, viable theories must (1) account for experimental results already severely passed and (2) show the significance of the experimental data for gravitational phenomena. Viable theories would have to be able to analyse and explore experiments about as well as GTR; ... One may regard these criteria as intertwined with the "pursuit" goals – that a theory should be useful for testing and learning more. (this volume, p. 48)

Experimental knowledge is the severely tested and, hence, reliable and stable part of scientific knowledge. Theories are a welcome part of science only insofar as they are useful instruments that aid the growth of experimental knowledge.

5 Reinstating Theory as a Component of Scientific Knowledge

One way to counter Mayo's emphasis on experimental, as opposed to theoretical, knowledge is to stress key roles for theory in science that she neglects or gives insufficient emphasis. Theories can be criticized, not by putting them to experimental test but by showing them to be incompatible with other theories. Some nineteenth-century theories of electromagnetism were criticised by showing them to be inconsistent with the conservation of energy,[1] whereas Einstein's major innovations in relativity stemmed from the difficulties he perceived in fitting together Newton's and Maxwell's theories, difficulties that we can now describe as following from the fact that Newton's theory is Galilean invariant whereas Maxwell's is Lorentz invariant. I would argue that Mayo's account of advances in relativistic gravitational theories gives insufficient emphasis to the extent to which theoretical considerations were a major driving force.

None of these kinds of observation strike deeply at Mayo's position, and I can foresee how she could respond to them by accommodating them into her view of theories as heuristic aids. Clashes between theories are significant just because, or insofar as, they imply clashes at the experimental level, and such problems are resolved at the experimental level. After all, the mere clash of a theory with the conservation of energy does not establish an inadequacy of the former any more than a clash between the kinetic theory and the second law of thermodynamics established the falsity of that theory. Exploring the consequences of theory clashes is productive insofar as it leads to new experimental knowledge, as the latter clash led to Perrin's experimental knowledge of Brownian motion – a manifestation of the falsity of the second law at the experimental level. A significant challenge to Mayo's position on the place of theory in science needs to find a role that cannot be shunted off into the "theory as heuristic aid" category. I attempt to pose such a challenge in the remainder of this section.

The position that ascribes a merely heuristic status to theory, as opposed to the severely tested, reliable status attributed to experimental knowledge,

---

[1] For an example, see Maxwell (1873, pp. 483–4).

presupposes that a clear dividing line can be drawn between the two. However, such a line cannot generally be drawn. There should be no doubt that two hundred years of development of Newtonian mechanics added significantly to the stock of scientific knowledge – knowledge that is reliable to the extent that any knowledge can be said to be so. A striking exemplification of this latter claim concerns the retrieval of a wayward, manned spacecraft a few decades ago. The spacecraft was on a course that was destined to take it past Earth, never to return. There was only sufficient fuel for one short firing of the spacecraft's rockets. The relevant data were fed into the appropriately programed computer that, within a few minutes, had provided the details of just how the rockets should be fired to ensure that the spacecraft would enter an orbit that would return it to Earth. The craft, along with its occupants, was safely retrieved. Newtonian theory was built into the program's computer and the salvage operation could not have been accomplished without it. Indeed, the development of Newtonian theory generally – which had led, for example, to a steady increase in the accuracy of knowledge of the motions of bodies in the solar system – presupposed and applied Newton's laws of motion as did Hooke's law and the phenomenological theory of elasticity developed in its wake.

The preceding point can be illustrated by an example from general relativity that Mayo has been keen to absorb into her new experimentalist framework. In recent decades the equivalence principle – the claim that all bodies fall at the same rate in a given gravitational field irrespective of their weight – has been put to the test for planet-sized bodies. More specifically, it has been tested for the case of the moon and the twenty-five times more massive Earth moving in the gravitational field of the sun. The separation of moon and Earth is measured by bouncing laser beams off reflectors placed on the former and measuring the time for the return journey. The separation can be measured to the nearest centimeter by modern techniques, and I am sure Mayo can give a nice account of this that does not involve an appeal to high-level theory. The modern experiments are well able to discriminate between the predictions of theories, such as Einstein's, that entail the equivalence principle and the Brans–Dicke theory, which does not; the two differ in their predictions of Earth–moon separation by about 120 centimetres. However, using the data to test the equivalence principle involves a comparison of the measurements with those expected were the equivalence principle true. And this necessarily involves theory. I quote C.M. Will, the authority to whom Mayo herself appeals on these matters: "These measurements [of moon–earth separation] are fit using a method of least squares to a theoretical model for the lunar motion that takes into account perturbations due to the Sun and other planets, tidal interactions, and post-Newtonian gravitational effects" (Will (2004), 3.6, p. 42). Experimental support for the strong equivalence principle in this example is not available to us without appeal to theory.

By the end of the nineteenth century, Maxwell's equations and the Lorentz force equation had become as indispensable in electromagnetism as Newton's laws were in mechanics. The experiments performed by J.J. Thomson in 1896 on cathode rays are another telling example of the indispensability of appeal to theory for the establishment of experimental knowledge. Thomson was able to show that cathode rays are beams of charged particles (later called electrons) and was able to estimate the charge-to-mass ratio, e/m, of those particles. In his calculation of the deflection of the beam in his discharge tube, Thomson wrote down expressions for the deflection in electric and magnetic fields that are, in effect, applications of the Lorentz force law. The quantitative measure of e/m simply cannot be accessed unless that law is presupposed.

A standard ploy is exploited by Mayo in situations such as this. She can insist that use of the Lorentz force law by Thomson does not require the assumption that that law is true in general. All that is required is the assumption that the law gets it right about the deflection of the beam in an experiment like Thomson's. However, this cannot be convincingly argued. If we raise the question of what warranted the assumption of the Lorentz force by Thomson, then it is simply not the case that Thomson had an argument for the truth of the law in the context of his experiment that was more localised and less general than a case for the truth of the law generally. The case for the Lorentz force law at the time rested on the way in which that law was compatible with well-established laws, such as those governing the deflection of current-carrying wires in magnetic fields, and the way in which that law, added to Maxwell's equations, enabled some deep-seated theoretical problems to be solved. Thomson's experiments in fact involved an extension of the law to a new type of situation in which the law had not been previously tested. The assumption that moving charged bodies constitutes field-producing currents in the same sense as currents through a wire, an assumption implicit in Lorentz's theory and taken for granted by Thomson, had not previously been directly tested. Macroscopic charged bodies cannot be made to move in a straight line fast enough or long enough for the magnetic fields they engender to be measured. Cathode rays provide a microscopic version of such a setup not available macroscopically. Elaboration of the historical detail would take us well beyond the scope of this chapter.[2] My basic point is that use of the Lorentz force equation by Thomson cannot be warranted in Mayo's strict sense and yet it must be warranted in some sense if the measurement of e/m for the electron is to be assumed to be part of experimental knowledge.

Mayo does allow a use of theory in science insofar as it is "preferred" or found "worthy of pursuit" or "rationally acceptable," but she distinguishes such notions from testing and as involving a goal distinct from the one she seeks. However, the roles played by Newton's laws in the Newtonian program or the Lorentz force in Thomson's experiments cannot be adequately comprehended by talk, for example, of Thomson's preference for, reason for believing or decision to pursue the Lorentz force law. Whatever Thomson's subjective attitude was to the Lorentz force law, he used it in his calculation of e/m for the electron and, unless it is used, the experiment does not yield that knowledge. The measurement of e/m is not warranted if the Lorentz force law is not warranted. Correspondingly, fruits of the Newtonian programme in astronomy are not warranted if Newton's laws are not warranted, and so on.

I have tried to show by means of examples that theory constitutes an integral component of scientific knowledge in such a way and to such a degree that it cannot be adequately construed merely as a heuristic guide leading us to the real thing – experimental knowledge – established independently of it. Significant experimental knowledge cannot be warranted unless theoretical knowledge can be warranted. To deny this would render impossible the development of those theories to the stage where their limitations could be appreciated. That means, as we have seen, that we need a sense of warrant that is weaker than one based on Mayo's notion of a severe test.

6 Confirmation, Underdetermination, and Reliability

Theories can be said to be confirmed to the extent that they are borne out by a wide variety of otherwise unconnected phenomena. I claim that some such notion, appropriately spelt out, can give us an account of theory confirmation that is strong enough to weed out pseudoscience and nonscience (e.g., creation theory or Boscovich's atomic theory) but not so strong that it prevents theories from being warranted in the way that Mayo's does. The

---

[2] For details of the path that led to the solution of deep theoretical problems within Maxwell's theory by the Lorentz electron theory, see Buchwald (1985) and Hunt (1991). The production of magnetic fields by convection currents had been measured in the case of rapidly rotating charged discs at the time of Thomson's experiment by Rowland and Hutchinson (1887).

general form of the argument is best grasped as an argument from coincidence. Would it not be an amazing coincidence that Newton's theory got it right, and in detail, about such a wide variety of phenomena, including the trajectory of the endangered spacecraft mentioned earlier, if it were false? The sense in which a theory is borne out by the phenomena needs to be demanding.[3] It is no coincidence that a theory fits the phenomena if the details of the theory are adjusted in the light of the phenomena to bring about the fit. Both Mayo and Worrell give overlapping but not identical accounts of why the mere compatibility of data with a theory is not sufficient as a test of it. My favored account of confirmation by arguments from coincidence would draw on Mayo (1996), Hacking (1983), and Worrall (2002) and would add some considerations of my own from Chalmers (2003 and 2009) concerning how seemingly circular arguments linking theories and evidence can in fact constitute arguments for both. The result, I anticipate, would be a suitably updated version of what Whewell aspired to with his "concilience of inductions" involving arguments that I denote with the phrase "arguments from coincidence."

Arguments from coincidence are less than compelling. Strong arguments from coincidence for a theory cannot guarantee that a theory will not break down in some unexpected ways when applied in novel domains or to new levels of accuracy. Newton's theory eventually broke down in relativistic and quantum domains in spite of the formidable arguments from coincidence made possible by two centuries of successful application. But it is important to note that the progression beyond Newtonian theory to relativity does not completely undermine the arguments for Newton's theory and leave us with unexplained coincidences. In the light of relativity, Newton's theory must be satisfied, provided the masses and speeds involved are small enough. A theory of warrant based on arguments from coincidence serves to support what has been termed the "general correspondence principle" by Post (1971) and Krajewski (1977). According to that principle, successful theories – theories that have been borne out by arguments from coincidence – must live on as limiting cases of their successors. If this were not the case, then replacement of previously successful theories would leave us with unexplained coincidences.

---

[3]   In Chapter 1, Mayo conflates my position with the "comparativist" view that she attributes to Larry Laudan. However, the mere fact that a theory has passed tests better than an available rival is insufficient to establish a strong argument from coincidence of the kind my position requires. Many of the examples of successful scientific theories cited by Laudan (1989) in his celebrated article critical of realism do not qualify as well-confirmed theories on my account. Here my position is in line with that of John Worrall (1989).

The considerations of the forgoing paragraph serve to indicate why the fact that arguments from coincidence are less than compelling is no serious threat to the reliability of science. Replaced theories continue to be reliably applicable in the domains in which they have been borne out by powerful arguments from coincidence. As I have written before (Chalmers, 2002, pp. 168–9), reliance on Newtonian theory, for example, to calculate the trajectory of a spacecraft is increased, not undermined, by the recognition that Newtonian theory is false in the relativistic domain. Our confidence in Newton's theory is enhanced by the understanding that it is indeed a limiting case of relativity theory and that the speeds involved are not sufficient to render relativistic effects appreciable. A sense exists that arguments from coincidence and theories warranted by them are unreliable; however strong the case for theories may be, they can break down when applied in new areas. But it is no fault of an account of confirmation that it renders theories unreliable in that sense. Science *is* unreliable in that sense and to demand otherwise is to demand the impossible. Inferring theories from evidence using strong arguments from coincidence, although "unreliable" in the sense that the inferred theory is liable to be replaced by a superior alternative, poses no threat to some reasonable interpretation of the claim that scientific knowledge can be reliably applied.

Because Mayo has chosen developments of general relativity as illustrative of her stance on the status of theory, I conclude this section by pointing out ways in which the development of that theory as construed by Will fits closely with my picture and poses problems for Mayo. In Will's view, "the time has come to take general relativity for granted" (1993, p. 242). The general theory of relativity, a product of Einstein's imagination, "turned out in the end to be so right" (1993, p. 259). Will justifies this by appealing to the degree to which the theory has passed a wide variety of tests. That is, Einstein's general theory of relativity is warranted by arguments from coincidence in my sense. Because it is warranted, it can be assumed to be a tool for investigating other phenomena, just as Newton's or Maxwell's theories were accepted and used as exploratory tools. "Just as Newtonian gravity is used as a tool for measuring astrophysical parameters of ordinary binary systems, so GTR is used as a tool for measuring astrophysical parameters in the binary pulsar" (Will, 2004, p. 2). This use of theories as tools is warranted by my arguments from coincidence but cannot be warranted in Mayo's more demanding sense. Note, however, that Will (2004, sec. 7) does not regard general relativity theory as confirmed beyond all doubt. One worry concerns the lack of fit between relativity and quantum theory.

All attempts to quantize gravity and to unify it with the other forces suggest that the standard general relativity of Einstein is not likely to be the last word.... Although it is remarkable that this theory, born 80 years ago out of almost pure thought, has managed to survive every test, the possibility of finding a discrepancy will continue to drive experiment for years to come. (Will, 1996, p. 280; 2004)

However, this fallibility and openness to revision of the theory is not a general threat to the reliability of its applicability, because any acceptable theory must yield the well-tested parts of current theory. Criteria for any viable gravity theory, according to Will, include the demands that (1) in the limit, as gravity is "turned off," the nongravitational laws of physics reduce to the laws of special relativity and (2) in the limit of weak gravitational fields and slow motions it must reproduce Newton's laws (1993, pp. 18–20). Here Will in effect adopts the general correspondence principle that is endorsed by arguments from coincidence. But note that it is well-confirmed *theories* that impose the constraint on successor theories. Mayo cannot endorse this insofar as those theories are not warranted. In Mayo's account, Will's constraints become "viable theories [that] would have to...analyze and explore experiments about as well as GTR" (this volume, p. 48). At best, she can list only established experimental results amongst the constraints. Even some of those, such as contemporary tests of the strong equivalence principle mentioned earlier, should be regarded as suspect on her view because of their reliance on theory.

7 Concluding Remarks

Deborah Mayo's version of the new experimentalism is a much needed response to excesses of theory-dominated accounts of science that portray experiment as theory- or paradigm-dependent to a degree that poses problems for the claim that science progresses in some clear, objective sense. She has codified the resources available to experimentalists for establishing experimental results in a way not dependent on or threatened by theory or paradigm dependence. For instance, she shows how Eddington established the bending of light rays by the sun by superimposing photographs of stars and how Perrin demonstrated that Brownian motion is random by tracing and analysing the changing position of a Brownian particle. She also makes valuable observations about work at the theoretical level, which makes it possible to partition a theory into the part of it responsible for some set of phenomena and the remaining part not implicated by those phenomena. A telling example is the work that led to the realisation that curved space-time in general relativity can be partitioned from accounts of its degree and causes, which made it possible to realise that the equivalence principle and the redshift are consequences of the former only; tests of those phenomena do not bear on the adequacy of general theories about the latter. Not only do I welcome and applaud these observations, but I have incorporated them into my own work in the history of atomism. My recent, controversial claim (Chalmers, 2005) that nineteenth-century chemistry offered little support to atomism because the chemical formulas responsible for much of the success of that chemistry can be understood as referring to combining portions, rather than the stronger assumption of combining atoms, shows the mark of my debt to Mayo's ideas.

Despite this recognition of the value of Mayo's work, I cannot accept it as *the* account of the status of scientific knowledge. I have argued that it cannot adequately accommodate the role and place of theory in science. This is the case if we restrict ourselves to the view of science as a reliable guide to our manipulations of the world. It is even more the case when we look to science as providing the best, objective answers to questions about the nature of the kind of world we live in. We need an objective account of how theories are warranted that captures the fallibility and revisability of theory while being able to capture some sense of the way science progresses and of how it is reliable to the extent that it is reliable. Mayo's severe testing does not provide an account of how theories are warranted because it is inappropriately demanding.

References

Buchwald, J. (1985), *From Maxwell to Microphysics*, University of Chicago Press, Chicago.

Chalmers, A. (2002), "Experiment and the Growth of Experimental Knowledge," pp. 157–69 in P. Gardenførs, J. Wolenski, and K. Kijania-Placek (eds.), *In the Scope of Logic, Methodology and Philosophy of Science*, Kluwer, Dordrecht.

Chalmers, A. (2003), "The Theory-Dependence of the Use of Theories in Science," *Philosophy of Science*, 70: 493–509.

Chalmers, A. (2005), "Transforming Atomic Chemistry into an Experimental Science: The Status of Dalton's Theory," *The Rutherford Journal*, 1, http://www.rutherfordjournal.org.

Chalmers, A. (2009), *The Scientist's Atom and the Philosopher's Stone: How Science Succeeded and Philosophy Failed to Gain Knowledge of Atoms*, Springer, Dordrecht.

Hacking, I. (1983), *Representing and Intervening*, Cambridge University Press, Cambridge.

Hunt, B. (1991), *The Maxwellians*, Cornell University Press, Ithaca, NY.

Krajewski, W. (1977), *Correspondence Principle and Growth of Science*, Reidel, Dordrecht.

Laudan, L. (1989), "A Confutation of Convergent Realism," in J. Leplin (ed.), *Scientific Realism*, University of California Press, Berkeley.

Maxwell, J.C. (1873), *A Treatise on Electricity and Magnetism*, Vol. 2, Dover, NY (reprinted 1953).

Mayo, D.G. (1996), *Error and the Growth of Experimental Knowledge*, University of Chicago Press, Chicago.

Mayo, D.G. (2002), "Theory Testing, Statistical Methodology and the Growth of Experimental Knowledge," pp. 171–90 in P. Gardenførs, J. Wolenski, and K. Kijania-Placek (eds.), *In the Scope of Logic, Methodology and Philosophy of Science*, Kluwer, Dordrecht.

Post, H.R. (1971), "Correspondence, Invariance and Heuristics," *Studies in History and Philosophy of Science*, 2: 213–55.

Rowland, A., and Hutchinson, C.T. (1887), "Electromagnetic Effects of Convection Currents," *Philosophical Magazine*, 27: 445–60.

Will, C.M. (1993), *Was Einstein Right? Putting General Relativity to the Test*, Basic Books, New York.

Will, C.M. (1996), "The Confrontation Between General Relativity and Experiment, a 1995 Update," pp. 239–80 in G.S. Hall and J.R. Pulham (eds.), *General Relativity: Proceedings of the Forty Sixth Scottish Universities Summer School in Physics*, SUSSP Publications, Edinburgh, and Institute of Physics, London.

Will, C.M. (2004), "The Confrontation Between General Relativity and Experiment," *Living Reviews in Relativity*, http://relativity.livingreviews.org/Articles/lrr-2001-4/title.html.

Worrall, J. (1989), "Structural Realism: The Best of Both Worlds," *Dialectica*, 43: 99–124.

Worrall, J. (2002), "New Evidence for Old," pp. 191–209 in P. Gardenfors, J. Wolenski, and K. Kijania-Placek (eds.), *In the Scope of Logic, Methodology and Philosophy of Science*, Kluwer, Dordrecht.

Can Scientific Theories Be Warranted with Severity? Exchanges with Alan Chalmers

Deborah G. Mayo

Reacting to Alan Chalmers's most serious challenges to the account of theories I have put forward provides an excellent springboard for dealing with general questions of how to characterize and justify inferences beyond the data (ampliative or inductive inference) and how to view the role of theory appraisal in characterizing scientific progress:

1. *Experimental Reasoning and Reliability*: Can generalizations and theoretical claims ever be warranted with severity? What is an argument from coincidence? Do experimental data so underdetermine general claims that warranted inferences are limited only to the specific confines in which the data have been collected?
2. *Objectivity and Rationality*: Must scientific progress and rationality be framed in terms of large-scale theory change? Does a negative answer entail a heuristic role for theories?
3. *Metaphilosophical Themes*: How do philosophical assumptions influence the interpretation of historical cases?

Although Chalmers appreciates that "the new experimentalism has brought philosophy of science down to earth in a valuable way" (1999), his decade-long call for us to supplement "the life of experiment" with a "life of theory," he argues, remains inadequately answered. In Section 9 ("Progress?"), however, I note how we may now be moving toward agreement on the central point at issue in this essay.

1 Argument from Coincidence

Chalmers denies that the account of the roles and appraisals of theories that emerges in the error-statistical account is adequate, because he thinks that requiring a theory to be severely tested (i.e., to pass severely) in my sense is too severe. According to Chalmers, scientists must invariably "accept" theories T as warranted "to the extent that they are borne out by a wide variety of otherwise unconnected phenomena" even though such "arguments from coincidence" fail to warrant T with severity (this volume, pp. 67–8). This acceptance, moreover, must be understood as accepting T as fully true; it must not be regarded as taking any of the weaker stances toward T that I consider, such as regarding T as a solution to key experimental problems, or employing T in pursuit of "learning" or "understanding," or in discovering rivals. Such stances, Chalmers thinks, regard theories as mere heuristics to be tossed away after being used to attain experimental knowledge.

Although all of T is not warranted with severity, all of T may be warranted by dint of an "argument from coincidence," wherein T correctly fits or predicts a variety of experimental results, for example, $E_1, E_2, \ldots, E_n$. Would it not be an amazing coincidence if theory T got it right, and in detail, about such a wide variety of phenomena, if T were false? We are not told how to determine when the agreement is appropriately amazing were T false, except to insist on a version of the novelty requirement: "The sense in which a theory is borne out by the phenomena needs to be demanding. It is no coincidence that a theory fits the phenomena if the details of the theory are adjusted in the light of the phenomena to bring about the fit" (p. 68). It is hard to see how this argument differs from an inference to a severely tested theory in my sense; if it is really an amazing coincidence that theory T has passed diverse tests $E_1, E_2, \ldots, E_n$, "if it were false," then we have T passing with severity. If Chalmers's "demanding" requirement is to be truly demanding, it would require the passing results to be very difficult to achieve if T were false. Oddly, Chalmers insists that an argument from coincidence can never be sufficiently strong to warrant an inference to a theory, or even to a generalization beyond the evidence!

2 Unreliable Methods: Rigged Hypotheses and Arguments from Conspiracy

According to Chalmers, "If scientific claims are warranted only by dint of surviving severe tests then theories cannot be warranted" (this volume, p. 60). That is because "the content of theories goes beyond the empirical evidence in their support and, in doing so, opens up various ways in which theories could be wrong" (p. 61). But high severity never demanded infallibility, and to charge as Chalmers does that I seek error-free knowledge is inconsistent with the centrality of statistical arguments in my account. But imagining such a charge to be a slip, we can consider his argument for denying that a general hypothesis H can ever pass severely on the basis of passing results $E_1, E_2, \ldots, E_n$. According to Chalmers, if we ask of a theory that has passed some experimental test "in what ways could it have passed this test if it were false," then one can answer, "if the theory got it right about the experiment in question but wrong in other contexts, or if some alternative theory, which says the same about the experimental claim in question, is true" (p. 61). Fortunately one's freedom to say this scarcely precludes arguing that it is very improbable for such passing results to occur, were H false.

But it is important to see that even without being able to argue in a particular case that H has passed severely, our error-statistical tester most certainly can condemn Chalmers's general underdetermination gambit: Although one can always claim that the passing results hold only for the observed cases, or that all the results are as if H is true but in fact it is false, such a ruse would be highly unreliable. No true hypothesis could be corroborated! One can always argue that any hypothesis H, however well probed, is actually false and some unknown (and unnamed) rival is responsible for our repeated ability to generate results that H passes!

Note, too, that this would also preclude Chalmers's arguments from coincidence: One can always deny that an agreement of results in the wide variety of cases suggested by Chalmers would be a coincidence were H false; thus, Chalmers's own argument against severity would preclude the argument from coincidence he wishes to uphold.

Chalmers's gambit comes under the heading of a "rigged alternative" (Mayo, 1996, p. 205):

Rigged hypothesis R: A (primary) alternative to H that, by definition, would be found to agree with any experimental evidence taken to pass H.

The problem is that even where H had repeatedly passed highly severe probes into the ways H could err, this general procedure would always sanction the argument that all existing experiments were affected in such a way as to systematically mask the falsity of H. Thus, such an argument has high if not maximal probability of erroneously failing to discern the correctness of H, even where H is true. Whenever it can be shown that such a stratagem is being used, it is discounted by the error statistician.

Thus it is unclear why Chalmers maintains that "it is not merely high-level theories such as Newton's that cannot be severely tested, Mayo style. Low-level experimental laws cannot be severely tested either" (p. 61). Aside from the argument from "rigging," Chalmers points to "one of Mayo's own examples":

Mayo rightly points out that, because a range of theories besides Einstein's predicts Einstein's law of gravity, testing the law of gravity by the eclipse experiments does not serve as a test of Einstein's theory as opposed to those alternatives. (Chalmers, this volume, p. 61)

True. However, the fact that Eddington's experiments failed to pass all of GTR with severity scarcely shows that no theory passes with severity. We have belied such skepticism, both in qualitative examples and in formal statistical ones. Having inferred a deflection effect in radio astronomy, experimental relativists can argue with severity that the deflection effect is approximately $L \pm \varepsilon$. Why? Because if this inference were false, then with high probability they would not have continually been able to reliably produce the results they did. Analogously, having studied the effects on blood vessels of high-level exposure to radioactive materials in well-designed studies, we argue with severity that there is evidence of such radioactive effects to any human similarly exposed. Although the inductive claim depends upon future experiments contingent on certain experimental conditions holding approximately, the fact that we can check those conditions enables the inference to pass. We could thereby discredit any attempt to dismiss the relevance of those studies for future cases as engaging in a highly unreliable, and hence unscientific, method of inference.

To hold as Chalmers does that "if we are to extract from science only those claims that have survived Mayo's version of a severe test, then we will be left with only some very low-level, and extremely qualified, statements" (this volume, p. 62) about *H* being consistent with past results. If we accept this position, it would be an extraordinary mystery that we have worked so hard at experimental design, and at developing procedures to avoid errors in generalizing beyond observed data. Why bother with elaborate experimental controls, randomized treatment control studies, and so on, if all one is ever able to do is announce the data already observed!

According to Chalmers, "the point that theories cannot be severely tested . . . is one that Mayo herself accepts. It is precisely because this is so that she herself has seen it necessary to take up the question of the role of theory in science" (p. 62).

I take up the question of the role of theory because I find that the error-statistical philosophy of science transforms the entire task of constructing an adequate philosophical account of the roles of high-level theories in science. The main point of my contribution was to show how progress is made in coming to learn more about theories by deliberately exploiting the knowledge of errors not yet ruled out and by building on aspects that are severely probed. It is incorrect to suppose that I deny theories can pass severely simply because I deny that passing the comparativist test suffices.

Chalmers finds my position on theories deeply problematic, and I shall spend the remainder of this exchange examining why. Note that even granting my arguments in Section 1 – that our experimentalist does not deny that theories pass severely – we are left with the argument of Chalmers I want to consider now. For even if he were to allow that some theories and hypotheses may be severely warranted, here Chalmers is keen to argue that scientists require a weaker notion wherein inseverely warranted theories may be accepted. So to engage Chalmers's position, we need to consider an argument from coincidence to theory T that is insevere; that is, we need to consider cases where evidence licenses an argument from coincidence to T, but T fails to be severely passed by dint of this evidence. Such an argument only warrants, with severity, some weaker variant of the full theory T, or proper subsets of T. The most familiar and yet the weakest form warrants merely an inference to a "real" or nonchance effect among the phenomena T seeks to explain. Allowing the move from severely passing a subset of T to all of T is easy enough; the question is why one would think it a good idea, much less required, as Chalmers does, to permit such an inference. I focus on what Chalmers deems his strongest arguments: those that he takes to preclude my getting away with claiming that only the severely corroborated portions or variants of a theory are warranted. The arguments, however, appear to land us immediately in contradictions and also conflict with the historical episodes on which he draws.

3 The Argument from Needing to Avoid Unexplained Coincidences

Chalmers argues that if we do not allow insevere arguments of coincidence to warrant all of theory T_1, then we will be stuck with unexplained coincidences when T_1 is replaced by theory T_2. He claims this would violate what he calls the "general correspondence principle," wherein "successful theories – theories that have been borne out by arguments from coincidence – must live on as limiting cases of their successors" (this volume, p. 68).

But we see at once that Chalmers's argument, to be consistent, must be construed as claiming that the argument from coincidence warrants not all of T_1 (as he supposes), but at most the severely corroborated subportions or variants of T_1.

For example, suppose that all of T_1 is warranted when T_1 is replaced by an incompatible theory T_2. Then the scientist would be accepting incompatible theories. Theory T_1 is replaced by T_2 when T_1 is determined to give an erroneous explanation of the results. If any portions of T_1 are retained, they could not comprise all of T_1, understood to hold in all the domains circumscribed by T_1, but at most those aspects that were and remain well corroborated. Nor are we then without an explanation of the passing results, because the replacement theory T_2 explains them. Therefore, Chalmers's appeal to the correspondence principle works against his position and he is faced with this dilemma: He must either deny that all of T_1 is retained when replaced or deny that accepting T_1 by an insevere argument from coincidence can really mean to accept all of T_1. Several of his own remarks suggest he adopts the former horn: "Replaced theories continue to be reliably applicable in the domains in which they have been borne out by powerful arguments from coincidence" (this volume, p. 69). But that would mean that the domains in which they are borne out by his argument from coincidence cannot be the full domain of the theory but only a truncated domain determined after the full theory is replaced.

To sum up this part, Chalmers is at pains to show that T_1 must be warranted in all of its domains, not just in those where it has passed severely. Yet he is implicitly forced to shift to the claim that what is borne out by powerful arguments from coincidence is T_1 *restricted to the domains in which it continues to be reliably applicable*, or some such T_1 variant. Moreover, when we add the falsifying case E_{n+1} to $E_1, E_2, \ldots, E_n$ that lead to T_1 being replaced by T_2, the full set of data no longer offers an argument from coincidence, because that would require accordant results for T_1. It would not be an amazing coincidence to observe $E_1, E_2, \ldots, E_n$ as well as E_{n+1}, were T_1 false (and T_2 true) – insofar as T_2 accords with the available data. His favorite example is retaining Newton's theory when replaced by Einstein's, but what would be retained would be a variant wherein relativistic effects (that falsify Newton) are imperceptible. Retaining the nonfalsified variants of a theory is scarcely to retain the full theory.

The only coherent position is that what remains as warranted by the data, if anything, are the portions severely corroborated, as we recommend. Granted, at any point in time one might not know which portions these are, and I concede Chalmers's point that we may later need to reinterpret or correct our understanding of what we are entitled to infer; that is why one deliberately sets out to explore implications on which theories disagree. But to suppose that hypothetically assuming a theory T for purposes of drawing out T's implications requires accepting T as true, leads to inconsistencies. This takes us to Chalmers's next argument.

4 The Argument from the Need to Explore New Domains

Chalmers's second argument is that scientists must assume T holds in domains other than those in which it has been probed with severity when they set out to explore new domains and phenomena – were I to deny this, my account "would render impossible the development of those theories to the stage where their limitations could be appreciated" (this volume, p. 67). For example, he claims, following Clifford Will, that experimental relativists assume GTR in probing areas beyond those for which it has been severely corroborated at any point in time. But do they? Such explorations fall into one of two types:

1. A theory is merely assumed hypothetically to derive predictions to find new data to discriminate between rivals that are thus far corroborated, or
2. Only aspects of T that have passed severely, or on which all viable rivals agree, are relied on to learn more about a phenomenon of interest.

The case of experimental relativity does provide examples of each of these, but in neither case is a theory regarded as warranted beyond the domains it has passed severely. Moreover, we seem forced to conclude once again that Chalmers's own arguments become incoherent if taken to warrant all of T.

5 Aims of Science: A Tension between Severity and Informativeness?

We may agree that a "tension" exists between severely passed and highly informative theories, but it is important to see that the aim of science for our error statistician is not severity but rather finding things out. One finds things out by means of experimental inquiries, which we define as inquiries where error probabilities or severity may be assessed, whether quantitatively or qualitatively. By distinguishing those hypotheses that have passed severely from those that have not, one gets ideas as to new hypotheses to test, and how to test them. It is the *learning goal* that drives us to consider implications where thus far well-corroborated theories and hypotheses may fail. Chalmers disparages such goals as merely heuristic uses of theories, but this seems to forfeit what is crucial to the most dynamic parts of the life of theory (see Glymour, Chapter 8, this volume).

5.1 The Contrast between Us

To make the contrast between us clear, for the error statistician, theories and hypotheses may be "used" in an inquiry so long as either (1) they have themselves been warranted with severity or, if not, (2) any inferences remain robust in the face of their falsity. Commonly, under condition 1, what is used is not the truth of theory *T* but a hypothesis of *how far from true T may be* with regard to some parameter (having been determined, say, as an upper or lower bound of a confidence interval or severity assessment). By contrast, Chalmers maintains that inseverely warranted aspects of *T* must be assumed in exploring *T*; thus, unless robustness could be shown separately, it would prevent warranted inferences about *T* – at odds with our account. If the validity of an inquiry depends on the truth or correctness of the theory being investigated, then the inquiry would be circular! By failing to consider condition 2, Chalmers views "using *T*" as assuming the truth of *T*. An inference to *T* where *T* was already assumed would earn minimal severity, and thus supply poor evidence for *T*. So if Chalmers's description of scientific practice were correct, it would invalidate our account of evidence as capturing what goes on in science. But if we are right to deny that scientists must circularly accept the theory they are appraising – the kind of position the Kuhnians advance – then Chalmers's account of the role of theory will not do.

6 The Case of GTR

One of the sins that besets philosophers is their tendency to take a historical episode, view it as exemplifying their preferred approach, and then regard our intuitive endorsement of that episode as an endorsement of that approach. To combat this tendency in the HPS literature was one of the goals of Laudan's 1983 project – the so-called VPI program[1] – to test philosophies of science naturalistically. In my contribution to that project, I proposed that severe testing principles be applied on the "metalevel" (Mayo, 1988). This approach enjoins us to examine carefully how historical episodes might "fit" the philosophical account while actually giving an erroneous construal of both what happened and its epistemological rationale. Such a metastatistical critique, alas, is rare; many today largely regard such HPS case studies as mere illustrations, however interesting in their own right.

---

[1] Laudan launched this program by means of a conference at Virginia Tech, then called Virginia Polytechnic Institute and State University (Donovan, et al., 1988, 1992).

Alluding to the example of experimental gravitation given by Clifford Will, Chalmers claims that the "ways in which the development of that theory as construed by Will fits closely with my picture and poses problems for Mayo" (this volume, p. 69). But Chalmers's construal fails to square with that episode in much the same way that the comparativist account fails. The fact that GTR accorded with a variety of results (deflection effect, Mercury, redshift – the three classical tests) would warrant inferring all of GTR for Chalmers; but far from taking this as warranting all of GTR, scientists instead wished to build the PPN parameterization, whose rationale was precisely to avoid accepting GTR prematurely. The PPN framework houses a host of theoretical models and parameters, but the secret to its successful use was that it allowed hypotheses about gravitation to be tested without having to assume any one full gravitational theory, as we already discussed at length (Chapter 1, this volume). Will calls this a "gravitation theory-free" approach, by which he means we do not have to assume any particular theory of gravity to proceed in probing those theories. Because this stage of testing GTR has already been discussed, I turn to Chalmers's appeals to the more contemporary arena, wherein experimental gravitation physicists test beyond the arenas in which viable gravity theories have thus far been severely tested (e.g., in learning about quasars, binary pulsars, and gravity waves).

Here Chalmers's position seems to find support in Will's remarks that "when complex astrophysical systems [are involved] a gravitation-theory independent approach is not useful. Instead, a more appropriate approach would be to assume, one by one, that individual theories are correct, then use the observations to make statements about the possible compatible physics underlying the system. The viability of a theory would then be called into question if the resulting 'available physics space' were squeezed into untenable, unreasonable, or ad hoc positions." (Will, 1993, p. 303)

However, rival theories T_1 and T_2 are not accepted as true when used conditionally in order to derive consequences; else in considering "one by one" the predictions of rival gravity theories, the scientist would be forced to regard as warranted both T_1 and T_2. So unless we are to imagine that Chalmers is endorsing arguments from coincidence to mutually inconsistent rivals, he cannot really mean to say, as he does, that in moving to unexplored domains scientists take the theory as warranted in the untested domain. Instead they deliberately use the gaps in existing tests to *stress* theories further – to identify hurdles for lopping off some that have thus far survived.

Clifford Will brings out a very important insight into the roles of background theories in testing a primary theory T; namely, where we enter domains involving uncertain physics we may not be able to test "cleanly" in his sense, or with severity in mine. What clean (severe) tests enable us to do is to *detach inferences* (in this case about gravity) and thereby shrink the possible alternative theories of gravity – in what he calls a "theory-independent way. The use of the PPN formalism was a clear example of this approach. The result was to squeeze theory space" (Will, 1993, p. 303). In cases where we cannot do this, we may at most hypothetically assume now this theory and then the other, in the hope that the predictions from some of the viable theories will be so qualitatively off that even so imprecise a "test" enables some to be killed off (which does not prevent their being used again for purposes of learning).

Chalmers is correct to note that in some cases GTR is used as a tool for measuring astrophysical parameters in the binary pulsar – we become "applied relativists" in Will's terminology. However, he overlooks the careful arguments that ensure the robustness of the resulting inferences. In particular, although "gravitational theory" is "used," it is used in such a way that avoids invalidating any inferred measurements. Several strategies exist to achieve this end. For instance, in using relativistic gravity to estimate the masses of the binary pulsar and its twin, one may rely on aspects for which all viable gravity theories agree, or they may conservatively take the values of parameters that represent the furthest a gravitation theory can disagree with GTR values. "The discovery of PSR 1913 + 16 caused considerable excitement in the relativity community . . . because it was realized that the system could provide a new laboratory for studying relativistic gravity"; in general, "the system appeared to be a 'clean' laboratory, unaffected by complex astrophysical processes" (Will, 1993, p. 284). Here, "relativistic gravitational theory" – but no one theory within the viable set – is used as a tool to estimate statistically such parameters as the mass of the pulsar. By obtaining knowledge of relativistic effects without assuming the truth of any one relativistic theory, we may opportunistically use the severely passed relativistic hypotheses to increase knowledge of relativistic phenomena such as gravity waves.

In particular, experimental relativists were able to contrast the predictions regarding the effects of gravity waves on the pulsar's orbit (in time for the centenary of Einstein's birth in 1979). Hypothetically assuming alternative theories of gravitation, they discover that one theory, Rosen's bimetric theory, "faces a killing test" by yielding a prediction qualitatively different – the orbit should slow down rather than speed up (Will, 1993, p. 287; Mayo, 2000a). The orbital decay that is estimated is in sync with GTR, but this is not regarded as providing reliable evidence for GTR; at most it provides indirect evidence for the existence of gravity waves. The adjustable parameter in Brans–Dicke theory prevents the binary results from discriminating between them: "the theoretical predictions are sufficiently close to those of general relativity, and the uncertainties in the physics still sufficiently large that the viability of the theory cannot be judged reliably" (Will, 2004, p. 307). In interpreting the results, in other words, there is a careful assessment to determine what is ruled out with severity. The techniques by which uncertainties are "subtracted out" are part of the day-to-day measurements in experimental gravity, and their properties need to be understood (and critically evaluated) by philosophers of science if they are to learn from the episode. Far from providing grounds that all of T must be accepted as true for this opportunistic learning, the implications of Chalmers's arguments reinforce our claim that "enough experimental knowledge will do" in making progress.

7 The Role of Theories and the Error-Statistical Perspective

The error-statistical conception of the roles of theory admittedly is at odds with some standard conceptions. These differences, or so our exchanges have seemed to show, may result in our meaning different things using the same terms. I consider three areas that still need to be developed:

1. *Not One by One Elimination.* In claiming large-scale theories may pass severely through piecemeal tests, theory testers (Laudan, Chalmers, Musgrave) suppose I mean we must plod through all the predictions and all the domains, and they conclude it cannot be done (Laudan, 1997: "there is little prospect of severity flowing up"). Some assume I would need an exhaustive account of all theories or all ways a given theory may fail and then eliminate them one by one. However, this process overlooks the ingenuity of experimental learning. By building up severely affirmed effects and employing robustness arguments, a single type of local inference – once severely corroborated – can yield the theory (along the lines of "A Big Shake-up Turns on a Small Result" in the case of Brownian motion; Mayo, 1996, p. 246). When Will asserts that "in fact, we conjecture that for a wide class of metric theories of gravity, the binary pulsar provides the *ultimate* test of relativistic gravity" (Will, 1993, p. 287), he is referring to the fact that a distinct type of gravity wave signature (reflecting its adherence to the strong equivalence principle), once found, could entail GTR. It matters not whether this is actually the case for this episode; what matters is the idea of large-scale appraisal turning on very local results.

2. *Understanding a Theory.* One often hears scientists make claims about having a correct or an incorrect *understanding* of a theory or of the way some processes of interest behave in some domain. This seems to be what is captured in talking about experimental knowledge in relation to theories. An example would be how experimental relativists claim that it is only when they began experimentally testing GTR that they really started to understand relativistic gravity. What is behind this type of claim is basically what I mean to capture in saying that what is learned with severity is "experimental knowledge" (certainly I did not mean to limit it to knowledge of "observables"). So even if it was somehow known in 1930 that GTR was true (despite the limited evidence), scientists could not be said to have correctly understood relativistic gravity – how it behaves in its variety of interactions and domains. (Work on explanation does not quite get at this goal; see Glymour, Chapter 8, this volume.)

Thus, what is learned with severity need not be well captured by writing down some hypotheses or equations that are a piece of the full theory, at least not as that is usually understood. It may be best captured by one or more hypotheses couched in one of the intermediate models linking data and theory. But even those hypotheses seem more like placeholders for the full comprehension of the mechanism or process involved.

The interest in deliberately probing the ways one may be wrong – deliberate criticism – stems from the goal of making progress in understanding and revealing how misunderstandings were concealed, which leads to my third point:

3. *Not Kuhnian Normal Science.* Those who champion accepting a large-scale theory, even knowing many domains have yet to be probed, often sound as if they see their task as providing the scientist with a valid excuse for holding onto *T*, as if scientists wished to have an alibi for adhering to the status quo. The Kuhnian "normal" scientist who will not rock the boat except when faced with a crisis of anomalies comes to mind.

Ironically, while these theory testers are hardly fans of Kuhn, the picture they paint is far too redolent of the Kuhnian normal scientist, at least to my tastes. Popper rejected ("as dangerous") the Kuhnian conception of normal science for just the kind of complacency it fostered, advocating instead the view of the scientist as continually looking for flaws and fallibilities (see Mayo, 1996, ch. 2). One may grant that Kuhnian normal scientists are being perfectly "rational" and perhaps earning their paychecks, while denying this mindset would spark the kind of probing and criticism that leads to extending the frontiers of knowledge. Why then does so much philosophical talk about theory appraisal (even among followers of Popper) seem to reflect the supposition that it is desirable to have things settled? Taking seriously the goal of "finding things out," the scientist in my view is driven to create novel effects and to compel the replacement of hypotheses and theories with ones that not only give a more correct understanding about more domains, but also teach us where our earlier understanding was in error.

Looking back, these three points hook up neatly with the experimental theses delineated in Chapter 1, Mayo (1996). Admittedly, I was unaware of these connections until I took up the challenge to supply the error-statistical account of experiment with an account of large-scale theories – thanks to Chalmers and our many exchanges.

8 A Word on the Task of Metamethodology

The error-statistical philosophy of science requires identification of the epistemological rationale behind strategies (e.g., varying the results, replication, novelty). It will not do to identify an impressive historical episode and note that it exemplifies the strategy. The method we supply for pinpointing the rationale is to consider how it might contribute to controlling error probabilities and making inferences severely. This task of "metamethodology" (Mayo, 1996, p. 455) requires "(a) articulating canonical models or paradigm cases of experimental arguments and errors, and (b) appraising and arriving at methodological rules by reference to these models. Historical cases, if handled correctly, provide a unique laboratory for these tasks... [however] the data we need do not consist of the full scientific episode, all finished and tidied up. The data we need are the experimental data that scientists have actually analyzed, debated, used, or discarded... Especially revealing are the processes and debates that take place before the case is settled and most of the learning is going on."

Chalmers's rule – require demanding arguments from coincidence – is clearly in the spirit of demanding severity, but by leaving it at a vague level we cannot determine its valid use or rationale. The most familiar canonical exemplar of an argument from coincidence is statistical null hypothesis testing for ruling out "mere chance" in inferring a "real" or reproducible effect. The null hypothesis, H_0, asserts that the effect or observed agreement is "due to coincidence" and the test is designed to ensure that if in fact H_0 can adequately account for experimental results, then with high probability, outcomes consistent with H_0 would occur (i.e., there is a low probability that the null hypothesis H_0 would be found false erroneously). In effect the nonnull hypothesis – that the effect is real or not chance – is given a "hard time" before data are regarded as evidence for it (see Mayo and Cox, Cox and Mayo, Chapter 7, this volume). The argument in the formal statistical arena provides an exemplar for the argument from coincidence more generally. However, to apply it more generally to cover cases without any explicit probability model, we erect various strategies for giving H_0 a "hard time." One such strategy is to require that the effect be repeatable in a "wide variety" of cases.

But not just any kind of intuitively varied results constitute relevant variety. The different trials should check each other, so that whatever causes an error in experiment E_1 would not also be causing an error in E_2. Otherwise the variability does not strengthen the severity of the inference. Consider my highly informal example (Chapter 1) of inferring that George had not gained weight by checking him with a variety of scales with known calibrations. This is best classified as a strategy for checking errors in measuring instruments or in underlying assumptions of statistical models. For example, if I tested George's weight using only a single scale, it would not severely warrant the hypothesis of no weight gain because it might be some property of this scale that is responsible. Of course, we would really like to check directly that our instruments are working or that the underlying assumption holds in the case at hand; it is precisely in cases where such checking is not feasible that arguments from variety are important. If the identical effect is seen despite deliberately varying the backgrounds that could mar any one result, then we can subtract out the effects of flawed assumptions. However, this would not be relevant variability for ruling out other claims such as explanations for his weight maintenance (e.g., a pill that acts as a thermostat causing more fat to burn with increased consumption – how I wish!). That sort of theory would require different kinds of variable results to be inferred with severity. My general point is that, without a clear understanding of the epistemic value of a given strategy, interpretations from cases in HPS may be illicit.

9 Progress?

In Chalmers's new book (2009), there are signs that our positions on theory testing are moving closer on a key point we have been discussing. Agreeing now that "theories can be partitioned into those parts that have been and those that have not been tested," at least in some cases, he is perhaps now prepared to concur that a theory is confirmed by an argument from coincidence only if "the successful tests cannot be accounted for by some specified sub-set of the theory." It will be interesting to see whether a weaker notion of theory testing is still thought to be needed.

References

Chalmers, A. (1999), *What is This Thing Called Science?* 3rd ed., Open University Press, and University of Queensland Press.

Chalmers, A. (2009), *The Scientist's Atom and the Philosopher's Stone: How Science Succeeded and Philosophy Failed to Gain Knowledge of Atoms*, Springer, Dordrecht.

Donovan, A., Laudan, L., and Laudan, R. (1988), *Scrutinizing Science*, Kluwer, Dordrecht (reprinted by Johns Hopkins University Press, 1992).

Kuhn, T. (1962), *The Structure of Scientific Revolutions*, University of Chicago Press, Chicago.

Laudan, L. (1997), "How About Bust? Factoring Explanatory Power Back into Theory Evaluation," *Philosophy of Science*, 64: 306–16.

Mayo, D. (1988), "Brownian Motion and the Appraisal of Theories," pp. 219–43 in A. Donovan, L. Laudan, and R. Laudan (eds.), *Scrutinizing Science*, Kluwer, Dordrecht (reprinted by Johns Hopkins University Press, 1992).

Mayo, D.G. (1996), *Error and the Growth of Experimental Knowledge*, University of Chicago Press, Chicago.

Mayo, D.G. (2000a), "Experimental Practice and an Error Statistical Account of Evidence," pp. S193–S207 in D. Howard (ed.), *Philosophy of Science*, 67 (Symposia Proceedings).

Will, C.M. (1993), *Theory and Experiment in Gravitational Physics*, Cambridge University Press, Cambridge (revised edition).

Will, C.M. (2004), "The Confrontation between General Relativity and Experiment," *Living Reviews in Relativity*, http://relativity.livingreviews.org/Articles/lrr-2001–4/title.html.

Related Exchanges

Chalmers, A. (2000), "'What Is This Thing Called Philosophy of Science?' Response to Reviewers of *What Is Thing Called Science?* 3rd Edition," *Metascience*, 9: 198–203.

Chalmers, A. (2002), "Experiment and the Growth of Scientific Knowledge," pp. 157–69 in P. Gardenfors, J. Wolenski, and K. Kijania-Placet (eds.), *In the Scope of Logic, Methodology and Philosophy of Science*, Vol. 1, Kluwer, Dordrecht.

Mayo, D.G. (2000b), "'What Is This Thing Called Philosophy of Science?' Review Symposium of A. Chalmers' *What Is This Thing Called Science?*" *Metascience*, 9: 179–88.

Mayo, D.G. (2002), "Theory Testing, Statistical Methodology, and the Growth of Experimental Knowledge," pp. 171–90 in P. Gardenfors, J.Wolenski, and K. Kijania-Placek (eds.), *In The Scope of Logic, Methodology and Philosophy of Science* (Vol. 1 of the 11th International Congress of Logic, Methodology, and Philosophy of Science, Cracow, August 1999), Kluwer, Dordrecht.

Staley, K. (2008). "Error-Statistical Elimination Of Alternative Hypotheses," *Synthese (Error and Methodology in Practice: Selected Papers from ERROR 2006)*, Vol. 163(3): 397–408.

Worrall, J. (2000), "'What Is This Thing Called Philosophy of Science?' Review Symposium of A. Chalmers' *What Is This Thing Called Science?*" *Metascience*, 9: 17–9.

THREE

Revisiting Critical Rationalism

Critical Rationalism, Explanation, and Severe Tests

Alan Musgrave

This chapter has three parts. First, I explain the version of critical rationalism that I defend. Second, I discuss explanation and defend critical rationalist versions of inference to the best explanation and its meta-instance, the Miracle Argument for Realism. Third, I ask whether critical rationalism is compatible with Deborah Mayo's account of severe testing. I answer that it is, contrary to Mayo's own view. I argue, further, that Mayo needs to become a critical rationalist – as do Chalmers and Laudan.

1 Critical Rationalism

Critical rationalism claims that the best method for trying to understand the world and our place in it is a critical method – propose views and try to criticise them. What do critical methods tell us about truth and belief? If we criticize a view and show it to be false, then obviously we should not believe it. But what if we try but fail to show that a view is false? That does not show it to be true. So should we still not believe it?

Notoriously, the term "belief" is ambiguous between the act of believing something (the believing) and the thing believed (the belief). Talk of "reasons for beliefs" inherits this ambiguity – do we mean reasons for believings or reasons for beliefs? Critical rationalists think we mean the former. They think there are reasons for believings that are not reasons for beliefs. In particular, failing to show that a view is false does not show it to be true, is not a reason for the belief. But it is a reason to think it true, for the time being anyway – it is a reason for the believing. Thus, it may be reasonable to believe a falsehood if you have sought but failed to find reasons to think it false. If you later find reason to think a view false, you should no longer believe it. Then we should say that what you previously believed was false – not that it was unreasonable for you ever to have believed it.

This is just common sense. The trouble is that philosophical tradition denies it. Philosophical tradition says that a reason for believing something must also be a reason for what is believed; it must show that what is believed is true, or at least more likely true than not. I call this "justificationism" and reject it. I think we can justify (give reasons for) believings without justifying the things believed.

This is not the usual reading of Popper's critical rationalism. Most of the so-called "Popperians" reject it. The High Priests of the Popper Church (I am thinking chiefly of David Miller) think there are no reasons for beliefs and, therefore, no reasons for believings either. They accept justification*ism* but reject all justification*s*. They think that critical rationalism has no need of a theory of justified or reasonable believing. They say they do not believe in belief, and fly instead to Popper's "Third World" of the objective contents of our thoughts. But you do not solve the problem of induction in that way. You do not avoid Hume's radical inductive scepticism – the thesis that all evidence-transcending believings are unreasonable – by agreeing with him. You can only avoid Hume's radical inductive scepticism by saying that some evidence-transcending believings are reasonable after all (and by proceeding to say which ones). Most philosophers object that Popper's answer to Hume smuggles inductive reasoning in somewhere. But this objection smuggles in precisely the justificationist assumption that Popper denies – the assumption that a reason for believing an evidence-transcending hypothesis must be a reason for that hypothesis.

Scepticism is underpinned by justificationism. Justificationism says that, to show that it is reasonable to believe something, you must show that what is believed is true or more likely true than not. If you invoke a reason for your belief, the sceptic asks for a reason for your reason, and an infinite regress opens up. To stop that regress, philosophers invoke certainly true "first principles" of some kind – "observation statements" if you are a classical empiricist, "self-evident axioms" if you are a classical rationalist. Rejecting justificationism enables critical rationalists to drive a wedge between scepticism about certainty (which is correct) and scepticism about rationality (which is not). Failure to show that a belief is false does not show it to be true, but does show that believing it is reasonable. Have sceptics not shown that our beliefs are false? No, sceptics produce no criticisms of our beliefs – they only produce excellent criticisms of attempts to prove them.

Justificationism also lies behind inductivism. If you accept justificationism, then you need inductive or ampliative reasoning to show that evidence-transcending beliefs are true or more likely true than not and, hence, reasonably believed. And you need inductive logic to show that inductive reasoning is valid or "cogent." Critical rationalists reject justificationism and have no need of inductive reasoning or inductive logic. Deductive reasoning is enough for them, and deductive logic is the only logic that they have or need. I illustrate this in the next section, which is about inference to the best explanation.

2 Explanationism

As I see it, critical rationalism and critical realism go hand-in-hand. What do I mean by critical realism? I mean that science is in the business of explaining the phenomena. Realists disagree with van Fraassen's constructive empiricist slogan – that the "name of the scientific game is saving the phenomena." For realists, the name of the scientific game is explaining phenomena, not just saving them. Realists typically invoke "inference to the best explanation" (IBE). But what sort of inference is IBE? Is it acceptable? And if it is, cannot constructive empiricists and other antirealists accept a version of it, too?

IBE is a pattern of argument that is ubiquitous in both science and everyday life. Van Fraassen has a homely example:

I hear scratching in the wall, the patter of little feet at midnight, my cheese disappears – and I infer that a mouse has come to live with me. Not merely that these apparent signs of mousely presence will continue, not merely that all the observable phenomena will be as if there is a mouse, but that there really is a mouse. (1980, pp. 19–20)

The mouse hypothesis is supposed to be the best explanation of the phenomena – the scratching in the wall, the patter of little feet, and the disappearing cheese.

What exactly is the *inference* in IBE, what are the premises, and what the conclusion? Van Fraassen says, "I infer that a mouse has come to live with me." This suggests that the conclusion is "A mouse has come to live with me" and that the premises are statements about the scratching in the wall and so forth. Generally, the premises are the things to be explained (the *explanandum*) and the conclusion is the thing that does the explaining (the *explanans*). But this suggestion is odd. Explanations are many and various, and it will be impossible to extract any general pattern of inference taking us from *explanandum* to *explanans*. Moreover, it is clear that inferences of this kind cannot be deductively valid ones, in which the truth of the premises guarantees the truth of the conclusion. For the conclusion, the *explanans*, goes beyond the premises, the *explanandum*. In the standard deductive model of explanation, we infer the *explanandum* from the *explanans*, not the other way around – we do not deduce the explanatory hypothesis from the phenomena, rather we deduce the phenomena from the explanatory hypothesis.

But did not the great Newton speak of "deducing theories from the phenomena?" Yes, he did. Moreover, he was right to speak of "deduction" here – not of induction, abduction, or anything fancy like that. He was wrong, however, to think that these deductions were from phenomena *alone.* They are deductions from phenomena *plus* general principles of one kind or another. As Peter Achinstein reminds us in Chapter 5, Mill thought that in some cases a single instance is "sufficient for a complete induction," whereas in other cases many instances are not. Why? Mill said, "Whoever can answer this question knows more of the philosophy of logic than the wisest of the ancients, and has solved the problem of induction" (*System of Logic,* Book III, Chapter 3, Section 3). The answer to Mill's question is plain. In the first case, we are assuming that what goes for one instance goes for all, whereas in the second case we are not. Achinstein writes that whether an inductive inference is a "good" or valid inference is an empirical issue. Referring to Mill's own examples, he says

> we may need only one observed instance of a chemical fact about a substance to validly generalize to all instances of that substance, whereas many observed instances of black crows are required to [validly?] generalize about all crows. Presumably this is due to the empirical fact that instances of chemical properties of substances tend to be uniform, whereas bird coloration, even in the same species, tends not to be. (this volume, p. 174)

Quite so. And if we write these empirical facts (or rather empirical assumptions or hypotheses) as explicit premises, the inferences become deductions. In the first case, we have a deductivly valid inference from premises we think true; in the second case, we have a deductively valid inference from premises one of which (that what goes for the color of one or many birds of a kind goes for them all) we think false. No special inductive logic is required, in which the "goodness" or "validity" or "cogency" of an argument depends on the way the world happens to be. The problem of induction is solved. But this is a long story, which I have told elsewhere.

The intellectual ancestor of IBE is Peirce's *abduction,* and here we find a different pattern:

> The surprising fact, C, is observed.
> But if A were true, C would be a matter of course.
> Hence, . . . A is true.
> (Peirce, 1931–1958, vol. 5, p. 189)

The second premise is a fancy way of saying "A explains C." Notice that explanatory hypothesis A figures in this second premise as well as in the conclusion. The argument as a whole does not generate the *explanans* out of the *explanandum*. Rather, it seeks to justify the explanatory hypothesis. Abduction belongs in the context of justification, not in the context of discovery. (This is a point of some importance. Peirce's abduction was once touted, chiefly by Norwood Russell Hanson, as a long-neglected contribution to the "logic of discovery." It is no such thing. Newtonian "deduction from the phenomena" has a better title to be a "logic of discovery." And its logic is deductive logic.)

Abduction is deductively invalid. Deductivists view it as a deductive *enthymeme* and supply its missing premise, "Any explanation of a surprising fact is true." But this missing premise is obviously false. Nor is any comfort to be derived from weakening the missing premise (and the conclusion) to "Any explanation of a surprising fact is *probably* true" or to "Any explanation of a surprising fact is *close to the* truth." It is a surprising fact that marine fossils are found on mountaintops. One explanation of this is that Martians came and put them there to surprise us. But this explanation is not true, probably true, or close to the truth.

IBE attempts to improve upon abduction by requiring that the explanation is the best explanation that we have. It goes like this:

> F is a fact.
> Hypothesis H explains F.
> No available competing hypothesis explains F as well as H does.
> Therefore, H is true.   (Lycan, 1985, p. 138)

This is better than abduction, but not much better. It is also deductively invalid. Deductivists view it as a deductive *enthymeme* and supply its missing premise, "The best available explanation of a (surprising) fact is true." But this missing premise is also obviously false. Nor, again, does going for probable truth, or closeness to the truth, help matters.

There is a way to rescue abduction and IBE. We can validate them *without* adding missing premises that are obviously false, so that we merely trade obvious unsoundness for obvious invalidity. Peirce provided the clue to this. Peirce's original abductive scheme was not quite what we have considered so far. Peirce's original scheme went like this:

> The surprising fact, C, is observed.
> But if A were true, C would be a matter of course.
> Hence, *there is reason to suspect that* A is true.
> (Peirce, 1931–1958, Vol. 5, p. 189)

This is obviously invalid, but to repair it we need the missing premise "*There is reason to suspect that* any explanation of a surprising fact is true." This missing premise is, I suggest, true. After all, the epistemic modifier "There is reason to suspect that . . ." weakens the claims considerably. In particular, "There is reason to suspect that A is true" can be true even though A is false. If the missing premise is true, then instances of the abductive scheme may be both deductively valid and sound.

IBE can be rescued in a similar way. I even suggest a stronger epistemic modifier, not "There is reason to suspect that . . ." but rather "There is reason to believe (tentatively) that . . ." or, equivalently, "It is reasonable to believe (tentatively) that. . . ." What results, with the missing premise spelled out, is the following:

> *It is reasonable to believe that* the best available explanation of any fact is true.
> *F* is a fact.
> Hypothesis *H* explains *F*.
> No available competing hypothesis explains *F* as well as *H* does.
> Therefore, *it is reasonable to believe that H* is true.

This scheme is valid and instances of it might well be sound. Inferences of this kind are employed in the common affairs of life, in detective stories, and in the sciences.

Of course, to establish that any such inference is sound, the "explanation-ist" owes us an account of when a hypothesis explains a fact and of when one hypothesis explains a fact better than another hypothesis. If one hypothesis yields only a circular explanation and another does not, the latter is better than the former. If one hypothesis has been tested and refuted and another has not, the latter is better than the former. These are controversial issues, to which I shall return. But they are not the most controversial issue – that concerns the major premise. Most philosophers think that the scheme is unsound because this major premise is false, whatever account we can give of explanation and of when one explanation is better than another. So let me assume that the explanationist can deliver on the promises just mentioned, and focus on this major objection.

People object that the best available explanation might be false. Quite so – and so what? It goes without saying that any explanation might be false in the sense that it is not necessarily true. It is absurd to suppose that the only things we can reasonably believe are necessary truths.

What if the best explanation not only might be false but actually is false. Can it ever be reasonable to believe a falsehood? Of course it can. Suppose van Fraassen's mouse explanation is false – that a mouse is not responsible for the scratching, the patter of little feet, and the disappearing cheese. Still, it is reasonable to believe it, given that it is our best explanation of those phenomena. Of course, if we *find out* that the mouse explanation is false, it is no longer reasonable to believe it. But what we find out is that what we believed was wrong, not that it was wrong or unreasonable for us to have believed it.

People object that being the best available explanation of a fact does not prove something to be true or even probable. Quite so – and again, so what? The explanationist principle – "It is reasonable to believe that the best available explanation of any fact is true" – means that it is reasonable to believe or think true things that have not been shown to be true or probable, more likely true than not. If you think that it can only be reasonable to believe what has been shown to be true or probable, and if you think that a good reason for believing something must be a good reason for what is believed, then you will reject abduction and IBE. Justificationism rules out explanationism. Critical rationalists reject justificationism and so they can be explanationists as well. The two go hand-in-hand.

But do explanationism and realism go hand-in-hand as well? Why cannot constructive empiricists also accept IBE and put their own gloss upon it? When it comes to van Fraassen's mouse hypothesis, truth and empirical adequacy coincide, because the mouse is an observable. But when it comes to hypotheses about unobservables, truth and empirical adequacy come apart. Why cannot the constructive empiricist accept IBE, but only as licensing acceptance of the best explanation as empirically adequate, not as true? As Howard Sankey puts it: "The question is why it is reasonable to accept the best explanation *as true*. Might it not be equally reasonable to accept the best explanation as empirically adequate . . . ?" (2006, p. 118)

My answer to this question is NO. Suppose that H is the best explanation we have of some phenomena. Remember the T-scheme: It is true that H if and only if H. Given the T-scheme, to believe that H and to believe that H is true are the same. Given the T-scheme, to accept that H and to accept that H is true are the same. So what is it to "accept that H is empirically adequate?" It is not to accept H, for this is the same as accepting that H is true. Rather, it is to accept a meta-claim about H, namely the meta-claim "H is empirically adequate" or equivalently "The observable phenomena are as if H were true." Call this meta-claim H^*. Now, and crucially, H^* is *no explanation at all of the phenomena*. The hypothesis that it is raining explains why the streets are wet, but "The phenomena are as if it were raining" does not. Ergo, H^* is not the best explanation – H is, or so we assumed. (Actually, all we need assume is that H is a better explanation than H^*.) So, given IBE, H^* should not be accepted as true. That is, given IBE, H should not be accepted as empirically adequate.

I wonder which part of this argument will be rejected by those who believe that there is a constructive empiricist version of IBE. Not IBE – at least, they are pretending to accept it. Not, presumably, the *T*-scheme. Not, presumably, its consequence, that to accept *H* and to accept *H* as true are the same thing. Not, presumably, the equivalence of "*H* is empirically adequate" and "The observable phenomena are as if *H* were true." Not, presumably, the claim that *H* is a better explanation of the phenomena than "The phenomena are as if *H* were true."

What this shows is that realism and explanation go hand-in-hand. If you try to recast IBE in terms of empirical adequacy rather than truth, you end up with something incoherent. You start off thinking that it is reasonable to accept the best explanation as empirically adequate, and you end up accepting something that is no explanation at all.

This does not refute constructive empiricism. It only refutes the idea that constructive empiricists can traffic in explanation, and in IBE, just as realists do. It is no accident that down the ages acute antirealists have pooh-poohed the idea that science explains things. Van Fraassen should join Duhem in this, as he already has in most other things.

So much for explanation, and for IBE, in science. The same considerations apply to IBE in metascience, to the so-called Miracle Argument for scientific realism. What is to be explained here is not a fact about the world, like the scratching in the wall or the disappearing cheese. What is to be explained is a fact about science – the fact that science is successful. The success in question is predictive success, the ability of a theory to yield true predictions about the observable, and the technological success that sometimes depends upon this. The key claim is that the best explanation of a theory's predictive success is that it is true. Given this claim, IBE licenses reasonable belief in the truth of that theory.

It is only *consistent* empirical success that can be explained in terms of truth – you cannot explain the *partial* success of a falsified theory in terms of its truth. (This is an important point to which I shall return.) So the *explanandum* is of the form " All *T*'s predictions about observable phenomena are true," or (putting it in van Fraassen's terminology) "*T* is empirically adequate," or (putting it in surrealist terminology, following Leplin, 1993) "The observable phenomena are as if *T* were true." The realist thought is that *T*'s actually being true is the best explanation of why all the observable phenomena are as if it were true.

To accept a theory as empirically adequate and set out to explain why is, of course, to make an "inductive leap." But it is no different in science itself. Scientists typically seek to explain general statements rather than statements of particular fact. If it is reasonable for scientists to seek to explain why sticks always look bent when dipped in water, then it might be reasonable for metascientists to seek to explain why some scientific theory is empirically adequate.

So, supposing that it makes sense to try to explain empirical adequacy, how exactly does truth do it? Suppose the theory in question asserts the existence of unobservable or theoretical entities. The theory will not be true unless these existence claims are true, unless the theoretical entities really exist, unless the theoretical terms really do refer to things. So part of the realist story is that T is observationally adequate because the unobservables it postulates really do exist. But this cannot be the whole realist story (as Larry Laudan has tirelessly pointed out). Reference may be a necessary condition for success, but it cannot be a sufficient condition. A theory may be referential yet false and unsuccessful. The other part of the realist story is that what the theory says about the unobservables it postulates is true.

The Miracle Argument says not just that truth explains empirical adequacy, but that it is the only explanation, or at least the best explanation. To evaluate this claim, we need to pit the realist explanation of success, in terms of successful reference and truth, against other possible antirealist explanations. What might such antirealist explanations be like? Van Fraassen replaces truth by empirical adequacy as an aim for science. But it is obvious that we cannot satisfactorily explain the empirical adequacy of a theory in terms of its empirical adequacy:

> T is empirically adequate.
> Therefore, T is empirically adequate.

This explanation is no good because it is blatantly circular.

Variants on empirical adequacy, like empirical puzzle-solving (Kuhn) or empirical problem-solving (Laudan), do no better. When we unpack the definitions, we are just explaining empirical adequacy by invoking empirical adequacy. Leplin's *surrealism* does no better. Surrealism arises by taking some theory T and forming its surrealism transform T^*: "The observed phenomena are *as if T* were true." It is clear that "The observed phenomena are as if T were true" is just a fancy way of saying that T is empirically adequate. That being so, we cannot satisfactorily explain the empirical adequacy of T by invoking the surrealist transform of T, because that is, once again, explaining empirical adequacy just by invoking empirical adequacy.

Kyle Stanford's "Antirealist explanation of the success of science" (Stanford, 2000) does no better. Jack Smart suggested long ago that the Copernican astronomer can explain the predictive success of Ptolemaic astronomy by showing that it generates the same predictions as the Copernican theory and by assuming the truth of the Copernican theory (Smart, 1968, p. 151). Stanford considers Smart's suggestion, and says of it:

Notice that the actual content of the Copernican hypothesis plays no role whatsoever in the explanation we get of the success of the Ptolemaic system: what matters is simply that there is some true theoretical account of the domain in question and that the predictions of the Ptolemaic system are sufficiently close to the predictions made by that true theoretical account. (Stanford, 2000, p. 274)

This is quite wrong. The detailed content of the Copernican theory, and the fact that some of the detail of the Ptolemaic theory is similar to it, is essential to the explanation of the success of Ptolemaic theory. For example, it is the fact that the annual periods of the epicycles of the superior planets mimic the true annual motion of the Earth around the Sun that explains why Ptolemy can correctly predict the retrogressions of the superior planets.

Stanford suggests that the predictive success of a theory can be explained by saying that it makes the same predictions as the true theory, whatever that is. But this is explaining "T is predictively successful" by saying "It is predictively as if T were true," or for short, "T is predictively successful." It is incredible that earlier in his paper (2000, pp. 268–9) Stanford accepts that we cannot satisfactorily explain empirical adequacy in terms of empirical adequacy, nor can we adequately explain it in the surrealist way. Yet what he ends up with is just a variant of the surrealist explanation.

Stanford says, in defense of his proposal, that – unlike the realist, constructive empiricist and surrealist proposals, which all appeal to some relationship between the theory and the world to explain its success – his proposal "does not appeal to a relationship between a theory and the world at all; instead it appeals to a relationship of predictive similarity *between two theories*" (2000, p. 276). This seems to be a double joke. First, there are not two theories here at all; there is one theory and an existential claim that there is some true theory somewhere predictively similar to it. Second, you hardly explain the success of T by saying "T is predictively similar to some other theory T^*," for T^* might be false and issue in false predictions. The truth of T^*, whether T^* is spelled out or just asserted to exist (as here), is essential to the explanation of T's success. The relation of T^* to the world is essential, in other words.

Stanford counts it a virtue of his proposal that it does not involve asserting the truth of any particular theory – all that is asserted is that *there is* some true theory T^* predictively similar to T. It might be thought that, simply by invoking the truth of some unspecified theory or other, Stanford's proposal remains a realist proposal. (This is suggested by Psillos [2001, p. 348].) Not so. I can satisfy Stanford by invoking the truth of the surrealist transform of

T. But then I end up saying that it is the truth of "The phenomena are as if *T* were true" that explains *T*'s success. I can also satisfy Stanford by invoking the truth of Berkeley's surrealist philosophy. It is the truth of "God creates experiences in our minds as if science were true" that explains why science is successful. Surrealist transforms are by design structurally similar to what they are transforms of. There is nothing realist about them. And, to repeat, Stanford previously conceded that the explanations of success they offer are no good.

In the Ptolemy-Copernicus case, the empirical success of a false theory (Ptolemy) is explained by invoking its similarity to a true theory (Copernicus). The similarity explains why the two theories make the same predictions – the truth of the second theory explains why the predictions of the first theory are true even though the first theory is false. The surrealist transform of Ptolemy's theory – "Observed planetary motions are *as if* Ptolemy's theory were true" – follows from Ptolemy's theory *and* from Copernicus. Realists about Copernicus become surrealists about Ptolemy to explain the empirical adequacy of Ptolemy. But Copernican realism, not Ptolemaic surrealism, is doing the explaining here. Copernicus tells us *why* the phenomena are as if Ptolemy were true.

The key premise of the Miracle Argument was that the truth of a theory is the best explanation of the empirical adequacy of that theory. So far, at least, that key premise seems to be correct. From which it follows, provided we accept IBE, that it is reasonable to believe that an empirically adequate theory is true.

We have assumed that truth explains empirical adequacy better than empirical adequacy does, because the latter "explanation" is completely circular. Normally, when we go for explanatory depth as opposed to circularity, we would like some independent evidence that the explanation is true. But this is a curious case. There can be no independent evidence favouring an explanation in terms of truth against a (circular) explanation in terms of empirical adequacy. The realist explanation is not circular. It tells us more than the antirealist explanation, but in the nature of the case there can be no evidence that the more it tells us is correct. My response to this is to bite the bullet: there are explanatory virtues that do not go hand-in-hand with evidential virtues. How could the two go hand-in-hand, when the explanatory rival is *by design* evidentially equivalent?

It should really be obvious that explanatory virtues do not always go hand-in-hand with evidential virtues. The ancients explained the motions of the fixed stars by saying that they were fixed on the surface of an invisible celestial sphere that rotates once a day around the central Earth. Compare that hypothesis with its surrealist transform, the hypothesis that the stars move *as if* they were fixed to such a sphere. The realist hypothesis is explanatory, and the surrealist hypothesis is not, despite the fact that the latter is expressly designed to be evidentially equivalent with the former. Similarly with the nineteenth-century geological theory of fossil formation, G, and Philip Gosse's surrealist transform G^*: God created the universe in 4004 BC *as if* G were true. There are quite different explanations here, but no geological evidence can decide between them – it was not on evidential grounds that nineteenth-century thinkers rejected G^* out of hand. Finally, and most generally, consider the realist explanation of the course of our experience proffered by common sense and science, R, with its Berkeleyan surrealist transform R^*: God causes our experiences *as if* R were true. Again, no experience can decide between R and R^*, since R^* is expressly designed to be experientially equivalent with R.

These examples are meant to show that realists should not be browbeaten by the fact that antirealists can come up with alternative hypotheses to the realist ones that empirical evidence cannot exclude. These alternatives can be excluded on explanatory grounds. Either they provide no explanations at all, or only incredible ones. It is the same with antirealist explanations of the success of scientific theories in terms of their empirical adequacy (however precisely formulated). Such explanations are either no explanations at all or are completely inadequate circular ones, and can be rejected as such.

If Larry Laudan were dead, he would now be turning in his grave. Happily, he is not dead; he is sitting here today and is probably squirming in his seat. How good an argument for realism is it, he will be asking, that the truth of a theory best explains its empirical adequacy? The argument refers only to the special case of empirically adequate theories. But empirical adequacy is an extreme case – rare, perhaps even nonexistent – in the history of science. Most, perhaps all, theories in the history of science enjoy only partial success (at best). It is the sum of these partial successes to which phrases like "the success of science" refer. You cannot explain the partial success of a falsified theory by invoking its truth. So the realist has no explanation of partial success. And if the "success of science" is a collection of partial successes, the realist has no explanation of the success of science either.

This is, in essence, Laudan's criticism of the Miracle Argument on historical grounds. He points out, first of all, that the global claim that science is successful is a hopeless exaggeration. Many scientific theories are spectacularly *un*successful. We must confine ourselves to successful theories rather than to science as a whole. But even among successful theories, most if not all enjoy some success but are not completely successful. What this means is that a theory yields some true observational consequences and some false ones, saves some regularities in the phenomena, but gets others wrong. Now assuming that the scientists involved have made no logical or experimental error, and assuming that the false predictions have actually been tested, a partially successful theory of this kind has been *falsified*. No sensible realist can invoke the truth of a falsified theory to explain its partial success! Laudan produces many examples of theories that were partially successful yet not true, nor even referential.

Actually, it was "referential realism," the idea that reference explains success, that was the chief target of Laudan's famous "confutation of convergent realism" (Laudan, 1981). To be fair to Laudan, this was the view that one could glean from incautious formulations found chiefly in Putnam's writings (e.g., Putnam, 1975). It is a view that spawned what I have called "entity realism" – the idea that realists need not believe in the truth or near truth of any theories, that it is enough just to believe in the theoretical entities postulated by those theories. It was not for nothing that Laudan attributed to the realist claims like "A theory whose central terms genuinely refer will be a successful theory." And he proceeded to refute this claim by giving examples of referential theories that were not successful and of successful theories that were not referential.

Referential realism or entity realism is a hopeless form of realism. There is no getting away from truth, at least for realists. To believe in an entity, while believing nothing else about that entity, is to believe nothing or next to nothing. I tell you that I believe in hobgoblins. "So," you say, "You think there are little people who creep into houses at night and do the housework." To which I reply that I do not believe that, or anything else about what hobgoblins do or what they are like – I just believe *in* them. It is clear, I think, that the bare belief in hobgoblins – or equivalently, the bare belief that the term "hobgoblin" genuinely refers – can explain nothing. It is equally clear, I think, that mere successful reference of its theoretical terms cannot explain the success of a theory. Laudan has an excellent argument to prove the point. Take a successful theory whose terms refer, and negate some of its claims, thereby producing a referential theory that will be unsuccessful. "George W. Bush is a fat, blonde, eloquent atheist" refers to Bush all right, but would not be much good at predicting Bush phenomena. Not that the ink spilled on reference was wasted ink – reference is (typically) a necessary condition for truth. A theory that asserts the existence of an entity will not be true unless that entity exists. But reference is not a sufficient condition for truth. A theory can be referential, yet false – and referential, yet quite unsuccessful.

Laudan's critique prompts a further antirealist worry. Partial success cannot be explained in terms of truth. So if it is explicable at all, it must be explicable in terms other than truth. So why can we not also explain total success in terms other than truth?

There is an obvious realist response to this. Just as total success is best explained in terms of truth, partial success is best explained in terms of partial truth. Return to the Ptolemy-Copernicus case. So far we have assumed that Ptolemaic astronomy was empirically adequate and Copernican astronomy true. Of course, neither of these assumptions is correct. What is really the case is that Ptolemy's explanation of retrograde motions shared a true part with Copernican theory. That true part, common to both theories, sets out the relative motions of Earth, Sun, and superior planets.

Partial truth is not the same as *verisimilitude*. Verisimilitude is closeness to the truth – the "whole truth" – of a false theory taken as a whole. Partial truth is just truth of parts. A simple example will make the difference clear. "All swans are white" is false because of the black swans in Australasia. (I had to get this baby example in – as some uncharitable soul once joked, having black swans in it is Australasia's chief contribution to the philosophy of science!) Despite its falsity, "All swans are white" is predictively successful in Europe, and bird-watchers find it useful to employ it there. I do not know how close to the (whole) truth is "All swans are white," and none of the captains of the verisimilitude industry can tell me in less than one hundred pages of complicated formulas. I do know that "All swans are white" has a true part (a true consequence), "All European swans are white," whose simple truth explains the success had by European bird-watchers.

The simple example with the swans can be generalized. A false theory T might be successful (issue nothing but true predictions) *in a certain domain D*. Explain this, not by saying that T is close to the truth but by saying that "In domain D, T" is true. A false theory T might be successful (issue nothing but true predictions) *when certain special conditions C are satisfied*. Explain this, not by saying that T is close to the truth but by saying that "Under conditions C, T" is true. A false theory T might be successful *as a limiting case*. Explain this, not by saying that T is close to the truth but by saying that "In the limit, T" is true. Notice that "In domain D, T" and "Under conditions C, T" and "In the limit, T" are all *parts* of T, that is, logical consequences of T. Of course, the conjunction S of the successes of T is also a logical consequence of T. But whereas S does not (satisfactorily) explain S, "In domain D, T" or "Under conditions C, T" or "In the limit, T" might explain S perfectly well. These restricted versions of T are not the same as its surrealist transform – restricted versions of T may be explanatory whereas its surrealist transform is not.

Of course, if we accept such an explanation, it immediately raises the question of *why* the restricted version of *T* is true while *T* is false. Typically, it is the successor theory to *T* that tells us why *T* is true in a certain domain, under certain special conditions, or as a limiting case. Still, that this further question can be asked and answered does not alter the fact that a true restricted version of *T* can explain *T*'s partial success while *T*'s surrealist transform does not.

It is the same with *approximate truth*, as when we say that "It is four o'clock" or "John is six feet tall" are only *approximately true*. What we mean is that the statements "It is *approximately* four o'clock" or "John is *approximately* six feet tall" are true. And if we want to be more precise, we can say that "It is four o'clock give or take five minutes" or "John is six feet tall give or take an inch" are true. Approximate truth is not to be explained by trying to measure the distance of a sentence from the (whole) truth. Approximate truth is truth of an approximation. Approximate truth is a species of partial truth, because the approximations in question are logical parts of what we began with. "It is four o'clock" logically implies "It is approximately four o'clock" as well as "It is four o'clock give or take five minutes," and "John is six feet tall" logically implies "John is approximately six feet tall" as well as "John is six feet tall give or take an inch."

I have come to believe that the entire verisimilitude project was a bad and unnecessary idea. Popper's definition of the notion of "closeness to the (whole) truth" did not work. The plethora of alternative definitions of "distance from the (whole) truth" that have taken its place are problematic in all kinds of ways. And what was the point of the verisimilitude project? It was precisely to explain how a false theory can have partial success. Now it is obvious that a true theory will be successful – after all, true premises yield true conclusions. But it is not obvious that a theory that is close to the truth will be successful, since near truths yield falsehoods as well as truths. We should eschew the near truth of false wholes in favour of the simple truth of their parts. We should explain partial success in terms of truth of parts. Whole truths are wholly successful, partial truths partially successful. Either way, it is simple truth, not verisimilitude, that is doing the explaining.

I am not saying that partial success can always be explained by partial truth in this way, nor am I saying that it need be so explained. There is a kind of partial predictive success that needs no explanation at all because it is no "miracle" at all – it is not even mildly surprising! Here is a simple schematic example to illustrate what I mean. Suppose a scientist has the hunch that one measurable quantity *P* might depend linearly on another measurable quantity *Q* – or perhaps the scientist does not even have this hunch, but just wants to try a linear relationship first to see if it will work. So she measures two pairs of values of the quantities P and Q. Suppose that when Q is 0, P is 3, and when Q is 1, P is 10. She then plots these as points on a graph and draws a straight line through them representing the linear relationship. She has performed a trivial *deduction*:

> $P = aQ + b$, for some a and b.
>
> When Q is 0, P is 3 (so that $b = 3$)
>
> When Q is 1, P is 10 (so that $a = 7$)
>
> Therefore $P = 7Q + 3$

Now the point to notice is that the hypothesis $P = 7Q + 3$ successfully predicts, or "postdicts," or at least entails that, when Q is 0, P is 3, and that when Q is 1, P is 10. Are these successes miraculous, or even mildly surprising? Of course not. Those facts were used to construct the hypothesis (they were premises in the deductive argument that led to the hypothesis). It is no surprise or miracle that the hypothesis gets these things right – they were used to get the hypothesis in the first place. (Of course, *given that* we have a linear relationship here, the results of the measurements are a conclusive reason for the hypothesis that $P = 7Q + 3$. But the measurements were not tests of that hypothesis, let alone severe tests.)

This trivial example illustrates a general point. Success in predicting, postdicting, or entailing facts used to construct a theory is no surprise. It is only *novel* predictive success that is surprising, where an observed fact is novel for a theory when it was not used to construct it.

Finally, a realist can say that accidents happen, some of them lucky accidents, in science as well as in everyday life. Even when a fact is not used to construct a theory, that theory might successfully predict that fact by lucky accident. It is not my claim that the correct explanation of predictive success is *always* in terms of truth or partial truth. My claim is that the best explanation of total predictive success is truth, and that the best explanation of partial predictive success (where it is not a lucky accident) is partial truth.

Nancy Cartwright argues that the predictive success of science is *always* a kind of lucky accident. It always arises from what Bishop Berkeley called the "compensation of errors." According to Cartwright, the laws or theories in science are always false (I shall come back to this). But scientists busy themselves to find other premises which, when combined with these false laws, generate true predictions. And, scientists being clever folk, it is no wonder that they succeed. A trivial example may make the point clear. Suppose the "phenomenological law" we want is "Humans are two-legged," and the false law of nature we have to work with is "Dogs are two-legged." What do we have to add to the false law to get the phenomenological law? Well, the auxiliary assumption "Humans are dogs" will do the trick. And two wrongs, carefully adjusted to each other, make a right.

Bishop Berkeley complained that the mathematicians of his day were only able to get correct results in their calculations because they systematically made mistakes that cancelled one another out. Berkeley observed that there was nothing so scandalous as this in the reasoning of theologians. Cartwright thinks the scandal is endemic in the reasoning of physicists: "Adjustments are made where literal correctness does not matter very much in order to get the correct effects where we want them; and very often . . . one distortion is put right by another" (Cartwright, 1983, p. 140).

In a case like this, one would be crazy to suppose that the best explanation of the theory's predictive success is its truth. A better explanation is that we get out what we put in to begin with. We use a known fact ("Humans are two-legged" in my trivial example) and a false theory we have ("Dogs are two-legged," in my trivial example) to generate an auxiliary theory ("Humans are dogs," in my trivial example) that will get us back to the known fact. It is no miracle that we get out what we put in. And our success in getting it is no argument for the truth of what we get it from.

Why does Cartwright think that the laws of physics lie, that is, are always false? The laws lie, she thinks, because they idealize or simplify things – they are false because they do not tell the *whole* truth. This is a mistake. "Nancy Cartwright is clever" is true, even though it does not tell the whole truth about Nancy Cartwright. Similarly, Newton's law of gravity is not false just because it does not tell the whole truth about the forces of nature.

But never mind this; the important point is that predictive success is no miracle if the predicted facts are used to construct the theory in the first place. What is miraculous is novel predictive success. And the best explanation of such "miracles" is truth – either truth of wholes or truth of parts.

So, the basic idea of critical rationalism is that it is reasonable to believe something if you have tried and failed to criticise it and show it to be false. And the basic idea of explanationism is that it is reasonable to believe, or adopt, or prefer the best available explanation we have. Obviously, an explanation is not the best we have if it has failed to withstand criticisms that other explanations have withstood. These views clearly need to be fleshed out with a theory of criticism – an account of the ways we can criticise explanatory hypotheses of various kinds. One way to criticise a hypothesis is to subject it to observational or experimental tests. Quite generally, criticism must be genuine or serious criticism, which could reveal that the hypothesis is false. So when it comes to empirical testing, the tests must be severe tests. Critical rationalism owes us a theory about what a serious or severe empirical test is. This brings me to my third and last section.

3 Severe Tests

The organizer of our conference is best known to the world for her resolute defence of a particular theory about when a test is a severe test of a hypothesis. And my question is – can we incorporate this theory into critical rationalism? Deborah Mayo does not think so. She rejects critical rationalism. She rejects it because she accepts justificationism. The upshot is that she abandons the project of appraising large-scale theories altogether. I want to entice her back into the critical rationalist camp.

First, I note an oddity. Mayo's entire focus is on what we learn when a test is *passed*. She seems to forget that if a theory *fails* a severe test, we learn that it is false – provided, of course, that we accept the particular experimental result. When we falsify a prediction, however "local" it is, we falsify whatever entails that prediction, however general or large-scale. There is, in this respect, no localization of the refuting process. The fact that we may try to find out which part of the refuted whole is to blame is another question – the Duhem question. But if a theory does not fail a severe test, Mayo does not want to say that the whole theory has passed the test. Why not? Because Mayo thinks "passing a severe test" means "certified by that test to be correct." She wants to know what passing a severe test proves to be true. Passing a severe test does not prove the theory tested to be true. Therefore, the theory tested has not passed the test. Folk used to think we test theories *by* testing their consequences. Mayo thinks that all that we really test *are* the consequences. What gets tested is not any "large-scale theory" but rather local and specific claims about particular experimental phenomena. And if a local and specific claim passes a severe test, this teaches us nothing whatsoever about the large-scale theory of which that claim might be a part.

Does it not teach us that the theory is preferable to a rival theory that was refuted by the test? No. Mayo is a justificationist – a reason for preferring or adopting or believing a theory must be a reason for the theory itself. That a theory has not been refuted by a severe test does not prove it to be true or probable, so it is no reason to believe or accept or prefer that theory to its known rivals. To the question "What does passing Mayo-severe tests teach us about the theory tested?" she answers, "Nothing." She writes:

The growth of knowledge has not to do with replacing or confirming or probabilify-ing or "rationally accepting" large-scale theories, but with testing specific hypotheses in such a way that there is a good chance of learning something – whatever theory it winds up as part of. (this volume, p. 28)

Think what an extraordinary statement that is. The growth of scientific knowledge has nothing to do with replacing geocentric astronomy with heliocentric astronomy, Aristotelian physics with Newtonian physics, creationism with Darwin, or anything like that!

Mayo is quite open about the fact that she has nothing to say about which "full or large-scale theories" should be believed or accepted or preferred: "Rather than asking, Given our evidence and theories, which theory of this domain is the best? We ask, Given our evidence and theories, what do we know about this phenomenon?" (Mayo, 2000, p. 187). Again, she described Alan Chalmers and others as having posed a "challenge" to her new experimentalist philosophy: "If scientific progress is thought to turn on appraising high-level theories, then this type of localized account of testing will be regarded as guilty of a serious omission, unless it is supplemented with an account of theory appraisal" (this volume, p. 29). Well, does Mayo rise to the challenge and give us an account of theory appraisal? No. Instead, she denies the antecedent here – denies that scientific progress turns on appraising high-level theories. She has nothing to say about the rationality or otherwise of "large-scale theory change."

I am not the first to worry about this. Alan Chalmers and Larry Laudan worry about it, too. As I said already, Mayo wants to know what a severe test proves to be true, not what theory it might refute. Chalmers correctly reports that "It is this focus on reliability that leads Mayo to reject comparativist accounts of confirmation according to which a theory is warranted if it has passed tests that its known rivals cannot pass. She is unhappy with this because being best-tested in this sense is no guarantee that a theory is free from error. Best tested does not imply well tested, as far as Mayo is concerned" (this volume, p. 60). Here "warranted" requires a "guarantee that a theory is free of error;" that is, shown to be true. Chalmers tries to bring back scientific theories as a legitimate part of scientific knowledge and as "indispensably implicated in science and its ... progress" (this volume, p. 59). But Chalmers's problem is that he, too, is a justificationist. The very title of his paper gives the game away – it asks, "Can scientific theories be warranted?" Chalmers thinks that explanatory theories can only be a legitimate part of scientific knowledge if they can be warranted. He accepts that theories cannot be warranted by severe tests as Mayo construes them but hopes to show that they can be "warranted in some sense," and wheels in Whewell's concilience of inductions to do the job instead (this volume, pp. 67–68). But if one Mayo-severe test does not prove, justify, or probabilify a theory, how do several different ones turn the trick? Mayo is more consistent – she chucks out the theories precisely because they cannot be warranted by severe tests as she construes them. The way out of this impasse is to adopt critical rationalism and reject justificationism. What needs "warranting," if we must use that trendy term, is not the theories but our adoption of them. Mayo is quite right that having passed tests that known rivals cannot pass does not show that a theory is true. But does it show that it is reasonable to believe that it is? Does it warrant believing the theory, though not the theory itself? Justificationism says that believing is warranted if and only if what is believed is warranted. Chalmers should reject justificationism and become a critical rationalist if he wants to warrant believing in a comparativist fashion.

Larry Laudan sees this. He also worries about what he aptly calls Mayo's "balkanization" of the testing process: "Mayo seems to have so balkanized the testing process that global or otherwise very general theories can rarely if ever be said to be well tested" (Laudan, 1997, p. 315). To resist this balkanization, Laudan wheels in "comparativism":

Within . . . a comparativist perspective, we can say that a theory has been severely tested provided that it has survived tests its known rivals have failed to pass (and not vice versa). . . . Now, on this analysis, when we ask if GTR can be rationally accepted, we are not asking whether it has passed tests which it would almost certainly fail if it were false. As Mayo herself acknowledges, we can rarely if ever make such judgements about most of the general theories of the sciences. But we can ask "Has GTR passed tests which none of its known rivals have passed, while failing none which those rivals have passed?" Answering such a question requires no herculean enumeration of all the possible hypotheses for explaining the events in a domain. . . . By relativizing severity to the class of extant theories, one can determine severity without stumbling over the problems of the catchall hypothesis. . . . [C]omparativism keeps the focus on general theories, without diverting it away, Mayo-fashion, onto their subordinate parts. The comparativist believes that, if a theory like GTR explains or predicts phenomena which its known rivals have not and apparently cannot, then we have good grounds for preferring GTR to its known rivals. He insists on the point, which Mayo explicitly denies, that testing or confirming one "part" of a general theory provides, defeasibly, an evaluation of all of it. Thus, the observation of Neptune provided grounds for a global claim about the superiority of Newtonian mechanics to its known rivals, not merely support for the Newtonian subhypothesis that there was a massive object in the vicinity of Uranus. (Laudan, 1997, pp. 314–5)

Notice that Laudan's question is whether the best-tested theory "can be rationally accepted," not whether it has been shown to be true, nearly true, or probable. Notice that his comparativist view is that we have "good grounds for preferring" the best-tested theory, not good grounds for the theory itself. Of course, for justificationists, good grounds for accepting or preferring a theory must be good grounds for the theory. Comparativism is at odds with justificationism. Comparativism is music to the ears of critical rationalists, for they are comparativists, too. Critical rationalism does not traffic in the undreamt-of possibilities of the catchall. It need not traffic in them because it does not seek to justify any hypothesis. Its question is this: Which of the *available* competing theories is it reasonable to adopt or prefer or believe? You cannot believe an undreamed-of hypothesis.

Despite his apparent rejection of justificationism, Laudan is no critical rationalist. He thinks that rational acceptance or preference have nothing to do with truth. He goes for problem-solving ability rather than truth. But problem-solving ability is out of the same stable as empirical adequacy, and those who think that science is in the business of explaining things will have no truck with it, as I already argued. For critical rationalists, to accept or adopt a theory is to accept or adopt it as true – that is, to believe it. And to prefer a theory when you are aiming at truth is also to believe it.

How far does Mayo take her balkanization of the testing process? Back in 1999, Chalmers had begun to suspect that the new experimentalists were throwing the baby out with the bathwater: "Some of the new experimentalists seem to wish to draw a line between well-established experimental knowledge on the one hand and high-level theory on the other.... Some have pushed this view to a point where only experimental laws are to be taken as making testable claims about the way the world is" (Chalmers, 1999, p. 209). Similarly, Laudan complained that the famous light-bending experiments "are a severe test (in Mayo's technical sense) not of GTR but of the hypothesis that 'there is a deflection of light approximately equal to that predicted by [GTR]'" (Laudan, 1997, p. 313) – that is, they are a severe test of Einstein's prediction, not of Einstein's theory. In his paper for our conference, Chalmers goes further: "It must be stressed that theories, not just high-level theories but theories in general, cannot be severely tested in Mayo's sense.... It is not merely high-level theories like Newton's that cannot be severely tested, Mayo style. Low-level experimental laws cannot be severely tested either" (this volume, pp. 60–61). Is this right? Mayo says, "It seems disingenuous to say all of theory T is well-tested... when it is known there are ways T can be wrong that have received no scrutiny or that there are regions of implication not checked at all" (this volume, pp. 39–40). Here T is meant to be a "large-scale" or "high-level" or "full-blown" theory. But any general hypothesis, however small-scale or low-level, has "regions of implication not checked at all." An experimental generalization says that in certain circumstances something will happen always and everywhere. Our experiments are confined to particular times and places. Suppose I check some experimental generalization in my laboratory in Baltimore on a Tuesday afternoon, and it passes the test. Has the generalization as a whole been well tested in Mayo's sense? After all, it has implications, hitherto untested, about what will happen if I do the same test on Wednesday afternoon, or in Buenos Aires instead of Baltimore. According to Chalmers, all that is Mayo-well-tested is the specific and local claim about what will happen in my laboratory in Baltimore on a Tuesday afternoon. Thus, the light-bending experiments were not Mayo-severe tests even of Einstein's general prediction about what will happen in *any* light-bending experiment, since that prediction or "sub-hypothesis" has untested implications about the results of other light-bending experiments. Mayo says we severely test "experimental hypotheses" (this volume, p. 28). But Chalmers says not – that all we severely test are particular predictions.

Mayo's book is called *Error and the Growth of Experimental Knowledge*. She writes as though experimental knowledge grows, where experimental knowledge consists of knowledge of experimental effects, what are called "experimental generalizations" or "experimental laws." But, according to Chalmers, she is not really entitled to this "inductive leap." Experimental knowledge, in Mayo's sense, does not even consist of experimental *laws* – it consists of particular experimental *facts*. Rutherford once said: "All science is either physics or stamp-collecting." According to Chalmers, Mayo wants to reduce physics to stamp-collecting as well.

I am pretty sure that, despite her rhetoric, Mayo has a different view. She thinks that experimental generalizations can pass severe tests as a whole, despite the fact that they have "regions of implication not checked at all." We are warranted in accepting well-tested experimental generalizations despite the fact that they are not guaranteed by the tests to be free from error. In other words, Mayo is a critical rationalist at the level of experimental or observational generalization. Whereupon my question is, what is to stop the critical rationalism creeping upwards to encompass "large-scale" or "high-level" or "full-blown" theory as well?

But how far upwards? There is a well-known problem here – the tacking problem. Suppose some large-scale theory T yields some experimental generalization G, which is Mayo-well-tested and corroborated. Critical rationalism says scientists are warranted in accepting G, and Mayo agrees. Critical rationalists also say that, if T is the best available explanation of G, scientists are also warranted in accepting T. But suppose we tack onto T some idle wheel X, to form a larger-scale theory, $T\&X$. Will critical rationalists say scientists are warranted in accepting $T\&X$ as well? They should not. But how can they avoid it?

One way to avoid it is to say that scientists are only warranted in accepting the smallest (logical) part of $T\&X$, or for that matter of T itself, from which G follows. But that is hopeless; G itself is the smallest logical part of $T\&X$, or of T, from which G follows. This way of avoiding the tacking problem takes us back to a radical "balkanization of the testing process." All that we really test, and are warranted in accepting if the tests are passed, are the consequences themselves.

A better way for critical rationalists is to invoke explanation. Science is in the business of explaining G; G is no explanation of G What about $T\&X$? This is an explanation of G because (by assumption) T is an explanation of G. But is $T\&X$ the best explanation of G? Is it a better explanation than T? No, it is not. We assumed that T was the best explanation, and tacked the idle wheel X onto it. You do not get a better explanation than T by tacking an idle wheel onto it, just as you do not get a better explanation than T by watering it down to "The observed phenomena are as if T."

The tacking problem is the obverse of the Duhem problem. The Duhem problem asks what part of a theoretical system we should blame if a prediction turns out wrong. The obverse problem asks what part of a theoretical system we should praise if a prediction turns out right. Had G turned out to be false, it would be no good blaming the idle wheel X rather than T – for G follows from T by itself. But what we cannot blame, we should not praise either, which is why $T\&X$ is neither a better-tested explanation nor a better explanation, than T.

Of course, if X is not an "idle wheel," and if $T\&X$ has regions of implication that T does not, which have not yet been checked, then critical rationalists will not advocate acceptance of $T\&X$ over T. As Mayo rightly insists, they require evidence that these excess implications are correct. If that evidence is forthcoming, then $T\&X$ will be preferable to T on explanatory grounds; it will be a well-tested explanation of broader scope.

So, what Mayo needs to do to "bring theories back in" is to become a critical rationalist, not just at the level of experimental generalisations but also at the level of explanatory theories. Critical rationalism proposes that it is reasonable to believe (adopt, prefer) that explanatory theory, if there is one, that has been best tested. Critical rationalists do not infer the theory, or the equivalent claim that the theory is true, from the fact that it is the best tested. They infer that it is reasonable to think that the theory is true, or to believe it. Mayo talks about "inferring a full theory severely" (this volume, p. 30) and prohibits us from doing it. Critical rationalists do not do it. Of course, if you assume justificationism, you will think that they must be doing it, for you will take for granted that a reason for believing something must be a reason for what is believed.

I like Mayo-severe testing and want to incorporate it into the critical rationalist point of view. I like it for all the reasons Alan Chalmers has pointed out:

> Implicit in the new experimentalist's approach is the denial that experimental results are invariably "theory" or "paradigm" dependent to the extent that they cannot be appealed to adjudicate between theories. The reasonableness of this stems from the focus on experimental practice, on how instruments are used, errors eliminated, cross-checks devised and specimens manipulated. It is the extent to which this experimental life is sustained in a way that is independent of speculative theory that enables the products of that life to act as major constraints on theory. Scientific revolutions can be "rational" to the extent that they are forced on us by experimental results. . . . Adopting the idea that the best theories are those that survive the severest tests, and understanding a severe experimental test of a claim as one that the claim is likely to fail if it is false, the new experimentalists can show how experiment can bear on the comparison of radically different theories. (1999, p. 205)

Why bother with the theories? Because science seeks to understand the world and to explain the experimental facts, and you need theories to do that. Theories are not just heuristic aids, "useful instruments that aid the growth of experimental knowledge," as Chalmers accuses Mayo of thinking (Chalmers, this volume, p. 64). What is the *point* of doing experiments? It is not *just* to accumulate reliable experimental knowledge. That is important, to be sure, but we seek reliable experimental knowledge to adjudicate between rival explanatory theories about the way the world is.

References

Cartwright, N. (1983), *How the Laws of Physics Lie*, Oxford University Press, Oxford.

Chalmers, A. (1999), *What Is This Thing Called Science?* 3rd ed., Open University Press, and University of Queensland Press.

Laudan, L. (1981), "A Confutation of Convergent Realism," *Philosophy of Science*, 48: 19–49.

Laudan, L. (1997), "How About Bust? Factoring Explanatory Power Back into Theory Evaluation," *Philosophy of Science*, 64: 303–16.

Leplin, J. (1993), "Surrealism," *Mind*, 97: 519–24.

Lycan, W. (1985), "Epistemic Value," *Synthese*, 64: 137–64.

Mayo, D.G. (1996), *Error and the Growth of Experimental Knowledge*, University of Chicago Press, Chicago.

Mayo, D.G. (2000), "Experimental Practice and an Error Statistical Account of Evidence," *Philosophy of Science*, 67: S193–S207.

Mayo, D.G. (2006), "Critical Rationalism and Its Failure to Withstand Critical Scrutiny,"
  pp. 63–96 in C. Cheyne and J. Worrall (eds.), *Rationality and Reality: Conversations
  with Alan Musgrave*, Springer.
Mill, J.S. (1888), *A System of Logic*, 8th edition, Harper and Bros., New York.
Musgrave, A.E. (1999), *Essays on Realism and Rationalism*, Rodopi, Amsterdam.
Musgrave, A.E. (2001), "Rationalitat und Zuverlassigkeit" ["Rationality and Reliabil-
  ity"], *Logos*, 7: 94–114.
Peirce, C.S. (1931–1958), *The Collected Papers of Charles Sanders Peirce*, C. Hartshorne
  and P. Weiss (eds.), Harvard University Press, Cambridge, MA.
Psillos, S. (2001), "Predictive Similarity and the Success of Science: A Reply to Sanford,"
  Philosophy of Science, 68: 346–55.
Putnam, H. (1975), *Mathematics, Matter and Method: Philosophical Papers*, Vol. 1, Cam-
  bridge University Press, Cambridge.
Sankey, H. (2006), "Why Is it Rational to Believe Scientific Theories Are True?"
  pp. 109–32 in C. Cheyne and J. Worrall (eds.), *Rationality and Reality: Conversa-
  tions with Alan Musgrave*, Springer, New York.
Smart, J.J.C. (1968), *Between Science and Philosophy*, Random House, New York.
Stanford, K.J. (2000), "An Antirealist Explanation of the Success of Science," *Philosophy
  of Science*, 67: 266–84.
Van Fraassen, B. (1980), *The Scientific Image*, Clarendon Press, Oxford.

Toward Progressive Critical Rationalism
Exchanges with Alan Musgrave

Deborah G. Mayo

Reacting to Alan Musgrave's challenges to my account of theories directs us to a distinct set of issues under the same broad headings as in my exchange with Chalmers in Chapter 2: how do we characterize and justify inferences beyond the data (ampliative or inductive inference), and how do we view the role of theory appraisal in characterizing scientific progress? Under the first, I wish to tackle head-on the problem of inductive or ampliative inference retained from Popper and other logical empiricists. Under the second, I explore the possibility that Musgrave's move toward a piecemeal account of explanation moves the Popperian critical rationalist to a more progressive standpoint. Hovering in the background is the metaphilosophical query: How do Musgrave's assumptions about inductive justification lead to his skepticism about induction?

1. *Experimental Reasoning and Reliability*: Can inductive or ampliative inference be warranted? Can we get beyond inductive skepticism by showing the existence of reliable test rules?
2. *Objectivity and Rationality*: Must scientific progress and rationality be framed in terms of large-scale theory change? Does a piecemeal account of explanation entail a piecemeal account of testing?
3. *Metaphilosophical Themes*: How do philosophical assumptions about the nature of reliable inference lead to skepticism about induction?

1 Brief Overview

I begin with a useful passage from Musgrave's contribution:

So, the basic idea of critical rationalism is that it is reasonable to believe something if you have tried and failed to criticise it and show it to be false.... [This view] clearly need[s] to be fleshed out with a theory of criticism – an account of the ways we can criticise explanatory hypotheses of various kinds. One way to criticise a hypothesis is to subject it to observational or experimental tests. Quite generally, criticism must be genuine or serious criticism, which could reveal that the hypothesis is false. So when it comes to empirical testing, the tests must be severe tests. Critical rationalism owes us a theory about what a serious or severe empirical test is. (Musgrave, pp. 104–5)

This is clearly very much in the spirit of my severe testing account. Note, however, that for a hypothesis H to have passed a "serious or severe empirical test" with test result E, it does not suffice that the test "could reveal that the hypothesis is false" by resulting in not-E. After all, this condition would be satisfied whenever H entails (or fits) E, and the test merely checks whether E results. Although such a weak falsificationist requirement is countenanced in a simple "hypothetical deductive" account, the key point of requiring severity is to demand more than that H entails E and that E occurs. If the test would very probably result in such a passing condition, even if H is false, then the test is insevere (by the minimal or weak severity requirement). Given that I view my account of testing as implementing and improving upon the Popperian idea of critical rationalism by cashing out the notion of a severe test, I have long been puzzled (and frustrated!) to find myself the object of Alan Musgrave's criticisms. As with all the essays on our exchanges, I focus on the arguments that have moved us forward, as well as those that highlight the problems that remain.

Although my focus is on Musgrave's discussion of severe tests, in fact it was his discussion of explanation in his contribution to this volume that finally allowed us to make real progress, as much as a year after the ERROR 06 conference. If one is prepared, as Musgrave is, to talk of inferring only parts of explanations, I reasoned, why stick to holistically inferring an entire theory rather than the portions that have passed stringent tests? And once the door is open to piecemeal testing in probing large-scale theories, much of the apparent disagreement evaporates. The final pages of Musgrave's contribution records the concordance that eventually grew between us, as regards accepting only certain pieces of a proposed explanation. A key question that remains *is whether his position on explanation can sit consistently with the comparativist testing account that he retains.*

Before getting to that, Musgrave raises key criticisms that usefully illuminate central issues. Some of Musgrave's criticisms were already taken up in my exchange with Chalmers, and, although I will not rehearse them, I will try to tease out their origins. Doing so reveals the deepest problem that has stood in the way of progress, not just by Popperians but by philosophers of science more generally; namely, the problem of showing the existence of reliable or severe test procedures. This is a contemporary version of the problem of induction, which we may understand as the problem of warranting ampliative inferences – ones that generalize or go beyond the data.

Alongside these fundamental issues are matters that seem to me largely to reflect equivocations on terms; but not having convinced Musgrave that a disambiguation of meaning scotches his concerns, it will be important to take up some of the main ones in the sections that follow.

2 The Aims and Goals of Science

One of the chief motivations to turn to the new experimentalism and to scientific practice is to uncover the real-life driving forces of specific inquiries. I wrote, for example:

The growth of knowledge has not to do with replacing or confirming or probabilify-ing or "rationally accepting" large-scale theories but with testing specific hypotheses in such a way that there is a good chance of learning something – whatever theory it winds up as part of. (This volume, p. 28)

Musgrave has a lot of fun construing my "not to do" as "nothing":

Mayo thinks that . . . if a local and specific claim passes a severe test, this teaches us nothing whatsoever about the large-scale theory of which that claim might be a part. . . . Think what an extraordinary statement that is. (Musgrave, pp. 105–6)

My statement, however, was that the growth of knowledge "is not a matter of" assigning probabilities to, or rationally accepting, large-scale theories. Moreover, philosophical accounts that take those forms serve *at most* as rational reconstructions of science and miss what is most interesting and important about scientific progress, especially at the cutting edge. For one thing, we have to gather up a lot of experimental knowledge to even arrive at relevant data and reasonably probative tests. For another, we may not even have a large-scale theory when we set out – and often enough succeed – in learning about the world. In my view, "In much of day-to-day scientific practice, and in the startling new discoveries we read about, scientists are just trying to find things out" (Mayo, 1996, p. 56). Finding things out, moreover, includes finding out how to improve our capacities to probe, model, and detect errors and to make reliable inferences despite lacking a well-tested full-blown theory.

But, contrary to Musgrave's allegation, I hold that we learn a lot about the theory and about the domain in question even when only piecemeal hypotheses are passed severely. It is for this reason that I distinguished testing a large-scale theory from learning about it (Mayo, 1996, pp. 189–92). In the case of experimental general relativity, for instance, scientists had to develop the parameterized post-Newtonian (PPN) framework, estimate lots of parameters in statistical models, and learn about instruments and about effects that must be accounted for, subtracted, or dismissed later. This is theoretical knowledge all right, but it does not amount to accepting an entire theory or background paradigm or the like. Quite the contrary, the goal is to set out a testing framework that does not require assuming the truth of any theories not yet well corroborated.

A supposition that seems to underlie Musgrave's, and perhaps also Chalmers's position, is that anyone who does not set out to build an account of theory acceptance must be an antirealist, viewing the aim of science as limited to "empirical adequacy" perhaps along the lines of van Fraassen (1980). That is why Musgrave claims I am a critical rationalist at the observable level and wonders how I can avoid moving up to the theoretical level – but I never made any such distinction. The models that link data to hypotheses and theories all contain a combination of observational and theoretical quantities and factors, and I am glad to move "up" insofar as stringent testing allows. The entire issue of scientific realism looms large in Musgrave's account; it is one about which the error statistician may remain agnostic, and, thus, I leave it open.

3 Justifying *H* versus Warranting Test Rules That Pass *H*

Among the most puzzling of Musgrave's statements has to be this:

My question is – can we incorporate [Mayo's] theory into critical rationalism? Deborah Mayo does not think so. She rejects critical rationalism. She rejects it because she accepts justificationism. The upshot is that she abandons the project of appraising large-scale theories altogether. I want to entice her back into the critical rationalist camp. (Musgrave, p. 105)

"Justificationism," as Musgrave uses that term, holds that to warrant a hypothesis requires either showing that it is true (error free) or that it is probable, in some sense which he does not specify. Musgrave seems to think that my denial of comparativist testing of theories is explicable only by my being a justificationist! But I too reject justificationism as he defines it. A hypothesis is warranted for an error statistician not by being justified as highly probable but rather by having been highly probed, that is, by passing a stringent test (Mayo, 2005). The basis for rejecting the comparativist account of theory testing, as argued at length in Chapter 1, is that to infer a theory simply because it is the best tested thus far is a highly unreliable rule. In other words, if we go along with Musgrave's use of "justifying *H*," then we agree that the task at hand is not to justify *H* but rather to warrant a rule that outputs hypothesis *H*. The difference between us – and it is a significant one – is that, for me, a warranted test rule must be reliable (i.e., it must be a reliable probe of relevant errors).

Once it is understood that to warrant the test that outputs *H* requires the test rule to be reliable or severe, the critical rationalists' insistence on banishing talk of justifying hypotheses appears silly. A hypothesis may be said to be justified (in our sense) by having passed a warranted (severe) test rule. The trouble for critical rationalists is that they deny it is possible to warrant a test rule as reliable; for that would involve an inductive justification, which they do not allow themselves. In keeping with this standpoint, Musgrave feels himself bound to reject the need for any kind of inductive or ampliative claim even at the "meta-level," as he calls it – that is, even as regards the reliability of test rules or methods of criticism. Yet showing meta-level reliability would seem to be at the very foundation of implementing critical rationalism.

3.1 Progressive Critical Rationalism

I wish to entice Musgrave to advance the critical rationalist program by actually developing rules for good and severe tests and for distinguishing them from illicit, unreliable, and unscientific test rules. Musgrave himself claims that "critical rationalism owes us a theory about what a serious or severe empirical test is." (p. 105) The time is ripe to provide it! To avoid confusion, we need to distinguish what critical rationalism has been since Popper – *Popperian critical rationalism* – and a forward-looking theory of criticism, which we may call *progressive critical rationalism*. (Note: I leave to one side the more radical Popperianism espoused by David Miller but soundly rejected by other followers of Popper.) Progressive critical rationalism would proceed by developing tools for severe tests. Such tools seek reliable probes of errors, and yet, in another verbal misconstrual, Musgrave imagines that I mean "proved" (with certainty) when I say "probed"!

But if a theory does not fail a severe test, Mayo does not want to say that the whole theory has passed the test. Why not? Because Mayo thinks "passing a severe test" means "certified by that test to be correct." She wants to know what passing a severe test proves to be true. Passing a severe test does not prove the theory tested to be true. Therefore, the theory tested has not passed the test. (Musgrave, p. 105)

To this indictment, I am left to repeat the same baffled cry I let out in considering Chalmers. I would not be appealing to statistical inference were I to think that H's passing a severe test "proves" H. It suffices that H passes despite being put to stringent probes of the ways in which H may be false. If, after a series of stringent probes and deliberately varied results, hypothesis H holds up, then we may say H has passed severely – this entitles us to infer H, even though we qualify the inference by means of the severity assessment of the test rule (formally or informally).[1] In some cases, the inferred claim may assert how far off H is from a correct description of the process in question. With Musgrave's interest in talking of approximate truth, this would seem just the ticket.

I am happy to use Popper's more succinct "H is corroborated" inter-changeably with "H has passed a severe test" so long as my notion of severity is understood.[2] Whole theories may be corroborated; my objection is only to regarding a theory as having passed severely when it is known that it makes claims that have been probed poorly or not probed at all. Now Musgrave makes clear, here and in an earlier exchange, that critical rationalism "proposes that it is reasonable to believe (adopt, prefer) that theory, if there is one, whose consequences have been best tested" (Musgrave, 2006, p. 307). If one adopts this rule for "reasonable believings" (as Musgrave sometimes refers to them), it would be reasonable to believe all of the General Theory of Relativity (GTR) in 1970, say, because its consequences about the deflection of light had held up to testing. Such a rule for "reasonable believings" would be highly unreliable.

Is Musgrave's critical rationalist not bothered by the fact that he advocates a rule for inferring or believing that is known to be unreliable? In Musgrave (1999, p. 346) he said No, but in his (2006) response he agrees that he would be bothered. He now clarifies his position as follows: Although "critical rationalists need not assert or prove that their method is reliable before they can rationally adopt it . . . if the method can be criticized by showing

---

[1] Note an error in the first (and last) sentence of his quote above, although perhaps this is just a slip. We distinguish between passing a test and passing it severely. If the test is severe for all of H – meaning that H as a whole has passed severely with the test and data in question – then all of H has passed severely. Yet even if the test is insevere, I am prepared to say that H has passed, just inseverely.

[2] Popper at one point allowed that one was free to call an inference to a severely tested hypothesis H an inductive inference to H, although he did not do so himself. C.S. Peirce talked this way long before.

it to be *un*reliable, then they should cease to employ it" (p. 308). This is an important point of progress. However, it only spells trouble for Musgrave's critical rationalist because it would indict the comparativist rule of testing as he defines it.

Musgrave could solve this problem by replacing "best tested" with severely tested in my sense. He would then have advanced to what I am calling progressive critical rationalism. In the wrap-up to this essay I consider whether he has moved at least partway to this standpoint. For now I need to continue with Musgrave's fairly strenuous resistance to buying into this account.

3.2 Can Inductive Generalizations Be Warranted?

The elephant in the room is the fear of endorsing any kind of "induction." Because of this fear, Musgrave seems ready to claim that no test based on finitely available evidence can ever severely pass a claim that goes beyond that finite evidence. It is the same stance that we find with Chalmers, whom Musgrave quotes, approvingly, as having pointed out that data collected in Baltimore only warrants with severity assertions about how well the claim held in Baltimore: "Suppose I check some experimental generalization in my laboratory in Baltimore on a Tuesday afternoon, and it passes the test." According to Musgrave, this passing result cannot warrant with severity the general claim because "it has implications, hitherto untested, about what will happen if I do the same test on Wednesday afternoon, or in Buenos Aires instead of Baltimore." (Musgrave, pp. 108–9)

Think what an extraordinary statement that is. The chemical shown to be toxic in the laboratory in Baltimore only teaches us what happened in Baltimore? Notice, one could strictly only "infer" the results already observed! This is vintage Popper. Popper made it clear that for him cor-roboration was merely a report of past success; there was no warrant for ever *relying* on claims thereby passed (since that would allude to trials other than those run). Although this may seem surprising, in fact this is the skeptical standpoint that the critical rationalist recommends. Rather than reconsider Popper's scepticism, critical rationalists seek notions of scientific rationality that can live alongside this scepticism.

3.3 The "Wedge" between Skepticism and Irrationalism

This takes us to the distinction Musgrave draws between warranting a rule (or a procedure) for "believings" or "inferrings" as opposed to warranting the claims believed or inferred. Musgrave regards this distinction as the secret by which Popper is able to avoid the problem of induction. According to Musgrave, "The way out of this impasse is to adopt critical rationalism and reject justificationism. What needs 'warranting,' if we must use that trendy term, is not the theories but our adoption of them." In this way, Musgrave allows, we can "warrant believing the theory, though not the theory itself" (p. 107). Many will wonder what it can mean to say:

We are warranted in believing hypothesis *H*, although belief in *H* is not warranted.

If someone says "I am warranted in believing that *X* is toxic to the liver" while denying that he regards the assertion "*X* is toxic to the liver" to be warranted, it would seem he is playing a verbal game. Yet, this is the kind of convoluted language Popperian critical rationalists seem forced to adopt because they assume warranting *H* would mean justifying *H* as probably true. Therefore, the way to parse the above sentence is to replace the second "warranted" with "justified" in the justificationist sense. Then the claim is as follows:

We can warrant a rule for accepting (or believing) *H* without claiming that *H* is proved true or probable.

The assertion now is perfectly reasonable; moreover, it is one the error statistician happily endorses. Error statistics, after all, is based on adapting the idea from Neyman-Pearson (N-P) testing that what we want are rules for "deciding" to classify hypotheses as "accepted or rejected," and probabilistic properties attach to the rules, not to the hypotheses inferred. For instance, in 7.1 of the introductory chapter, we considered a test rule *T* that maps data into the claim "there is evidence for a discrepancy from μ_0." Where we differ from Musgrave concerns what is required to warrant a test rule (or, as he might prefer, a procedure for believing). We require the rule be sufficiently reliable or stringent or the like; whereas the Popperian critical rationalist deems this goal beyond reach. Fittingly, Lakatos once said that N-P tests are excellent examples of Popperian "methodological falsification-ism" (Lakatos, 1978, p. 109, note 6). Unfortunately, the Popperians never made use of the fact that N-P tests, in the realm of statistics, provide rules with low probabilities of erroneously rejecting or erroneously accepting hypotheses. If they had, they might have seen how to adapt those statistical rules to their purposes. That is essentially what I have tried to do.

A reformulation of the strict (behavioristic) construal of N-P tests is required because tests may have good long-run error probabilities while doing a poor job of ruling out errors in the case at hand (see, e.g., Chapter 7, this volume). Still, the formal determination of test rules with good error probabilities is extremely important because it demonstrates that stringent or reliable test rules exist, which lets us check putative rules for reliability. Numerous examples of how error properties of test rules relate to the severity of particular inferences are discussed throughout this volume.

To make some quick points, consider a formal test rule R that rejects a null hypothesis:

H_0: drug X is *not* toxic (e.g., to rats),

just when a randomized controlled trial yields an increase, $d(\mathbf{x}_0)$, in risk rates that is statistically significant at the .01 level. In those cases, $d(\mathbf{x}_0)$ is taken as grounds for rejecting H_0 and inferring H^*, where

H^*: there is evidence of the increased risk.

Then H^* has passed a severe test because

Prob(rule R would not infer H^*, when in fact H_0 is true) = .99.

Musgrave's random sample of rats in Baltimore is effectively a random sample of all rats with respect to this risk; the error probabilities apply for subsequent applications so long as the statistical model assumptions hold sufficiently, and this is testable. Now someone may charge that, although the generalization about X's toxicity holds for all rats today, X may not be harmful at some future date – but this is not problematic for our account. It is precisely because we can vouchsafe that the test rule continues to operate in a stable way that we can discover the effect has changed!

It is not clear why Musgrave charges that experiments on GTR fail to warrant general deflection hypotheses. I chose to discuss such "null tests" in physics precisely to apply the error-statistical account to the home turf of the high-level theory philosophers. Here the null hypothesis is set as a precise claim – one from which it would be very easy to detect departures:

H_0: deflection effect = 1.75(GTR prediction)

Having observed $d(\mathbf{x}_0)$ that accords with H_0, a test rule R infers that

H^*: the deflection effect differs no more from the GTR value than $\pm\varepsilon$,

where, say, ε is three standard deviations away from $d(\mathbf{x}_0)$. If one wishes to speak only in terms of falsifications, there is no problem in doing so: inferring or corroborating H^* is the same as falsifying the denial of H^*. We have

Prob(rule R corroborates H^*, when H^* is false) = .001.

Therefore, when R corroborates H^*, H^* passes the test with severity .999 (see also Introduction, p. 22, and Chapter 7, this volume). Thus, our warrant of the reliability of rule R is no different than warranting the rule for rejecting the zero-risk-increase hypothesis. Unless Popperians wish to discard the logic of falsification, they cannot object to corroborating claims based on falsifying their denials!

Popperian critical rationalists make the success of science a mystery (indeed, Popper calls it "miraculous"). It is no great mystery that statistical (and other) predictions hold in particular cases of a type of experiment because if we had failed to bring about the experimental conditions for testing the prediction, then we would discover our errors, at least with high probability. Scientists build their repertoires for checking instruments and models so that errors not detected by one tool will be picked up by another.

3.4 Beyond Formal Inductive Logics

Popper spoke of corroborating claims that passed sincere attempts to falsify them, but he was at a loss to operationalize "sincere attempts to falsify." Merely "trying" to falsify, even if we could measure effort, would be completely beside the point: one must deliver the goods. It is the properties of tools that have to be demonstrated (i.e., that were a specified discrepancy present, it would have been detected with high probability). Combining different tests strengthens and fortifies the needed claims.

One can perhaps understand why Popper remained skeptical of identifying reliable rules: at the time, philosophers thought that one needed a formal inductive logic that would be context-free, a priori and not empirical. Pursuing the path of formal inductive logics failed to yield applicable, useful rules, so the whole project of developing reliable rules and methods was abandoned. The time is ripe to get beyond this. We invite Musgrave and other critical rationalists to turn to the project of identifying reliable rules of criticism!

4 A Constructive Ending: Explanation and Testing

By the end of his chapter, Musgrave seems prepared to adopt what I call progressive critical rationalism, if only because it emerges naturally out of his discussion of inference to the best explanation (IBE) in earlier sections. For one thing, in his discussion of IBE, it is required that the predictive success to be explained by H be surprising or novel: First, if it is very easy to achieve explanatory success *H* would not pass severely. Second, when it comes to inferring explanations, Musgrave is prepared to infer pieces or aspects of theories. Here inference to parts is not only allowed but is welcomed. Indeed, Musgrave recommends replacing Popper's search for a measure of verisimilitude, understood as closeness to the truth, with partial truth understood as truth of the parts:

Just as total success is best explained in terms of truth, partial success is best explained in terms of partial truth. . . . That true part, common to both theories, sets out the relative motions of Earth, Sun and superior planets. (Musgrave, p. 101)

The position is clearly in sync with our piecemeal theory of severe testing: we may infer those aspects or variants or subparts of a theory that hold up to stringent scrutiny.

In his last section Musgrave hints at renouncing his advocacy of inferring the entire theory when only parts have been severely corroborated. He considers the addition to a hypothesis *T* of another hypothesis *X* that broadens or extends the domain of *T* − *X* is not what he calls an "idle wheel" or irrelevant conjunct. (The "tacking paradox" returns in Chapter 9.)

Of course, if *X* is not an "idle wheel," and if *T&X* has regions of implication that *T* does not, which have not yet been checked, then critical rationalists will not advocate acceptance of *T&X* over *T*. As Mayo rightly insists, they require evidence that these excess implications are correct. (Musgrave, p. 110)

Musgrave seems to be leaving off with the promise that he is ready to make the move to progressive critical rationalism, thereby attesting to the fruitfulness of multiple exchanges!

References

Lakatos, I. (1978), *The Methodology of Scientific Research Programmes*, J. Worrall and G. Currie (eds.), Vol. 1 of *Philosophical Papers*, Cambridge University Press, Cambridge.

Mayo, D.G. (1996), *Error and the Growth of Experimental Knowledge*, University of Chicago Press, Chicago.

Mayo, D.G. (2005), "Evidence as Passing Severe Tests: Highly Probed vs. Highly Proved," pp. 95–127 in P. Achinstein (ed.), *Scientific Evidence*, Johns Hopkins University Press, Baltimore.

Mayo, D.G. (2006), "Critical Rationalism and Its Failure to Withstand Critical Scrutiny," pp. 63–99 in C. Cheyne and J. Worrall (eds.), *Rationality and Reality: Conversations with Alan Musgrave*, Kluwer Studies in the History and Philosophy of Science, Springer, The Netherlands.

Mayo, D.G. and D.R. Cox (2006), "Frequentist Statistics as a Theory of Inductive Inference," pp. 77–97 in J. Rojo (ed.), *Optimality: The Second Erich L. Lehmann Symposium*, Lecture Notes-Monograph Series, Institute of Mathematical Statistics (IMS), Vol. 49.

Musgrave, A. (1999), *Essays in Realism and Rationalism*, Rodopi, Amsterdam.

Musgrave, A. (2006), "Responses," pp. 293–333 in C. Cheyne and J. Worrall (eds.), *Rationality and Reality: Conversations with Alan Musgrave*, Kluwer Studies in the History and Philosophy of Science, Springer, The Netherlands.

van Fraassen, B. (1980), *The Scientific Image*, Clarendon Press, Oxford.

Related Exchanges

Chalmers, A. (2006), "Why Alan Musgrave Should Become an Essentialist," pp. 165–82 in C. Cheyne and J. Worrall (eds.), *Rationality and Reality: Conversations with Alan Musgrave*, Kluwer Studies in the History and Philosophy of Science, Springer, The Netherlands.

Worrall, J. (2006), "Theory-Confirmation and History," pp. 31–62 in C. Cheyne and J. Worrall (eds.), *Rationality and Reality: Conversations with Alan Musgrave*, Kluwer Studies in the History and Philosophy of Science, Springer, The Netherlands.

FOUR

Theory Confirmation and Novel Evidence

Error, Tests, and Theory Confirmation

John Worrall

In this chapter I address what seems to be a sharp difference of opinion between myself and Mayo concerning a fundamental problem in the theory of confirmation.[1] Not surprisingly, I argue that I am right and she is (interestingly) wrong. But first I need to outline the background carefully – because seeing clearly what the problem is (and what it is not) takes us a good way towards its correct solution.

1 The Duhem Problem and the "UN" Charter

So far as the issue about confirmation that I want to raise here is concerned: in the beginning was the "Duhem problem." But this problem has often been misrepresented. No sensible argument exists in Duhem (or elsewhere) to the effect that the "whole of our knowledge" is involved in any attempt to test any part of our knowledge. Indeed, I doubt that that claim makes any sense. No sensible argument exists in Duhem (or elsewhere) to the effect that we can never test any particular part of some overall theory or theoretical system, only the "whole" of it. If, for example, a theory falls "naturally" into five axioms, then there is – and can be – no reason why it should be impossible that some directly testable consequence follows from, say, four of those axioms – in which case only those four axioms and not the whole of the theory are what is tested.

What Duhem *did* successfully argue is that if we take what is normally considered a "single" scientific theory – such as Newton's theory (of mechanics plus gravitation) or Maxwell's theory of electromagnetism or the wave theory of light – and carefully analyze any attempt to test it empirically by deducing from it some directly empirically checkable consequence, then

---

[1] See, e.g., Worrall (2006) and Mayo (1996).

the inference is revealed to be valid only *modulo* some further set of auxiliary theories (theories about the circumstances of the experiment, about the instruments used and so on). For example, as became clear during the famous dispute between Newton and Flamsteed, to deduce from Newton's theory of gravitation a consequence that can be directly tested against telescopic sightings of planetary positions, we need to invoke an assumption about the amount of refraction that a light beam undergoes in passing into the Earth's atmosphere.

Moreover, Duhem pointed out that at least in many cases the "single" theory that we test itself involves a "central" claim together with some set of more specific assumptions; and in such cases, so long as the central claim is retained, we tend to describe changes in the specific assumptions as producing "different versions of the same theory" rather than a new, different theory. An example of this, analyzed of course at some length by Duhem himself, is "the" (classical) wave theory of light. This "theory" was in fact an evolving entity with a central or core assumption – that light is some sort of wave in some sort of mechanical medium – an assumption that remained fixed throughout, with a changing set of more specific assumptions about, for example, the kind of wave and the kind of medium through which the waves travel. For example, one celebrated (and relatively large) change was that effected by Fresnel when he abandoned the idea that the "luminiferous ether" that carries the light waves is a highly attenuated fluid and the waves, therefore, longitudinal, and hypothesized instead that the ether is an (of course still highly attenuated) elastic solid that transmits *transverse* waves.

These facts about the deductive structure of tests of "single" scientific theories of course have the trivial consequence that no experimental result can refute such a theory. Even assuming that we can unproblematically and directly establish the truth value of some observation statement O on the basis of experience, if this observation statement follows, not from the core theory T alone but instead only from that core, plus specific assumptions, plus auxiliaries, then if O turns out in fact to be false, all that follows deductively is that *at least one* of the assumptions in the "theoretical system" involving core, plus specific, plus auxiliary assumptions is false.

Kuhn's account of "scientific revolutions" is to a large extent a rediscovery – of course unwitting – of Duhem's point (along with a great number of historical examples).[2] Perhaps the claim in Kuhn that most strikingly challenged the idea that theory change in science is a rational process is that in "revolutions" the old-guard (or "hold-outs" as he calls them) were no

---

[2] This is argued in Worrall (2003).

less rational than the "revolutionaries" – there being "some good reasons for each possible choice" (sticking to the older theory or accepting the newer one).[3] This claim in turn is at least largely based on Kuhn's observation that the evidence that the "revolutionaries" regard as crucial extra empirical support for their new paradigm-forming theory can in fact also be "shoved into the box provided by the older paradigm."[4] Exactly as Duhem's analysis of theory-testing shows, it is always logically possible to hold onto the basic (central or core) idea of the older theoretical framework by rejecting some other – either "specific" or auxiliary – assumption.

For example, results such as that of the two-slit experiment that were certainly correctly predicted by the wave theory of light are often cited by later accounts as crucial experiments that unambiguously refuted that theory's corpuscular rival. But in fact plenty of suggestions existed within the early nineteenth-century corpuscularist literature for how to accommodate those experimental results. Some corpuscularists, for example, conjectured that, alongside the reflecting and refracting forces to which they were already committed, results such as that of the two-slit experiment showed that there was also a "diffracting force" that emanates from the edges of "ordinary, gross" opaque matter and affects the paths of the light particles as they pass. Those corpuscularists laid down the project of working out the details of this diffracting force on the basis of the "interference" results. (Of course, because this force needs to be taken to pull some particles into the geometrical shadow and push others away from places outside the shadow that they would otherwise reach, the corpuscularists thereby denied that the fringe phenomena are in fact the result of interference.)

Similarly, and as is well known, Copernicus (and following him Kepler and Galileo) was especially impressed by his theory's "natural" account of the phenomenon of planetary stations and retrogressions – despite the fact that it had long been recognized by Copernicus' time that this phenomenon could be accommodated within the Ptolemaic geostatic system: although they are certainly inconsistent with the simplest Ptolemaic theory, which has all planets describing simple circular orbits around the Earth, stations and retrogressions could be accommodated within the Ptolemaic framework by adding epicycles and making suitable assumptions about their sizes and about how quickly the planet moved around the epicycle compared to how quickly the center of that epicycle moved around the basic "deferent" circle.

---

[3] Kuhn (1962, pp. 151–2).
[4] Kuhn (1977, p. 328).

Kuhn seems to presume that the fact that such phenomena can be accommodated within the older "paradigm" means that the phenomena cannot unambiguously be regarded as providing extra support for the newer theoretical framework and as therefore providing part of the reason why the theory shift that occurred was rationally justified. But this presumption is surely wrong. It is, instead, an important part of any acceptable account of theory confirmation that merely "accommodating" some phenomenon within a given theoretical framework in an ad hoc way does *not* balance the evidential scales: the theory underlying the framework that predicted the phenomenon continues to receive greater empirical support from it, even if it can be accommodated within the older system (as Duhem's analysis shows is always possible). The wave theory continues to derive more support from the result of the two-slit experiment even once it is conceded that it is *possible* to give an account of the phenomenon, though in an entirely post hoc way, within the corpuscular framework. Planetary stations and retrogressions give more (rational) support to the Copernican theory even though the Ptolemaic theory can accommodate them (indeed even though the Ptolemaic theory had, of course, long *pre*-accommodated them).

A suspicion of ad hoc explanations has guided science from its beginning and is widely held and deeply felt. Take another (this time noncomparative) example. Immanuel Velikovsky conjectured that in biblical times a giant comet had somehow or other broken away from the planet Jupiter and somehow or other made three separate series of orbits close to the Earth (before eventually settling down to a quieter life as the planet Venus). It was these "close encounters" that were responsible for such biblically reported "phenomena" as the fall of the walls of Jericho and the parting of the Red Sea. Velikovsky recognized that, if his theory were correct, it is entirely implausible that such cataclysmic events would have been restricted to the particular part of the Middle East that concerned the authors of the Bible. He accordingly set about looking for records of similar events from other record-keeping cultures of that time. He found records from *some* cultures that, so he (rather loosely) argued, fit the bill, but he also found some embarrassing gaps: the apparently fairly full records we have inherited from some other cultures make no mention of appropriately dated events that were even remotely on a par with the ones alleged to have occurred in the Bible. Velikovsky – completely in line with Duhem's point – held on to his favoured central theory (there really had been these close encounters and widespread associated cataclysms) and rejected instead an auxiliary assumption. Suppose that similar cataclysms *had* in fact occurred in the homeland of culture *C*. To predict that *C*'s scribes would have recorded such events (which were, after all, one would have thought, well worth a line in anyone's diary!) it must of course be assumed that those scribes were able to bear accurate witness. But what if, in some cultures, the events associated with the close encounters with this "incredible chunk" proved *so* cataclysmic that all of the culture's scribes were afflicted by "collective amnesia?" Velikovsky conjectured that collective amnesia had indeed afflicted certain cultures and proceeded to read off which exact cultures those were from the (lack of) records. Those cultures that recorded cataclysms that he had been able to argue were analogous to the biblical ones did *not* suffer from this unpleasant complaint; those cultures that would otherwise have been expected to but did not in fact record any remotely comparable events *did* suffer from it. Clearly although this modified theory now entails correctly which cultures would and which would not have appropriate records, this can hardly be said to supply any empirical support to Velikovsky's cometary hypothesis – a hypothesis that has been augmented exactly so as to yield the already known data.

These intuitive judgments need to be underwritten by some general principle that in turn will underwrite the rejection of Kuhn's implicit claim that the fact that evidence can be forced into the "box provided by the older paradigm" means that that evidence cannot be significant extra support for the newer theoretical framework. It was this idea that led some of us to sign up to the "UN Charter."[5] This phrase, in slogan form, has been interpreted as ruling that "you can't use the same fact twice, once in the construction of a theory and then again in its support." According to this "use novelty criterion" or "no-double-use rule" as it has generally been understood, theories are empirically supported by phenomena that they correctly *predict* (where prediction is understood, as it invariably is in science, not in the temporal sense but in the sense of "falling out" of the theory without having had to be worked into that theory "by hand")[6] and *not* by phenomena that have to

---

[5] For history and references see Worrall (2002).

[6] Notice that, despite the fact that UN stands for "use novelty," the UN charter in fact gives no role to novelty of evidence in itself. Those who had argued that evidence that was, as a matter of historical fact, discovered only as a result of its being predicted by some theory carried greater confirmational weight were missing the real issue. This issue is one of accommodation versus prediction, where the latter is used in the proper sense, just meaning "not accommodated." Some but not all predictions are of hitherto unknown phenomena (although all accommodations must of course have been of known phenomena). This sense of prediction is accurately reflected in the following passage from French's textbook, *Newtonian Mechanics*: "[L]ike every other good theory in physics, [the theory of universal gravitation] had predictive value; that is, it could be applied to situations besides the ones from which it was deduced [i.e., the phenomena that had been deliberately accommodated within it]. Investigating the predictions of a theory may involve looking

be "accommodated within," or "written into" the theory post hoc. Thus, it yields the judgment that the (amended) corpuscular theory gets no support from the two-slit experiment because the details of the "diffracting force" had to be "read off" from already-given experimental results such as that of the two-slit experiment itself, whereas the wave theory, which predicted this experimental outcome in a way that is entirely independent of that outcome, *does* get support from the result. Similarly Velikovsky's (amended) theory gets no support from the empirical fact that no records of cataclysms in culture *C* have been preserved, because the facts about which cultures have or have not left records of appropriately timed cataclysms were used in constructing the particular form of his overall theory that he defended.

It is a central purpose of this chapter to clarify further and defend the UN rule under a somewhat different interpretation than the one it has often been given and in a way that clashes with Mayo's (partial) defence of that rule. Many philosophers have, however, claimed that the view is indefensible in any form – a crucial part of the clarification consists in showing how exactly these opponents of the view go astray.

2 "Refutations" of the UN Rule

Allan Franklin once gave a seminar at the London School of Economics under the title "*Ad Hoc* Is Not a Four Letter Word." Beneath the (multiple) surface literal correctness of this title is a substantive claim that is undeniably correct; namely, it is entirely normal scientific procedure to use particular data in the construction of theories, without any hint of this being in any way scientifically questionable, let alone outright intellectually reprehensible.

Suppose, for (a multiply realized) example, that a scientist is facing a general theory in which theoretical considerations leave the value of some parameter free; the theory does, however, entail that the parameter value is a function of some set of observable quantities. A particular example of this kind that I like to use is that of the wave theory of light, which leaves as an open question, so far as basic theoretical considerations are concerned, what is the wavelength of the light from any particular monochromatic source? Because the theory provides no account of the atomic vibrations within luminous objects that produce the light, it does not dictate from first principles the wavelength of light from a particular source. The theory

---

for hitherto unsuspected phenomena, or it may involve recognising that an already existing phenomenon must fit into the new framework. In either case the theory is subjected to searching tests, by which it must stand or fall" (French, 1971, pp. 5–6).

does, however, entail that that value, whatever it is, is a function of the slit distances, the distance from the slits to the screen and the fringe distances in the two-slit experiment. In such a situation, the scientist will *of course* not make "bold conjectures" about the value of the wavelength of light from some particular source and then test those conjectures. Instead, she will "measure the wavelength;" that is, she will perform the experiment, record the appropriate observable values, and infer the wavelength of light from that particular source from the formula entailed by the theory. She has then *deduced* a particular, more powerful theory (wave theory complete with a specific value of this particular theoretical parameter) from her general (parameter-free) theory plus observational results.

Clearly using observational data as a premise in the deduction of some particular version of a theory is a paradigmatic example of "using data in the construction of a theory." And yet this is an entirely sensible, entirely kosher scientific procedure. Moreover, if asked why she holds the particular version of the theory that she does – that is, if she is asked why, given that she holds the general wave theory of light, she also attributes this particular wavelength to light from this particular source – she will surely cite the observations that she has used in that deduction. What then of the "rule" that you can't use the same fact twice, once in the construction of a theory and then again in its support?

Nor do the apparent problems for the UN rule end there. Colin Howson, for example, likes to emphasize a different general case – standard statistical examples such as the following (see Howson, 1990). We are given that an urn contains only red and white balls though in an unknown (but fixed) proportion; we are prevented from looking inside the urn but can draw balls one at a time from it. Suppose that a sample of size n has been taken (with replacement) and k of the balls have been found to be white. Standard statistical estimation theory then suggests the hypothesis that the proportion of white balls in the urn is $k/n \pm \varepsilon$, where ε is calculated as a function of n by standard confidence-interval techniques. The sample evidence is the basis here of the construction of the particular hypothesis and surely, Howson suggests, also supports that particular hypothesis at least to some degree – the (initial) evidence for the hypothesis just *is* that a proportion k/n of the balls drawn were white. For this reason (and others) Howson dismisses the UN rule as "entirely bogus."[7]

Mayo cites and analyzes in more detail similar statistical cases that seem to count against the "no-double-use idea" and also cites the following "trivial

---

[7] See Howson (1990).

but instructive example" (1996, p. 271). Suppose one wanted to arrive at what she characterizes as a "hypothesis H" about the average SAT score of the students in her logic class. She points out that the "obvious" (in fact uniquely sensible) way to arrive at H is by summing all the individual scores of the *N* students in the class and dividing that sum by *N*. The "hypothesis" arrived at in this way would clearly be "use constructed." Suppose the constructed "hypothesis" is that the average SAT score for these students is 1121. It would clearly be madness to suppose that the data used in the construction of the "hypothesis" that the average SAT score is 1121 fail to support that hypothesis. On the contrary, as she writes,

Surely the data on my students are excellent grounds for my hypothesis about their average SAT scores. It would be absurd to suppose that further tests would give better support. (1996, p. 271)

Exactly so: the data provide not just excellent, but, short of some trivial error, entirely *conclusive* grounds for the "hypothesis"– further "tests" are irrelevant. (This is precisely why it seems extremely odd to talk of a "hypothesis" at all in these circumstances – a point to which I return in my criticism of Mayo's views.)

How in the light of apparently straightforward counterexamples such as these can I continue to defend (a version of) the UN "rule"? Well, first we need to get a clearer picture of the underlying nature of all these "counterexamples." They all are (more or less clear-cut) instances of an inference pattern sometimes called "demonstrative induction" or, better, "deduction from the phenomena." The importance of this inference pattern to science was emphasized long ago by Newton and, after some years of neglect, has been increasingly reappreciated in recent philosophy of science.[8]

Of course, general theories are invariably logically stronger than any finite set of observational data and so a "deduction from the phenomena," if it is to be valid, must in fact implicitly involve extra premises. The idea is that certain very general principles are, somehow or other, legitimately taken for granted (as "background knowledge") and some more specific theory is deduced from those general principles plus experimental and observational data. Newton, in a complicated way that involves generalizing from models known to be (strictly) inaccurate, deduced his theory of universal gravitation from Kepler's "phenomena" plus background assumptions that included conservation of momentum. The statistical case cited by Howson, exactly because it is statistical, does not of course exactly fit the pattern – but

---

[8] See Worrall (2000) and the references to the literature therein.

something very similar applies. We are somehow given (or it seems reasonable to assume) that drawing balls from an urn (with replacement) is a Bernoulli process, with a fixed probability p – we then "quasi-deduce" from the fact that the sample of draws has produced a proportion k/n of white balls that the population frequency is $k/n \pm \varepsilon$. In Mayo's case, we deduce her "hypothesis" about the average SAT score of her logic students from background principles (basically the analytic principles that specify what an average is) plus the "observed" individual student scores. (The fact that the background principles in this last case are analytic is another reflection of the oddness of characterising the resultant claim as a "hypothesis.")

Of the cases cited, the most direct instance of the type of reasoning that is at issue (and that plays an important role in physics) is the wave theory case. The scientist starts with a theory, $T(\lambda)$, in which the theoretical parameter (in this case wavelength) is left free. However, the theory entails that λ is a determinate function of quantities that are measurable. Here the wave theory, for example, entails (subject to a couple of idealisations) that, in the case of the famous two-slit experiment performed using light from a monochromatic source – say, a sodium arc – the (observable) distance X from the fringe at the center of the pattern to the first fringe on either side is related to (theoretical) wavelength λ, via the equation $X/(X^2 + D^2)^{1/2} = \lambda/d$ (where d is the distance between the two slits and D the distance from the two-slit screen to the observation screen – both, of course, observable quantities). It follows analytically that $\lambda = dX/(X^2 + D^2)^{1/2}$. But all the terms on the right-hand side of this last equation are measurable. Hence, particular observed values e' determine the wavelength of the light (within some small margin of experimental error, of course) and so determine the more specific theory $T' = T(\lambda_0)$, with the parameter that had been free in T now given a definite value, λ_0 – again within a margin of error.

As always, "deduction from the phenomena" here really means "deduction from the phenomena plus general 'background' principles." In this case, the general wave theory with free parameter is given, and we proceed, against that given background, to deduce the more specific version with the parameter value fixed from the experimental data.

This case is clear and illustrative but rather mundane. More impressive cases exist such as Newton's deduction of his theory of universal gravitation from the phenomena or the much more recent attempt, outlined by Will and analysed by Earman,[9] to deduce a relativistic account of gravitation from phenomena. These cases involve background principles of extreme

---

[9] See Earman (1992, pp. 173–80); see also the discussion in Mayo (1996) and this volume.

generality that seem natural (even arguably "unavoidable"). These general principles delineate a space of possible – more specific – theories. Taking those principles as implicit premises, the data, by a process that can be characterized either as "deduction from the phenomena" or equivalently as "demonstrative induction," gradually cut down that possibility space until, it is hoped, just one possible general theory remains. Taking the simple case where the background principles specify a finite list of alternatives $T_1, \ldots, T_n$, each piece of data falsifies some T_i until we are left with just one theory, T_j – which, because the inference from $(T_1 \vee T_2 \vee \ldots \vee T_n)$ and $\neg T_1, \neg T_{j-1}, \neg T_{j+1}, \ldots, \neg T_n$ to T_j is of course deductively valid – is thus "deduced from the phenomena."

Clearly such a deduction, if available, is very powerful – it shows, if fully successful, that *the* representative of the very general background assumptions at issue is dictated by data to be one general but particular theory T_j. The data e in such a case therefore provide powerful support for T_j in a very clear and significant sense: the data *dictate* that if any theory that satisfies these natural assumptions can work then it must be T_j.

In the less exciting but more straightforward wave theory case, the data from the two-slit experiment uniquely pick out (modulo some small error interval) the more particular theory T' (with precise value of λ_0 for the wavelength of light from the sodium arc) as the more specific representative of the general wave theory. If you hold the *general* wave theory already, then data dictate that you hold T'.

In Mayo's still simpler case the general background principles are analytic – stating in effect just what the notion of an average *means*. And, hence, the data from her students *dictate* that the average SAT score is 1121 and, therefore (in a very stretched sense), support (maximally, of course) the "hypothesis" that the SAT average is 1121.

Again, because of its statistical character, Howson's standard statistical estimation case does not *quite* fit, but essentially the same situation holds. The basic model is again treated – or so we suppose – as a given: it is taken that this is a Bernoulli process with fixed probability p. Of course, in this case the interval estimate for the proportion of white balls to red balls is not *deduced* from the data provided by the sample, but it might be said that it is "quasi-deduced" in line with standard statistical procedure.

In all these cases, then, a clear sense exists in which the theory is "deduced from the phenomena e" and yet is given strong support by e. In the wave theory case, for example, the result of the two-slit experiment using light from the sodium arc deductively entails $T(\lambda_0)$, the specific version of the wave theory with the wavelength of that light fixed, and what better support or confirmation could there be than deductive entailment? The "no-double-use rule" seems therefore to be entirely refuted.

3 Two Qualitatively Distinct Kinds of "Confirmation" or "Empirical Support": How to Get the Best of Both Worlds

The "UN" or "no-double-use" rule is not, in fact, refuted by the support judgments elicited in the cases discussed in Section 2; instead it simply needs a little elaboration. The principal step towards seeing this is to recognize just how conditional (and *ineliminably* conditional) the support at issue is in all these cases.

In the wave theory case, for example, the judgment that the result of the two-slit experiment with sodium light strongly supports the specific version of the theory $T(\lambda_0)$ is entirely dependent on the prior acceptance of the general wave theory $T(\lambda)$. Insofar as we already have good empirical reason to "accept" that general theory (whatever exactly that means!), the deduction from the phenomena outlined in Section 2 shows that we have exactly the same reason to accept the more specific theory, $T(\lambda_0)$. The "deduction from the phenomena" *transfers* whatever empirical support the general theory already had to the more specific theory that is the conclusion from that deduction. But it surely does not add anything to the support for the more general theory – which was not in any sense tested by this experiment. Of course, so long as the experimental results (that is, the slit and fringe distances) satisfy the general functional formula entailed by that general theory, then *any* particular outcome – any distance between the central bright fringe and the first dark fringe to either side, for example – is consistent with the general theory. A set of fringe distances different from those actually observed (assuming again that the set had the same functional features c – central bright band, symmetrically placed dark bands on either side of that central band, etc.) would not, of course, have led to the rejection of $T(\lambda)$ but simply to the construction or deduction of a *different* specific version – say, $T(\lambda_1)$ – of that same general theory. The fact, then, that $T(\lambda_0)$ entails the correct fringe, slit, and screen distances in the two-slit experiment with sodium light from which it was constructed provides no *extra* empirical reason at all for holding the general theory $T(\lambda)$.

The conditional nature of this sort of empirical support for some relatively specific theory – conditional, that is, on independent empirical support for its underlying general theory being already present – is further underlined by the fact that the sort of theoretical maneuvers that give *ad hocness* a bad name fit the model of "deduction from the phenomena." Consider, for example, the Velikovsky dodge outlined earlier. We can readily reconstruct Velikovsky's overall general theoretical framework (involving not just his assumptions about Jupiter but also about how the (alleged) subsequent terrestrial cataclysms would be reported by appropriate scribes) as employing a free (functional) parameter indicating whether or not the scribes in society S were afflicted by collective amnesia. And then his more specific theory involving claims about which particular societies were, and which were not, afflicted by collective amnesia follows deductively from his general theory plus the "phenomena" (here of course the records, or lack thereof, of appropriate cataclysms). And the deduction proceeds in exactly the same way – both the wave theory and the Velikovsky cases are then instances of "parameter-fixing."

The difference between the wave theory and Velikovsky cases is simply that, in the former but not the latter, *independent support* existed for the general theory ahead of the deduction from the phenomena. But that aside, the logic is identical: in both cases the deduction does no more and no less than *transfer* the empirical support enjoyed by the general theory to the specific deduced theory; it is just that in the Velikovsky case there is no such empirical support for the general theory that could be transferred.[10]

Again there is no question of the underlying theory getting any support from the data at issue and for exactly the same reason as in the wave theory case. The data of records from some cultures, and lack of them from others, do nothing to support the general idea of cataclysms associated with close encounters with the alleged massive comet, because that general theory (once equipped with a "collective amnesia parameter") is not tested by any such data – different data would not have led to the rejection of Velikovsky's general theory but instead simply to a specific version different from the one that Velikovsky actually endorsed given the actual data he had. (This different version would, of course, have simply had a different series of values for the "collective amnesia parameter.")

The sort of confirmation or empirical support involved in these cases is what might be called "purely intra-framework" or "purely intra-research program support." The lack of records in cultures $C_1, \ldots, C_n$ and their (arguable) presence in $C'_1, \ldots, C'_m$ gives very good reason for holding the

---

[10] I am assuming throughout this discussion that the *only way* in which Velikovsky could reconcile the lack of records of suitable cataclysms from some record-keeping cultures within his general theory was via the collective amnesia dodge. Because of the relative laxity of his theory, this is of course far from true. I am therefore idealizing somewhat to make it a crisp case of deduction from the phenomena (it is not really as good as that!). But I believe that all the methodological points stand in spite of this slightly idealizing move.

specific collective-amnesia version of Velikovsky's theory that he proposed *if* you already hold Velikovsky's general theory, *but* (and this is where the initial UN intuitions were aimed) those data give you absolutely no reason at all for holding that general theory in the first place. (Although there might, of course, have been other empirical reasons for doing so, it is just that as a matter of fact in this case there were not.) The data from the two-slit experiment give you very good (in fact, to all intents and purposes, *conclusive*) reason to hold the specific version of the wave theory with the particular value of the wavelength for light from a sodium arc *if* you already hold the general wave theory, *but* the data give you absolutely no reason at all for holding that general theory in the first place (although of course there may have been – and in this case actually were – other empirical reasons for doing so).

Not all empirical confirmation or support can have this ineliminably conditional and ineliminably intra-program character. After all, as we just saw, it seems clear that the difference between the general wave theory and the general Velikovskian theory is that the former has empirical support, which the latter lacks. *Some* general theories – the wave theory of light, but not the general Velikovsky theory – have independent empirical support; that is, empirical reasons exist for holding those general theories ahead of the sort of conditional confirmation (or demonstration) of some particular version of them from data. How can this be, especially in view of the fact that the Duhem thesis implies that all confirmation is of general theories plus extra assumptions? The answer must be that cases exist in which, in contrast to the cases of confirmation we have just considered, confirmation "spreads" from the theoretical framework (central theory plus specific assumptions) to the central theory of the framework – that is, some empirical results must exist which – rather than giving us good reason to accept some specific version of a general theory, given that we have already accepted the general theory – in fact give us good reason to accept the underlying general theory itself (and this despite the fact that the result, in line with Duhem's point, only follows deductively from some specific version of the theory *plus* auxiliaries).

Two kinds of case seem to exist where this occurs. The first is easy to describe. Having used data to fix the value of some parameter in a general theory – that new specific theory complete with parameter value, as well of course as giving you back what you gave to it by entailing the "used" data – may go on to make further predictions that are *independent* of the used data. Thus, for example, the general wave theory entails not only a general functional relationship between wavelengths and quantities measurable in the two-slit experiment but also another general functional relationship between wavelengths and quantities measurable in other experiments – for example, the one-slit diffraction experiment. Thus, having gone from $T(\lambda)$ with free parameter λ plus evidence e about slit separations and fringe distances in the two-slit experiment to the "specific" theory $T(\lambda_0)$, $T(\lambda_0)$ not only entails the original two-slit data e (of course it does!), it also makes an independently testable prediction about the fringe distances produced by light from the same source in the entirely different one-slit experiment. Moreover, this prediction turns out (of course, entirely nontrivially) to be correct. Similarly – in another much-discussed case – Adams and Leverrier, having used evidence e about the Uranian "irregularities" to deduce the existence of a further planet produced a modified Newtonian framework that not only gets e – that is, Uranus's orbit – correct (of course it is bound to) but also makes independent predictions e' about the existence and orbit of Neptune, predictions that again turn out to be correct.

The independent evidence e' – the one-slit result in the case of the wave theory and the observations of Neptune in the Adams–Leverrier case – surely gives *unconditional* support to the general underlying theory: not just support for the wave theory made more specific by fixing parameter λ conditional on the general theory that light consists of waves through a medium, but support for that general theory itself; not just support to the Newtonian system that is committed to a particular assumption about the number of planets, conditional on the basic Newtonian theory, but to the fundamental Newtonian theory itself. So alongside the conditional intra-research program confirmation that is obtained in all the cases discussed in Section 2, a second, more-powerful kind of confirmation exists that provides support for the general theory, or research program, itself. What the UN rule was saying all along, and saying correctly, is that this unconditional kind of support for the underlying general theory involved cannot (of course!) be obtained when the evidence concerned was used in the construction of the specific theory out of that general framework.

Given that in both wave theory and Newtonian cases, the specific theory constructed using evidence e turns out to be independently tested and confirmed by evidence e' (in contrast, of course, to the Velikovsky case of no independent testability), it might seem reasonable to count the used evidence as itself supportive. Given that Adams–Leverrier-amended Newtonian theory makes correct predictions about Neptune, the evidence about Uranus's orbit from which it itself was "deduced" can count as evidence for it, too; given that the wave theory complete with wavelength for sodium light deduced from the two-slit result is independently confirmed by its prediction of the one-slit result with light from the same source, the two-slit result can also count as (unconditional) support for the general wave theory. But this seems to me prejudicial as well as unnecessary and misleading. If Velikovsky is to get only conditional support from the lack of records in culture *C*, then, because the logic is exactly the same, so should the amended Newton theory from the evidence concerning Uranus. The difference between the two is simply, to repeat, that Newton's theory garnered lots of the unconditional kind of support, whereas Velikovskian-specific theories have *only* support conditional on a framework that itself has no support. Two quite different sorts of scientific reasoning seem to be involved after all – obtaining support for a general theory from data and *using* data to construct specific versions of that general theory.

There is at least one respect in which matters are sometimes slightly more complicated. Not perhaps invariably, but certainly quite often, the value of a parameter within a powerful general theory is *overdetermined* by the data. Indeed this is bound to be held whenever, as in the wave theory case discussed earlier, the fixing of a parameter via one experimental result leads to a theory that is (successfully) independently testable via a further experimental result. The general wave theory entails not just one but a *number* of functional relationships between wavelength – clearly a theoretical parameter – and measurable quantities in a range of *different* experiments. So, for example, a mid-nineteenth-century wave theorist could just as well have used the results from the *one-slit* diffraction experiment to fix the value of the wavelength of monochromatic light from some particular source and then have gone on to predict the outcome of the two-slit experiment performed using that same light. This, in the end, would be equivalent to the converse process that I just described in which the theorist uses the results from the two-slit experiment to fix the parameter and then goes on to predict the one-slit result. In general, there may be a series of experimental results $e_1, \ldots, e_n$, any (proper) subset of which of some size r can be used to fix parameter values; then the underlying general theory with these fixed parameter values predicts the remaining $n - r$ pieces of evidence. There is clearly no a priori guarantee that the set of data $e_1, \ldots, e_n$ admits any consistent assignment of values for the theoretical parameter at issue – it will do so if and only if the results of the $(n - r)$ independent tests of the theory once the parameter has been measured using r of the results are positive.

Clearly, in cases where this does indeed happen, the data set $e_1, \ldots, e_n$ tells us something positive about the underlying theory. It would not seem unreasonable to say, as I believe Mayo would, that this data set is *both* used in the construction of the theory *and at the same time* "severely" tests it. And this judgement would again seem to be in clear conflict with the "no-double-use rule." However, this judgment is surely coarse-grained. What really (and, once you think about it, pretty obviously) ought to be said is that *part* of the evidence set fixes parameters in the underlying general theory and then *part* of that set tests the resulting, more specific version of the theory. It is just that in such a case it does not matter which particular subset of size r you think of as doing the parameter-fixing and which remaining subset of size $n - r$ you think of as doing the testing. Nonetheless, two separate things *are* going on that are dependent on different bits of data: genuine *tests* of a theory and *application* of a theory to data to produce more specific theoretical claims.

This may seem an unnecessary quibble – why not just agree that the "no-double-use rule" fails in such cases? The evidence set is used *both* in the construction of the specific theory involved *and* in its (unconditional) support. One reason is as follows: suppose we had two theories, T and T', one of which, say T, has no relevant free parameters and entails $e_1, \ldots, e_n$ straight off, whereas T' involves parameters that are left free by theoretical considerations and need to be fixed using r of the evidential results e_i. Surely we would want to say in such a circumstance that the evidential set $e_1, \ldots, e_n$ supports T *more* than it does T'? If so, then there must be some confirmational "discount" for parameter-fixing: speaking intuitively, in such a case, T gets n lots of (unconditional) confirmation from the data set, whereas T' gets only $n - r$ lots. How much of the data set is needed to fix parameters plays a role in the judgment of how much (unconditional) support the theory gets from the data set. And this condition holds even when the choice is arbitrary of which particular subset (of a certain size) is used to fix parameters and which is used to genuinely test and, hence, (possibly) supply genuine "unconditional" support. In this sense, although the set as a whole, if you like, both fixes parameter values and (unconditionally) supports, *no particular element of the data set does both.*

I said that two kinds of case exist where support is unconditional – two kinds of case in which support "spreads" from the specific theory that entails the evidence to the underlying general theory. The first of these is the case of independent testability that we have just considered. The second type is equally important though somewhat trickier to describe precisely. This sort of confirmation (again, of the general underlying theory, rather than of some specific theory, *given* the general underlying theory) is provided in cases in which, roughly speaking, some prediction "drops out of the basic idea" of the theory. Here is an example.

The explanation of the phenomena of planetary stations and retrogressions within the Ptolemaic geocentric theory is often cited as a classic case of an ad hoc move. The initial geocentric model of a planet, say Mars, travelling on a single circular orbit around a stationary Earth, predicts that we will observe constant eastward motion of the planet around the sky (superimposed, of course, on a constant apparent diurnal westward rotation with the fixed stars); this prediction is directly refuted by the fact that the generally eastward (apparent) motion of Mars is periodically interrupted by occasions when it gradually slows to a momentary halt and then begins briefly to move "backwards" in a westward direction, before again slowing and turning back towards the east (remember that it never moves or even seems to move backwards on any particular night because the diurnal movement is always superimposed). The introduction of an epicycle of suitable size and the assumption that Mars moves around the center of that epicycle at a suitable velocity while the whole epicycle itself is carried around the main circular orbit (now called the deferent) leads to the correct prediction that Mars will exhibit these stations and retrogressions. Although not as straightforward as normally thought, this case surely is one that fits our first, entirely conditional, kind of confirmation – if you *already accept* the general geocentric view, then the phenomena of stations and retrogressions give you very good reason to accept (and in that sense they strongly confirm) the particular version of geocentricism involving the epicycles.[11] However, the fact that stations and retrogressions are "predicted" (or better, entailed) by the specific version of geocentricism with suitable epicyclic assumptions gives absolutely no further reason to accept (and so no support for, or confirmation of) the underlying basic geocentric (geostatic) claim.

The situation with Copernican heliocentric (or rather heliostatic) theory and planetary stations and retrogressions is, I suggest, entirely different.[12] According to the Copernican theory we are, of course, making our observations from a moving observatory. As the Earth and Mars

---

[11] This is often thought of as the archetypically ad hoc move (epicycles are almost synonymous with *ad hoccery*). However, the Ptolemaic move does produce an independent test (and indeed an independent confirmation) but not one that, so far as I can tell, was ever recognized by any Ptolemaist. It follows from the epicycle-deferent construction that the planet must be at the "bottom" of its epicycle and, hence, at its closest point to the Earth exactly at retrogression. But this, along with other natural assumptions, entails that the planet will be at its brightest at retrogression – a real fact that can be reasonably confirmed for some planets with the naked eye. (Of course, even had it been recognized, this test would not have been reason to continue to prefer Ptolemy over Copernicus because, as will immediately become apparent, the Copernican theory also entails, in an entirely non–ad hoc way, that the planet is at its nearest point to the Earth at retrogression.)

[12] See the treatment in Lakatos and Zahar (1976).

both proceed steadily eastward around the sun, the Earth, moving relatively quickly around its smaller orbit, will periodically overtake Mars. At the point of overtaking, although both are in fact moving consistently eastward around the Sun, Mars will naturally *appear*, as observed from the Earth, to move backwards against the background of the fixed stars. Planetary stations and retrogressions, rather than needing to be explained *via* specially tailored assumptions (having to be "put in by hand" as scientists sometimes say), drop out naturally from the heliocentric hypothesis. Copernican theory, in my view, genuinely *predicts* stations and retrogressions even though the phenomena had been known for centuries before Copernicus developed his theory. (Here I am talking about the qualitative phenomenon, not the quantitative details which, as is well known, need to a large extent to be "put in by hand" by both theories – and courtesy of multiple epicycles in Copernicus no less than in Ptolemy.[13])

The way that Copernican theory yields stations and retrogressions may, indeed, seem to be *so* direct that it challenges Duhem's thesis: doesn't the basic heliocentric hypothesis on its own, "in isolation," entail those phenomena? This is a general feature of the sort of case I am trying to characterize: the way that the confirming phenomenon "drops out" of the basic theory appears to be so direct that scientists are inclined to talk of it as a direct test of just the basic theory, in contradiction to Duhem's thesis. But we can see that, however tempting this judgment might seem (and I *am*, remember, endorsing the view that especially direct or strong support is yielded in such cases), it cannot be literally correct.

First of all, there must be assumptions linking actual planetary positions (as alleged by the theory) to our observations of them – no less so, or not much less so, with naked-eye observations as with telescopic ones. (Remember that the Flamsteed–Newton dispute revealed the inevitable existence of an assumption about the amount of refraction undergone by the light reflected from any given planet as that light enters the Earth's atmosphere.) But even laying this aside, no theory T, taken "in isolation," can deductively entail any result e if there is an assumption A that is both self-consistent and consistent with T and yet which together with T entails not-e. Therefore, in the case we are considering, if the basic Copernican theory alone entailed stations and retrogressions, then there would have to be *no possible* assumption consistent with that basic heliocentric claim that, together with it, entailed that there would be no stations or retrogressions.

---

[13] See, for example, Kuhn (1957).

But such possible assumptions *do* exist. Suppose, for example, that the Earth and Mars are orbiting the Sun in accordance with Copernicus' basic theory. Mars happens, though, to "sit" on an epicycle but only starts to move around on that epicycle when the Earth is overtaking Mars and does so in such a way that exactly cancels out what would otherwise be the effects of the overtaking (that is, the station and retrogression). Of course, this assumption is a monstrous one, but it is both internally consistent and consistent with the basic heliocentric view. The existence of this assumption implies that, contrary perhaps to first impressions, Duhem's thesis is not challenged in this case; the heliocentric hypothesis *alone* does not entail the phenomena (even if we lay aside the dependence on assumptions linking planetary positions with our observations of them).

However, those first impressions and the monstrousness of the auxiliary necessary to "prevent" the entailment of stations and retrogressions both reflect just how "natural" the extra assumptions are that are necessary for heliocentricism to entail the phenomena. All that needs to be assumed, in addition to the basic idea that Mars and the Earth are both orbiting the sun, is that they both do so in relatively regular ways (no sudden pirouettes and the like) and that the Earth (which has an observably smaller average period) moves relatively quickly around its smaller orbit and, hence, periodically "laps" Mars.

Let me, then, sum up this section of the chapter. It seems obvious on reflection, or so I claim, that two quite different precise ways exist for using data in science, each of which falls under the vague notion of data providing "empirical support" for a theory. Using empirical data *e* to construct a specific theory T' within an already accepted general framework T leads to a T' that is indeed (generally maximally) supported by *e*; but *e* will not, in such a case, supply any support at all for the underlying general theory T. The second and stronger type of empirical support involves a genuine test of, and therefore the possibility of real confirmation for, not just the specific theory that entails some observational result *e* but also the underlying general theory. And, as we have just been seeing, there are in turn two separate ways in which this stronger kind of support can be achieved. The "UN" or "no-double-use" rule was aimed at distinguishing general theoretical frameworks or research program that are "degenerating" from those that are "progressive." At, in other words, systematically underwriting the intuitive judgment that, when some piece of evidence *e* is predicted by some specific theory within general program *P*, but only accommodated post hoc by some specific theory within general rival program P', this does not, contrary to what Kuhn seemed to suppose, balance the evidential scales – *e* continues to provide *a* reason for preferring *P* to *P'* (of course, this fact does not rule out other reasons for the opposite preference). The defenders of the rule were therefore pointing (correctly) at the importance of the "second," "stronger" unconditional type of support described earlier and (correctly) emphasizing that the conditional type of confirmation provides no support at all that "spreads" to the underlying general theory. What those who thought that they were criticising the "UN" or "no-double-use" rule were really doing was pointing out that the same manoeuvre – of using data to fix parameter values or particular theories within a given general framework – that is correctly regarded with suspicion when performed as a defensive, "degenerating" move when two general frameworks are vying for acceptance is often also used positively within general theoretical frameworks. The manoeuvre will seem positive when the general framework that is being presupposed is supported independently of the particular data being used. And it will look more positive the more such independent empirical support exists for the general framework. But, however positive the manoeuvre looks, the evidence involved does not – cannot! – supply any further support for the general framework. Instead that evidence simply (though importantly) transfers the support enjoyed by the general framework theory to the particular theory thus deduced from that evidence plus the general theory.

Mayo challenged me to be more explicit about the underlying *justification* for the two-type confirmation theory that I defend here. Well it is, I trust, clear that the justification for the conditional type (where the "no-double-use" rule *allegedly* fails) is deductive (or a close substitute): we already (we assume) have good reasons for holding some general theory; the relevant data then, *within that context*, support the specific version by deductively entailing it. As for the "stronger" "unconditional" type of confirmation, the underlying justification is exactly the same as that cited by Mayo in favour of her own approach (see next section) – a theory *T* is supported in this sense by some evidence *e* only if (and to the extent that) *e* is the outcome (positive so far as *T* is concerned) of some severe test of *T*. This, in turn – as Popper resisted recognizing – is underpinned by the intuitions that are often taken to be captured by the "No Miracles argument": it seems in some clear but (I argue[14]) elimininably intuitive sense very unlikely that a theory would survive a severe test of it if the theory were not somehow "along the right lines."

[14] See Worrall (n.d.).

4 Mayo's Alternative: Confirmation Is All about "Severe Tests"

How do these views on confirmation compare with Mayo's influential and more highly developed views? Some striking similarities certainly exist. Deborah starts, just as I do, with the "UN rule" and by emphasising the fact that the rule delivers judgments that accord with intuition in many cases; and she insists, just as I do, that the rule also seems to contradict what seem to be clearly valid intuitive judgments about support in other cases. Unlike me, however, she sees the UN rule as definitely refuted by these latter judgments and therefore as needing to be replaced, rather than, as I have argued, clarified.

Mayo's bold and challenging idea is in fact that *all* cases, both those that satisfy the UN rule and those that seem to conflict with it, are captured by one single underlying notion that is at once simple and powerful: the notion of a *severe test*. Confirmation of a theory for her *always* results from that theory's surviving a severe test. Echoing Popper, of course, she holds that hypotheses gain empirical credit only from passing genuine tests; the more severe the test, the higher the confirmation or support, if the theory passes it. This simple idea, when analysed from her own distinctive perspective, reveals – so she argues – *both* the rationale for the UN rule in the cases where it does correctly apply *and* the reason why that rule delivers incorrect judgments in other cases.

The defenders of the use-novelty account hold in effect that evidence used in the construction of a hypothesis cannot provide a genuine test of it and, hence, cannot supply genuine confirmation. Underlying their view, on Mayo's analysis, is the initially plausible-sounding claim that a severe test is one that a theory has a high probability of failing. Hence, the UN rule must, it seems, be correct because evidence *e* used in the construction of *T* cannot possibly test *T*, because there is no chance of *T*'s failing the "test" whose outcome is *e* – that outcome was instead "written into" *T*. No matter how plausible this may sound, argues Mayo, it in fact misidentifies the probability that we should be concerned to maximize so that we might maximize severity and, hence, it misidentifies the real notion of a severe test. It is easier to understand her characterization if we accentuate the negative: a *non*severe test is *not* one that has a high probability of being passed by a theory (in the limit, of course, is a test *certain* to be passed by the theory), but rather one that has a high probability of being passed by the theory, *even though the theory is false*. As she puts it, "what matters is not whether passing is assured but whether erroneous passing is" (1996, pp. 274–5).

In cases where the "no-double-use" rule delivers the correct answer (she cites "gellerized hypotheses,"[15] but would surely accept the Velikovsky case cited earlier as identical in the relevant respects), the "test" at issue was indeed nonsevere: the modified Velikovsky theory would have a good chance of passing the test of no records of suitable cataclysms in culture C even though that theory were false. On the other hand, in the cases where the no-double-use rule goes wrong, such as her SAT score example, although admittedly there was no chance of the "hypothesis" that the average score of her class is 1121 not passing the "test" arrived at by adding the N individual scores and dividing by N, the "test" was nonetheless genuine and severe, indeed maximally severe, because there would have been no chance of the "hypothesis" passing the test *if it were false*. Similarly in standard statistical estimation cases, such as the one cited by Howson and developed in much more detail by Mayo, assuming that we have some reason to think that the general model being applied really does apply to the real situation, then using the observed result of k out of n balls drawn being white to construct the hypothesis that the proportion of white balls overall in the urn is $k/n \pm \varepsilon$ (where ε is calculated as a function of n and the chosen significance level by standard confidence-interval techniques) does *not* preclude the sample relative frequency (e) of k/n red balls being good evidence for our hypothesis. Even though e was thus used in the construction of h, e still constitutes a severe test of h because there was little chance of h passing the test resulting in e if it were false.

Despite being a colleague of Nancy Cartwright, there are few bigger fans of unity than I. And Mayo here offers a unified alternative to my "two kinds of confirmation" view – two kinds of confirmation do not exist, only one: that supplied by a theory's surviving a severe test. It would seem churlish of me to turn this offer down and thus reject the call to join the "error paradigm." Moreover, so Deborah assures me, were I to join then I could avail myself of precise characterisations of notions such as that of an empirical prediction "falling out" of a theory, which are important to my view of confirmation but

---

[15] Uri Geller asserted (indeed no doubt still asserts) that he has genuine psychokinetic powers; when, unbeknownst to him, professional magicians controlled the situation in which he was to exhibit these powers, for example, by bending spoons at a distance, he proved impotent; however, Geller responded by claiming that his "special" powers were very delicate and had been affected by the presence of skeptics in the audience. Obviously he could only identify whether or not skepticism was playing this obstructive role post hoc: if he was able to bend spoons by the "power of pure thought" then no skeptics were around (and also, of course, no hindrance to his employing standard magicians' tricks); if he was unable to bend them "supernaturally" then clearly there *was* skepticism in the air.

are left as merely suggestive notions within it (though I hope with clear-cut illustrations from particular scientific cases).

Despite these enticements, I must turn this kind offer down. I do so for two interrelated reasons:

1. There seem to be a number of unclarities in or outright significant difficulties with Mayo's position; and more fundamentally
2. It just seems to be true – and plainly true – that, as I explained in the preceding section, there are two quite separate uses of evidence within science: using evidence in the construction of a theory is a quite different matter from using evidence to test it by "probing for errors;" Mayo's attempt to construct a one-size-fits-all account where all (positive) uses of evidence in science are regarded as the passing of a severe test is itself an error. (Einstein is reported as having said that physics should be as simple as possible, but not more so! The same surely applies to meta-science.)[16]

I begin with the already much-discussed SAT score example. As remarked, it does seem extraordinary to call the assertion arrived at about the average SAT score of Mayo's students a "hypothesis" and at least equally extraordinary to call the process of adding the individual scores and dividing by the number of students a "test" of that claim. Of course, had someone made a "bold conjecture" about the average score, then one might talk of the systematic process of working out the real average as a test of that conjecture. But boldly conjecturing would clearly be a silly way to proceed in this case, and, as already remarked, not one that would ever be used in more realistic cases in science. The process of adding the individual scores and dividing by the number of students surely is a *demonstration that* the average score is 1121, not a "*test*" of the "hypothesis" that this is the average score.[17] We construct the "theory" by deducing it from data (indeed the "theory" just encapsulates a feature of the data).

More important, because we all agree that the evidence here is conclusive for the "hypothesis" and it might be felt that it does not really matter

---

[16] The problems involved in reason 1 are in fact, as I shall explain, all produced by the fact that reason 2 is true.

[17] In fact this and some of the other cases that Mayo analyzes – such as the identification of the car that hit her own car's fender or the technique of "genetic fingerprinting" – seem altogether more naturally categorized as *applications* of already accepted theories (or "theories" in the case of the average SAT score) to particular circumstances rather than as any sort of empirical support for theories. We *apply* our theories of genetics to work out the probability that the match we have observed between the crime scene blood and that of the defendant would have occurred if he or she were innocent.

how we choose to express it, the case seems to me to highlight a problem with applying Mayo's central justification for all confirmation judgments. In the circumstances (and assuming that both the data on the individual students and the arithmetic have been carefully checked) there is *no* chance that the average SAT score is *not* 1121. If, as seems natural, this claim is interpreted as one about a conditional probability – namely, p (T passes the test with outcome e/T is false) $= 0$ – we are being asked to make sense of a conditional probability where the conditioning event (the claim's being false) has probability zero. Indeed we are asked not only to make sense of it but to agree that the conditional probability at issue is itself zero. It is well known, however, that – at any rate in all standard systems – $p(A/B)$ is not defined when $p(B) = 0$. Perhaps we are meant to operate with some more "intuitive" sense of chance and probability in this context. But I confess that I have examined my intuitions minutely and still have no idea what it might mean in this case to imagine that the average score is *not* 1121, when the individual scores have been added and divided by N and the result *is* 1121!

In correspondence Mayo tells me that I *should have* such intuitions "because the next time you set out to use your estimation tool it may NOT BE 1121." Well, maybe she can give me more hints on how to develop better intuitions, but this one certainly doesn't work for me: surely – again short of making some trivial arithmetical error – applying the "estimation tool" would just *have* to yield 1121 again with this particular group of students; if you were to arrive at any other figure you simply would not be taking the average. And if she means that the average score might not be 1121 for some *different* group of students then of course this is (trivially) true but whatever number you arrived at (assuming again that you arrived at it correctly without trivial error) would still be the group average for that new group!

It could, perhaps, be argued that this is simply a problem for this admittedly extreme case. But there are other, related problems with Mayo's account of severity and the associated intuitive probability judgments that underpin it that surface in other more standard, scientific cases.

One way that she likes now to put the partial connection, and partial disconnection, between "no double use" and severity is something like this:

It would seem that if hypothesis H is use-constructed then a successful fit (between H and [data] x) is assured, *no matter what* (and, hence, that in the case of use-construction the data always represent a nonsevere test). However, the "no matter what" here may refer to two quite different conditions:

1. no matter what the *data are*, or
2. no matter whether H is true or false.

In cases where the "no-double-use" rule gives the *wrong* answer, condition 1 is true alright (as always with double-use cases), *but condition 2 is false.* In cases in which "no double use" *correctly* applies, on the other hand, condition 2 is also true – that is, *both* conditions 1 and 2 must be met if x is to *fail* to represent a severe test. The no-double-users have mistakenly held that if condition 1 alone is met then the test is automatically nonsevere.

But this way of putting things, while sounding very neat, in fact leads simply to a new way of putting the previous objection. I have already pointed out that it seems impossible to me to make sense of condition 2, at least in the SAT score case (and, as I will shortly argue, more generally). But problems exist with condition 1, too – problems that apply across the board. The collective-amnesia–ized version of Velikovsky, V', is "use-constructed" from the data e concerning the cultures from which we have or have not appropriate records of suitably dated "catastrophes." This is surely a case where the "no-double-use" idea ought to apply in some way or another: V', because of its method of construction, is not really tested by data e and, hence, is not supported by it – not, at least, in any sense that makes it rationally more credible. Does Mayo's account deliver this judgment of nonseverity? In particular, does her condition 1 apply? That is, is it true that a successful fit between V' and data x is "assured no matter what the data are?"

Well if the data were anything other than they in fact are, say they were e' (that is, Velikovksy faced a different list of otherwise record-keeping cultures who have left no records of suitable cataclysms), then V' (that is, the particular version of the general Velikovskian theory actually constructed from the real data e) would of course conflict with this supposed data e': some cultures alleged by V' to have suffered from collective amnesia would have records of catastrophes and/or some cultures having no such records would not be alleged by V' to have suffered collective amnesia. Had the data been different then the Velikovskian would not have been proposing V' but instead some rival V'' – a specific version of the same general "cometary" theory that evaluated the "collective-amnesia parameter" differently.

It seems, then, to be straightforwardly untrue that a successful fit between V' and e is assured no matter what e is. Nor can we rescue the situation by allowing (as of course Mayo explicitly and elaborately does) grades of severity and, hence, the intuitive "probabilities" discussed earlier. It again makes no sense to me to say that the test of V' that turns out to have outcome e is nonsevere because there is a high probability of V' fitting the data no matter what those data are. Within the context of the general Velikovsky theory with free "collective-amnesia parameter," we can in effect derive the biconditional V' if and only if e: that is, V' would definitely not have fit the data had the data been different than they in fact were.

It is not the successful fit of a particular hypothesis with the data that is guaranteed in these sorts of case but rather the fit of *some particular* hypothesis developed within the "given" underlying general framework. We again need to recognize that, as my account entails, two separate issues exist – the "confirmation" of a theory within a general framework (*e* maximally confirms V' given *e and given* V) and "confirmation" of a specific theory within a general framework that "spreads" to the underlying general theory. This condition, as pointed out earlier, is not satisfied in the Velikovsky case and, hence, *e* gives no "unconditional" support to V' of the sort that would spread to the underlying V, - exactly because the general theory places no constraints on the relevant parameter, the value of which can be "read off" whatever the data turn out to be.

Mayo may reply that her account does yield the result that the evidence *e* does not support the general Velikovskian theory, V, because the latter is not tested (or, as she often says, "probed") by that evidence. Of course I agree with this judgment, but that is not the problem – her account of the support lent to V' is at issue. We surely want to say that *e* provides no good reason to take V' seriously. If she were to deliver this judgment directly, it would have to be that *e* fails to be any sort of severe test of V', which requires that conditions 1 and 2 of her latest formulation of (non)severity be satisfied; but, as we just saw, condition 1 is *not* in fact satisfied. If Mayo were tempted, in response, to rule that a specific theory like V' is only "probed" by a test with outcome *e* if that same test also probes (severely tests) its general version V, then it would mean that she had failed to capture the alleged exceptions to the UN rule. These (apparent) exceptions, as explained earlier, all involve deductions from the phenomena that all presuppose, and therefore cannot "probe," the underlying theory. And the attempt to capture these alleged exceptions is of course an important part of the motivation for her overall account. Adding the SAT scores of her N logic students and dividing by N does not probe the underlying definition of an average score! Adams and Leverrier's use of the anomalous data from Uranus to construct a version of Newton's theory (complete with a postulated "new" planet) did not probe the underlying Newtonian theory (three laws plus the principle of universal gravitation) but, on the contrary, presupposed it. It is again surely clear why her account meets these difficulties. We are dealing, in accordance with my own account, with *two quite different uses of evidence* relative to theories; her attempt to cover these two different cases with one set of criteria is bound to fail.

To see that these difficulties for Mayo's account are not simply artifacts of the strange, clearly pseudoscientific case of Velikovsky's theory, nor are they restricted to problems with her first condition for nonseverity, let us return to the case of the wave theory that I discussed earlier. This example, although deliberately a very simple one, nonetheless exemplifies an important and recurrent pattern of reasoning in real science. The simplicity of the case allows us to concentrate on the pattern of reasoning and not become sidetracked by scientific details and complexities.

Remember, the case involves the general wave theory of light, call it W. Theory W leaves the wavelengths of light from particular monochromatic sources as free parameters; it does, however, entail (a series of) functional relationships between such wavelengths and experimentally measurable quantities. In particular, subject to a couple of idealisations (which nonetheless clearly approximate the real situation), W implies that, in the case of the two-slit experiment, the (observable) distance X from the fringe at the center of the pattern to the first fringe on either side is related to (theoretical) wavelength λ via the equation $X/(X^2 + D^2)^{1/2} = \lambda/d$ (where d is the distance between the two slits and D the distance from the two-slit screen to the observation screen – both observable quantities). It follows analytically that $\lambda = dX/(X^2 + D^2)^{1/2}$. But all the terms on the right-hand side of this last equation are measurable. Hence, particular observed values of these terms, call their conjunction e, will determine the wavelength (of course within some small margin of experimental error) and so determine the more specific theory W', with the parameter that had been free in W now given a definite value – again within a margin of error.

This scenario is a paradigmatic case of "deduction from the phenomena" – exactly the sort of case, so critics of the UN rule have alleged, in which that rule clashes with educated intuition. We do want to say that e "supports" W' in some quite strong sense; and yet clearly e was used in the construction of W' and, hence, W' was guaranteed to pass the "test" whose outcome was e. Mayo's claim here is that, whenever "no double use" goes astray, it is because condition 2 has been ignored. A test of theory T may be maximally severe even if T is guaranteed to pass it, so long as T is not guaranteed to pass it *even if it is false*. (Remember: "what matters is not whether passing is assured but whether *erroneous* passing is.") But in fact Mayo's condition 2 for *nonseverity* is *met* here: whether or not W' is true the fit with e is assured, because the value of λ specified by W' has been calculated precisely to yield e. It is true that, for exactly the same reasons as we saw in the case of Velikovsky, it can be argued that Mayo's condition 1 fails to hold in this case. In fact an ambiguity exists over what "it" is in the condition (for nonseverity, remember) – that "it" would have passed the test concerned whatever that test's outcome. The particular W' that was in fact constructed from data e would certainly *not* have passed the test of measuring the fringe distances and so forth in the two-slit experiment with sodium light had those measurements produced results other than those expressed in e. What was bound to pass again *is some version or other* of W, with some value or other for the wavelength of sodium light. It might be argued, therefore, on Mayo's behalf that because it is not true that both conditions for nonseverity hold in this case, the test may be regarded as at least somewhat severe. But clearly what Mayo intended was that condition 1 *should* in fact hold in "use-constructed" cases and that it is the failure of condition 2 to hold in certain particular cases (despite condition 1 holding) that explains why the UN rule delivers incorrect verdicts in those cases.

It seems, therefore, to be at best unclear whether Mayo's scheme, when analysed precisely, can explain the judgment that e does at least something positive concerning the credentials of W' in the case we are considering. On the other hand, this judgment *is* captured by my account: e definitely supports W' in the conditional sense in that it establishes W' as *the* representative of the general theory W if that theory is to work at all; hence, one might say, the construction transfers to W' all the unconditional empirical support that W had already accrued (and in this case there was plenty of such support).

Mayo has claimed (personal correspondence) that this analysis entirely misrepresents her real view. She would *not* in fact want to say in this wave-theory case that e tests W', because it does not "probe the underlying [W]." Of course this is indeed true (and importantly true), though it is unclear how this relates to the issue of whether W' and e satisfy condition 2 of her latest account. But even supposing we go along with this view of what her account entails here, how then does that account deliver the (conditional but nonetheless positive) verdict concerning the support that e lends to W' that intuition does seem to require? Moreover, this interpretation of her account takes us back to the problem mentioned earlier in connection with Velikovsky – namely, that it then seems hard to understand how it delivers the judgments that she highlights as "refutations" of UN. How, if "probing the underlying theory" is also required for a test of a specific theory to be severe, can it be that the data in the SAT score case severely test the hypothesis about the average score for her class? Or that estimates of some parameter (such as the proportion of red to white balls in Howson's urn case) arrived at *via* standard statistical techniques severely test the hypothesis about that parameter? In neither case is the underlying theory "probed," but it is instead taken for granted (indeed in the SAT course case no option exists but to take the underlying theory for granted because it is analytic). No result that you could get from averaging SAT scores could challenge the definition of an average; no sample relative frequency of red and white balls could challenge the idea that the urn contains some unknown but fixed ratio of such balls and that the draws are independent.

If the Mayo account could be defended at all here, then it would have to be, so it seems to me, by reinterpreting her second condition for nonseverity. Her account would have to be understood as saying that a test of *T* is nonsevere if the test's outcome is bound to fail to refute *T* (condition 1) *and* if the general theory underlying *T* is not itself empirically supported by *other* tests. But this would be in effect just to rewrite my own account in something approximating Mayo's terms. Moreover, as I have suggested earlier more than once, by thus writing my analysis into one account of severe versus nonsevere tests, the important and qualitative difference between the two uses of evidence as related to theories would be obscured.

The attempt to see everything in terms of severe testing, and probing for error, seems to lead either to error or at best to a confusing reformulation of the view that I defended. It surely is just the case that science makes use of two separate roles for evidence: a role in the construction of theories ("observation as theory-development by other means" as I believe van Fraassen says somewhere) and a role in testing theories, in probing them for errors. The latter is of course a vastly important use of evidence in science but it is not, as Mayo has tried to suggest, everything.

References

Earman, J. (1992), *Bayes or Bust? A Critical Examination of Bayesian Confirmation Theory*, MIT Press, Cambridge, MA.

French, A. (1971), *Newtonian Mechanics*, MIT Press, Cambridge, MA.

Hitchcock, C., and Sober, E. (2004), "Prediction Versus Accommodation and the Risk of Overfitting," *British Journal for the Philosophy of Science*, 55: 1–34.

Howson, C. (1990), "Fitting Theory to the Facts: Probably Not Such a Bad Idea After All," in C. Wade Savage (ed.), *Scientific Theories*, University of Minnesota Press, Minneapolis.

Kuhn, T.S. (1957), *The Copernican Revolution*, Princeton University Press, Princeton.

Kuhn, T.S. (1962), *The Structure of Scientific Revolutions* (2nd enlarged ed., 1970), University of Chicago Press, Chicago.

Kuhn, T.S,. (1977), *The Essential Tension*, University of Chicago Press, Chicago.

Lakatos, I., and Zahar, E.G. (1976) "Why Did Copernicus's Programme Supersede Ptolemy's," Chapter 4 in I. Lakatos (ed.), *The Methodology of Scientific Research Programmes*, Cambridge University Press (reprinted 1978).

Mayo, D.G. (1996), *Error and the Growth of Experimental Knowledge*, University of Chicago Press, Chicago.

Worrall, J. (2000), "The Scope, Limits and Distinctiveness of the Method of "Deduction from the Phenomena": Some Lessons from Newton's "Demonstrations" in Optics," *British Journal for the Philosophy of Science*, 51: 45–80.

Worrall, J. (2002), "New Evidence for Old," in P. Gardenførs et al. (eds.), *In the Scope of Logic, Methodology and Philosophy of Science*, Kluwer, Dordrecht.

Worrall, J. (2003), "Normal Science and Dogmatism, Paradigms and Progress: Kuhn "versus" Popper and Lakatos," in T. Nickles (ed.), *Thomas Kuhn*, Cambridge University Press, Cambridge.

Worrall, J. (2006), "Theory Confirmation and History," in C. Cheyne and J. Worrall (eds.), *Rationality and Reality*, Springer, Dordrecht.

Worrall, J. (n.d.) "Miracles, Pessimism and Scientific Realism," forthcoming.

An Ad Hoc Save of a Theory of Adhocness?
Exchanges with John Worrall

Deborah G. Mayo

In large part, the development of my concept of severity arose to deal with long-standing debates in philosophy of science about whether to require or prefer (and even how to define) novel evidence (Musgrave, 1974, 1989; Worrall 1989). Worrall's contribution represents the latest twists on our long-running exchange on the issue of novel evidence, beginning approximately twenty years ago; discussions with Musgrave around that time were also pivotal to my account.[1] I consider the following questions:

1. *Experimental Reasoning and Reliability*: Do distinct uses of data in science require distinct accounts of evidence, inference, or testing?
2. *Objectivity and Rationality*: Is it unscientific (ad hoc, degenerating) to use data in both constructing and testing hypotheses? Is double counting problematic only because and only when it leads to unreliable methods?
3. *Metaphilosophy*: How should we treat counterexamples in philosophical arguments?

I have argued that the actual rationale underlying preferring or requiring novel evidence is the intuition that it is too easy to arrive at an accordance between nonnovel data and a hypothesis (or model) even if *H* is false: in short the underlying rationale for requiring novelty is severity. Various impediments to severity do correlate with the double use of data, but this correlation is imperfect. As I put it in Mayo (1996): "Novelty and severity do not always go hand in hand: there are novel tests that are not severe and severe tests that are not novel. As such, criteria for good tests that are couched in terms of novelty wind up being either too weak or too strong, countenancing poor tests and condemning excellent ones" (p. 253). This

[1] Mayo (1991; 1996, p. xv).

malady is suffered by the version of novelty championed by Worrall, or so I argue. His chapter turns us directly to the problem of metamethodology with respect to a principle much debated in both philosophy of science and statistical practice.

The UN requirement – or, as he playfully calls it, the UN Charter – is this:

1.1 Use-Novelty Requirement (UN Charter): For data **x** to support hypothesis *H* (or for **x** to be a good test of *H*), *H* should not only agree with or "fit" the evidence **x**, **x** must itself *not have been used* in *H*'s construction.

For example, if we find data **x** anomalous for a theory or model, and we use **x** to arrive at a hypothesized explanation for the anomaly *H*(**x**), it would violate the UN Charter to also regard **x** as evidence for *H*(**x**). Much as with the rationale for varying the evidence (see also Chapter 2, Exchanges with Chalmers), use-novelty matters just to the extent that its violation inhibits or alters the reliability or stringency of the test in question. We can write the severity requirement in parallel to the UN Charter:

1.2 Severity Requirement: For data **x** to support hypothesis *H* (or for **x** to be a good test of *H*), *H* should not only agree with or "fit" the evidence **x**, *H* must have passed a stringent or severe test with **x**.

If UN violations alter a test's probativeness for the inference in question, the severity assessment must be adjusted accordingly.

However, as I have argued, some cases of "use-constructed" hypotheses succeed in being well tested by the same data used in their construction. To allude to a case discussed in Mayo (1996, p. 284), an apparent anomaly for the General Theory of Relativity (GTR) from the 1919 Sobral eclipse results was shown to be caused by a mirror distortion from the sun's heat. Although the eclipse data were used both to arrive at and to test *A*(**x**) – the hypothesized mirror distortion explanation – *A*(**x**) passed severely because it was constructed by a reliable or stringent rule. Because Worrall agrees with the severity goal, these cases stand as counterexamples to the UN requirement. Worrall's chapter discusses his recent attempts to accommodate these anomalies; surprisingly, Worrall is prepared to substantially adjust his account of confirmation to do so.

In particular, he allows my counterexamples to stand, but regards them as involving a distinct kind of support or corroboration, a kind he has developed to accomodate UN violations. For instance, if *A*(**x**) is use-constructed to account for **x** which is anomalous for theory *T*, then the inference to *A*(**x**) gets what Worrall calls "conditional support." By this he means that *A*(**x**) is legitimately inferred only conditional on already assuming *T*, the theory to be "saved." So for Worrall the UN requirement still stands as necessary (and sufficient) for full-bodied support, but data **x** may still count as evidence for use-constructed $A(\mathbf{x})$ so long as we add "conditional on accepting an overarching theory T."

But I do not see how Worrall's attempt to save the UN Charter can adequately accommodate the counterexamples I have raised. It is false to suppose it is necessarily (or even commonly) the case that use-constructed hypotheses assume the truth of some large-scale theory, whether in the case of blocking an anomaly for theory T or in any of the other use-constructions I have delineated (Mayo, 1996, 2008). Certainly using the eclipse data to pinpoint the source of the GTR anomaly did not involve assuming the truth of GTR. In general, a use-constructed save of theory T takes the form of a hypothesis designed to block the alleged anomaly for T.

1.3 A use-constructed block of an anomaly for T:

$A(\mathbf{x})$: the anomalous data **x** are due to factor F, not the falsity of T, or
$A(\mathbf{x})$ explains why the data **x** did not accord with the predictions from T.

By lumping together all cases that follow this logical pattern, Worrall's account lacks the machinery to distinguish reliable from unreliable use-constructions. As a result, I argue, Worrall's account comes up short when it comes to real experimental inferences:

Any philosophy of experimental testing adequate to real experiments must come to grips with the fact that the relationship between theory and experiment is not direct but is mediated along the lines of the hierarchy of models and theories.... At various stages of filling in the links, it is standard to utilize the same data to arrive at as well as warrant hypotheses.... As a matter of course, then, the inferences involved violate even the best construals of the novelty requirement. (Mayo, 1996, p. 253)

By maintaining that all such use-constructions are conditional on already assuming the truth of some overarching theory or "research program," Worrall's philosophy is redolent of the image of scientists as locked into "theory-laden" paradigms (Lakatos, Kuhn). Conversely, by regarding UN as sufficient for warranted inference, Worrall overlooks the fact "that there is as much opportunity for unreliability to arise in reporting or interpreting (novel) results given knowledge of theoretical predictions as there is...in arriving at hypotheses given knowledge of (non-novel) results" (ibid, p. 254). By recognizing that what matters is the overall severity with which a claim may be inferred, we have a desideratum that allows us to discriminate, on a case-by-case basis, whether UN violations matter and, if so, how we might correct for them.

I begin by discussing Worrall's treatment of use-construction in blocking anomalies and then turn to some confusions and errors that lead Worrall to forfeit a discriminatory tool that would seem to fulfill the (Popperian) testing philosophy that he himself endorses.

2 Use-Constructing in Blocking Anomalies: Must All of *T* Be Assumed?

We are all familiar with a variety of "rigging" procedures so as to accommodate data while protecting pet hypotheses rather than subjecting them to scrutiny. One of Worrall's favorite examples is Velikovsky's method for use-constructing hypotheses to save his theory when confronted with any anomalous data **x**:

> The lack of records in cultures $C_1, \ldots, C_n$ and their (arguable) presence in $C_1', \ldots, C_m'$ gives very good reason for holding the specific collective-amnesia version of Velikovsky's theory that he proposed *if* you already hold Velikovsky's general theory, *but* (and this is where the initial UN intuitions were aimed) those data give you absolutely no reason at all for holding that general theory in the first place. (Worrall, this chapter, pp. 136–7)

But why suppose that the inference to blocking a *T* anomaly assumes all of *T*? We know that, even if it is warranted to deny there is evidence against *T*, this fact alone would not provide evidence *for T*, and there is no reason to saddle every use-constructed save with committing so flagrant a fallacy (circularity). Obviously any method that assumes *T* in order to save *T* is minimally severe, but it is false to suppose that in use-constructing $A(\mathbf{x})$, *T* is assumed. It is not even clear why accepting Velikovsky means that any lack of records counts as evidence for the amnesia hypothesis, unless it is given that no other explanation can exist for the anomaly, as I take it Worrall does (note 10, p. 136). But are we to always imagine this? I put this aside. Even a proponent of Velikovsky's dodge could thwart Worrall's charge as follows.

V-dodger: I am not claiming that lack of records of the cataclysms described in my theory *T* is itself evidence for *T* (other records and considerations provide that); I am simply saying that I have a perfectly sound excuse, $A(\mathbf{x})$, for discounting the apparent anomaly for my theory.

Despite the ability to escape Worrall's charge, the flaw in the V-dodger's inference seems intuitively obvious. The severity account simply provides

some systematic tools for the less obvious cases. We are directed to consider the use-construction rule R leading from $\mathbf{x}$ to the inference $A(\mathbf{x})$
and the associated threats of error that could render the inference unwarranted. Here, we can characterize the rule R in something like the following
manner.

Rule R (Velikovsky's scotoma dodge): For each possible set of data $\mathbf{x}^i$
indicating that culture C^i has no records of the appropriate cataclysmic
events, infer $A^i(\mathbf{x}^i)$: culture C^i had amnesia with regard to these events.

The blocking hypothesis $A^i(\mathbf{x}^i)$ is use-constructed to fit data $\mathbf{x}^i$ to save
Velikovsky from anomaly.

Clearly, rule R prevents any observed anomaly of this form to threaten
Velikovsky's theory, even if the culture in question had not suffered amnesia
in the least. If one wanted to put this probabilistically, the probability of
outputting a Velikovsky dodge in the face of anomaly is maximal, even if the
amnesia explanation is false (a case of "gellerization") – therefore, severity
is minimal. Because rule R scarcely guards against the threat of erroneously
explaining away anomalies, we would say *of any particular output of rule R*
that the observed fit fails to provide evidence for the truth of $A(\mathbf{x}_0^i)$.

Notice that one need not rule out legitimately finding evidence that a
given culture had failed to record events that actually occurred, whether
due to memory lapses, sloppy records, or perhaps enforced by political
will. For example, we could discern that all the textbooks in a given era were
rewritten to expunge a given event, whose occurrence we can independently
check. But with Velikovky's rule R there is no chance that an erroneous
attribution of scotoma (collective amnesia) would be detected; nothing has
been done that could have revealed this fact, at least by dint of applying
rule R.

Although we condemn inferences from tests that suffer from a low probability of uncovering errors, it is useful to have what I call "canonical
errors" that stand as extreme cases for comparison (cases of zero severity). Velikovsky's case gives one. We utterly discredit any inference to $A(\mathbf{x})$
resulting from Velikovsky's use-construction rule, as seems proper. It is surprising, then, that Worrall's account appears to construe Velikovsky's gambit
as no worse off than any other use-constructed saves, including those that
we would consider altogether warranted.

The detailed data analysis of eclipse plates in 1919 warranted the inference
that "the results of these (Sobral Astrographic) plates are due to systematic
distortion by the sun and not to the deflection of light" (Mayo, 1996,
p. 284). To warrant this explanation is to successfully block an interpretation

of those data as anomalous for GTR. In Worrall's account, however, all use-constructed saves of theory T are conditional on assuming T; the only way they can avoid being treated identically to the case of Velikovsky's dodge is if *other*, independent support arises for accepting T.

As a matter of fact, however, the data-analytic methods, well-known even in 1919, did not assume the underlying theory, GTR, nor is it correct to imagine Eddington arguing that, provided you accept GTR, then the mirror distortion due to the Sun's heat explains why the 1919 Sobral eclipse results were in conflict with GTR's predicted deflection (and in agreement with the Newtonian prediction). GTR does not speak about mirror distortions. Nor were even the staunchest Newtonians unable to agree (not that it was immediately obvious) that the detailed data analysis showed that unequal expansion of the mirror caused the distortion. It was clear the plates, on which the purported GTR anomaly rested, were ruined; accepting GTR had nothing to do with it. Nor could one point to GTR's enjoying more independent support than Newton at the time – quite the opposite. (Two data sets from the same eclipse afforded highly imprecise accordance with GTR, whereas Newton enjoyed vast support.) Nor would it make sense to suppose that vouchsafing the mirror distortion depended on waiting decades until GTR was warranted, as Worrall would seem to require.

Thus, I remain perplexed by Worrall's claim that we need "to recognise just how conditional (and *ineliminably* conditional) the support at issue is in all these cases" (this volume, p. 135). By this, he means not that there are assumptions – because that is always true, and Worrall is quite clear he does not wish to label all cases as giving merely conditional support. He means, rather, that the entire underlying theory is assumed. We have seen this to be false.

My goal (e.g., in Mayo, 1996, sec. 8.6) was to illustrate these counterexamples to the UN requirement, at several stages of testing:

> The arguments and counterarguments [from 1919 to ~1921] on both sides involved violating UN. What made the debate possible, and finally resolvable, was that all...were held to shared criteria for acceptable and unacceptable use-constructions. It was acceptable to use any evidence to construct and test a hypothesis...so long as it could be shown that the argument procedure was reliable or severe. (p. 289)

Although the inferences, on both sides of the debate, strictly violated UN, they were deliberately constrained to reflect what is correct, at least approximately, regarding the cause of the anomalous data.

These kinds of cases are what led me to abandon the UN Charter, and Worrall has yet to address them. Here the "same" data are used both to identify and to test the source of such things as a mirror distortion, a plane crash, skewed data, a DNA match, and so on – without threats from uncertain background theories ("clean tests"). In statistical contexts, the stringency of such rules may be quantitatively argued:

A Stringent Use-Construction Rule (R-α): The probability is very small, $1 - \alpha$, that rule R would output $H(\mathbf{x})$ unless $H(\mathbf{x})$ were true or approximately true of the procedure generating data $\mathbf{x}$. (Mayo, 1996, p. 276)

Once the construction rule is applied and a particular $H(\mathbf{x}_0)$ is in front of us, we evaluate the severity with which $H(\mathbf{x}_0)$ has passed by considering the stringency of the rule R by which it was constructed, taking into account the particular data achieved. What matters is not whether H was deliberately constructed to accommodate $\mathbf{x}$; what matters is how well the data, together with background information, rule out ways in which an inference to H can be in error.

2.1 Deducing a Version (or Instantiation) of a Theory

At several junctures, it appears that Worrall is taking as the exemplar of a UN violation "using observational data as a premise in the deduction of some particular version of a theory" (p. 131) so that there is virtually no threat of error. True, whenever one is in the context wherein all of the givens of Worrall's inference to "the representative or variant of the theory" are met, we have before us a maximally severe use-construction rule (α would equal 1). We can agree with his claim that "[w]e do want to say that $\mathbf{x}$ supports $T(\mathbf{x})$ in some quite strong sense," (see p. 151), where $T(\mathbf{x})$ is what he regards as the variant of theory T that would be instantiated from the data $\mathbf{x}$. Confronted with a particular $T(\mathbf{x}_0)$, it would receive maximal support – provided this is understood as inferring that $T(\mathbf{x}_0)$ is the variant of T that would result if T were accepted and $\mathbf{x}_0$ observed. Instantiating for the wave theory, W, Worrall asserts: "$\mathbf{x}_0$ definitely supports $W(\mathbf{x}_0)$ in the conditional sense in that it establishes $W(\mathbf{x}_0)$ as *the* representative of the general theory W if that theory is to work at all" (replacing e with $\mathbf{x}_0$).[2] (p. 152) Although this

---

[2] An example might be to take the results from one of the GTR experiments, fix the parameter of the Brans-Dicke theory, and infer something like: if one were to hold the B-D theory, then the adjustable constant would have to be such-and-such value, for example, $q = 500$. (See Chapter 1, Section 5.2.)

inference is not especially interesting, and I certainly did not have this in mind in waging the counterexamples for the UN Charter, handling them presents no difficulty. If the assumptions of the data are met, the "inference" to the instantiation or application of theory *T* is nearly tautological.

The question is why Worrall would take this activity as his exemplar for use-constructed inferences in science. Certainly I would never have bothered about it if that was the sort of example on which the debate turned. Nor would there be a long-running debate in methodological practice over when to disallow or make adjustments because of UN violations and why. Yet, by logical fiat – construing all UN violations as virtually error-free inferences that aspire to do no more than report a specific variant of a theory that would fit observed data – the debate is settled, if entirely trivialized. If philosophers of science are to have anything useful to say about such actual methodological debates, the first rule of order might be to avoid interpreting them so that they may be settled by a logical wand.

Worrall claims to have given us good reasons for accepting his account of confirmation in the face of anomalies – where the anomalies are counterexamples to his view that UN is necessary for full-bodied confirmation. We might concur, in a bit of teasing reflexivity, that Worrall has given reason to support his handling of anomalies if you already hold his account of conditional support! But I doubt he would welcome such self-affirmation as redounding to his credit. This point takes me to a cluster of issues I place under "metaphilosophy."

3 Metaphilosophy: The Philosophical Role of Counterexamples

To a large extent, "the dispute between those who do and those who do not accept some version of the novelty principle emerges as a dispute about whether severity – or, more generally, error characteristics of a testing process – matters" (Mayo, 1996, p. 254). If it is assumed that whether *H* is warranted by evidence is just a function of statements of evidence and hypotheses, then it is irrelevant how hypotheses are constructed or selected for testing (I call these evidential-relation accounts). What then about the disagreement even among philosophers who endorse something like the severity requirement (as in the case of Worrall)? Here the source of disagreement is less obvious, and is often hidden: to dig it up and bring it to the surface requires appealing to the philosopher's toolkit of counterexamples and logical analysis. However, "philosopher's examples" are anything but typical, so one needs to be careful not to take them out of their intended context – as counterexamples!

3.1 Counterexamples Should Not Be Considered Typical Examples: The SAT Test

Now Worrall agrees with the general severity rationale: "the underlying justification is exactly the same as that cited by Mayo in favour of her own approach . . . a theory *T* is supported in this [strong] sense by some evidence *e* only if (and to the extent that) *e* is the outcome (positive so far as *T* is concerned) of some severe test of *T*" (Worrall, this volume, p. 144). We concur that, for a passing result to count as severe, it must, first of all, *be a passing result*; that is, the data must fit or accord with hypothesis *H* (where *H* can be any claim). Although Worrall often states this fit requirement as entailment, he allows that statistical fits are also to be covered. The key difference regards *what more* is required to warrant the inference to *H*. Should it be Worrall's UN criterion, or my severity criterion?

To argue for the latter, my task is to show how UN could be violated while intuitively severity is satisfied. In this I turned to the usual weapon of the philosopher: counterexamples. Observe what happens in cases where it is intuitively, and blatantly, obvious that a use-constructed hypothesis is warranted: the method that uses the data to output $H(\mathbf{x})$ is constrained in such a way that $H(\mathbf{x})$ is a product of what is truly the case in bringing about data $\mathbf{x}$. Worrall and like-minded use-novelists often talk as if an accordance between data and hypothesis can be explained in three ways: it is due to (1) chance, (2) the "blueprint of the universe" (i.e., truth or approximate truth of *H*), or (3) the ingenuity of the constructor (Worrall, 1989, p. 155). That hypotheses can be use-constructed reliably is precisely what is overlooked. In my attempts to lead them to the "aha" moment, I – following the philosopher's craft – sought extreme cases that show how use-constructed hypotheses can pass with high or even maximal severity; hence, the highly artificial example of using the data on the SAT scores to arrive at the mean SAT score. As I made clear, "the extreme represented by my SAT example was just intended to set the mood for generating counterexamples" (Mayo, 1996, p. 272), after which I turn to several realistic examples. Worrall (who is not alone) focuses on the former and gives little or no attention to the latter, realistic cases.

Ironically, it was Musgrave's reaction long ago to such flagrant cases that convinced me the Popperians had erred in this manner: "An older debt recalled in developing the key concept of severe tests is to Alan Musgrave" (Mayo, 1996, p. xv). Actually, as Musgrave reminds me, the example that convinced him was the incident that first convinced me that UN is not necessary for a good test: using data on the dent in my Camaro to hunt for a car with a tail fin that would match the dent, to infer that "it is practically impossible for the dent to have the features it has unless it was created by a specific type of car tail fin" (p. 276). The point is that counterexamples serve as this kind of tool in philosophy, and no one would think the user of the counterexamples intended them as typical examples. Yet some charge that I must be regarding the SAT averaging as representative of scientific hypotheses, forgetting that it arises only in the service of getting past an apparent blind spot.

We should clear up a problem Worrall has with the probabilistic statement we make. He considers the example of deducing $H(\mathbf{x})$ from data $\mathbf{x}$ (e.g., deducing the average SAT score from data on their scores). The probability $H(\mathbf{x})$ would be constructed, if in fact the data came from a population where $H(\mathbf{x})$ is false, is zero. (Because this is true for any $\mathbf{x}$, it is also true for any instance $\mathbf{x}_0$). But Worrall claims it would be undefined because the denominator of a conditional probability of a false claim is zero. Now the correct way to view an error-probabilistic statement, for example,

$$P(\text{test } T \text{ outputs } H(\mathbf{x}); \ H(\mathbf{x}) \text{ is false}),$$

is *not* as a conditional probability but rather a probability *calculated under the assumption that* $\mathbf{x}$ *came from a population where* $H(\mathbf{x})$ *is false*. The probability that a maximally severe use-construction rule outputs $H(\mathbf{x}_0)$, calculated under the assumption that $H(\mathbf{x}_0)$ is false, is zero – not undefined. Moreover, if we bar conditional probabilities on false hypotheses, then Bayesians could never get their favorite theorem going because they must exhaust the space of hypotheses.

3.2 Equivocations and Logical Flaws

If counterexamples will not suffice (in this case, to deny UN is necessary for severity), a second philosophical gambit is to identify flaws and equivocations responsible for leading astray even those who profess to share the goal (severity). *But one can never be sure one has exhausted the sources of confusions!* Worse is that the analytic labors carefully crafted to reveal the logical slip can give birth to yet new, unintended confusions. This seems to have happened here, and I hope to scotch it once and for all.

Everything starts out fine: Worrall correctly notes that I identify, as a possible explanation for the common supposition that UN is necessary for severity, a slippery slide from a true assertion – call it (a) – to a very different assertion (b), which need not be true:

(a) A use-constructed procedure is guaranteed to output an $H(\mathbf{x})$ that fits $\mathbf{x}$, "no matter what the data are."

(b) A use-constructed procedure is guaranteed to output an $H(\mathbf{x})$ that fits $\mathbf{x}$, "no matter whether the use-constructed $H(\mathbf{x})$ is true or false" (Mayo, 1996, p. 270; Worrall, this volume, p. 148).

Giere, for example, describes a scientist unwilling to consider any model that did not yield a prediction in sync with an observed effect $\mathbf{x}$. "Thus we know that the probability of any model he put forward yielding [the correct effect $\mathbf{x}$] was near unity, independently of the general correctness of that model" (Giere, 1983, p. 282). It is this type of multiply ambiguous statement, I argue, that leads many philosophers to erroneously suppose that use-constructed hypotheses violate severity. Pointing up the slide from (true) assertion (a) to (false) assertion (b) was intended to reveal the equivocation. Let me explain.

A use-constructed test procedure has the following skeletal form:

Use-Constructed Test Procedure: Construct $H(\mathbf{x})$ to fit data $\mathbf{x}$; infer that the accordance between $H(\mathbf{x})$ and $\mathbf{x}$ is evidence for inferring $H(\mathbf{x})$.

We write this with the variable $\mathbf{x}$, because we are stating its general characterization. So, *by definition*, insofar as a use-constructed procedure is successfully applied, it uses $\mathbf{x_0}$ to construct and infer $H(\mathbf{x_0})$, where $\mathbf{x_0}$ fits $H(\mathbf{x_0})$. This is captured in assertion (a). But assertion (a) alone need not yield the minimally severe test described in assertion (b); it need not even lead to one with low severity. The construction rule may ensure that false outputs are rare. We may know, for example, that anyone prosecuted for killing JonBenet Ramsey will have to have matched the DNA from the murder scene; but this is a reliable procedure for outputting claims of form:

The DNA belongs to Mr. X.

In any specific application it outputs $H(\mathbf{x_0})$, which may be true or false about the source of the data, but the probability that it outputs false claims is low. The familiar argument that use-constructed tests are invariably minimally severe, I suggest, plays on a (fallacious) slide from assertion (a) to assertion (b).

Having gotten so used to hearing the Popperian call for falsification, it is sometimes forgotten that his call was, strictly speaking, for falsifying hypotheses, *if false*. Admittedly, Popper never adequately cashed out his severity idea, but I would surmise that, if he were here today, he would agree that some construction procedures, although guaranteed to output some

$H(\mathbf{x})$ *or other*, whatever the data, nevertheless ensure false outputs are rare or even impossible.

Worrall sets out my argument with admirable clarity. Then something goes wrong that numerous exchanges have been unable to resolve. His trouble is mainly as regards claim (a). Now claim (a) was intended to merely capture what is generally assumed (*by definition*) for any use-constructed procedure. So, by instantiation, claim (a) holds for the examples I give where a use-constructed procedure yields a nonsevere test. From this, Worrall supposes that claim (a) is necessary for nonseverity, but this makes no sense. Were claim (a) required for inseverity, then violations of claim (a) would automatically yield severity. Then hypotheses that do not even fit the data would automatically count as severe! But I put this error aside. More egregiously, for current purposes, he argues that claim (a) is false! Here is where Worrall's logic goes on holiday.

He considers Velikovsky's rule for blocking anomalies by inferring that the culture in question suffered amnesia $A(\mathbf{x})$. Worrall says, consider a specific example of a culture – to have a concrete name, suppose it is the Thoh culture – and suppose no records are found of Velikovsky-type events. Velikovsky conveniently infers that the apparent anomaly for his theory is explained by amnesia:

$A(\text{Thoh})$: Thoh culture suffered from amnesia (hence no records).

Says Worrall, "It seems, then, to be straightforwardly untrue that a successful fit between" $A(\text{Thoh})$ and $\mathbf{x}$ "is assured no matter what [$\mathbf{x}$] is" (p. 149). Quite so! (For instance, the procedure would not output $A(\text{Thoh})$, or any claims about the Thoh culture, if the observation was on some other culture). But this does not show that assertion (a) is false. It could only show that assertion (a) is false by an erroneous instantiation of the universal claim in assertion (a). The assertion in (a) is true because every anomalous outcome will fit *some Velikovsky dodge or other*. It does not assert that all anomalous cultures fit a *particular* instantiation of the Velikovsky dodge, e.g., $A(\text{Thoh})$.[3]

---

[3] Worrall seems to reason as follows:

1. According to assertion (a), for any data $\mathbf{x}$, if $\mathbf{x}$ is used to construct $A(\mathbf{x})$, then $\mathbf{x}$ fits $A(\mathbf{x})$.
2. But suppose the data from the Thoh culture is used to construct $A(\text{Thoh})$.
3. (From assertion (a) it follows that) all data $\mathbf{x}$ would fit $A(\text{Thoh})$ (i.e., a successful fit between $A(\text{Thoh})$ and $\mathbf{x}$ "is assured no matter what $\mathbf{x}$ is").

Then from the falsity of premise 3, Worrall reasons that premise (a) is false. But premise 3 is an invalid instantiation of the universal generalization in premise (a)! It is unclear whether Worrall also takes this supposed denial of assertion (a) as denying assertion (b), but to do so is to slip into the fallacy that my efforts were designed to avoid. (For further discussion of variations on this fallacy, e.g., in Hitchcock and Sober, 2004, see Mayo, 2008).

Sometimes a gambit that a philosopher is sure will reveal a logical flaw instead creates others. Pointing up the faulty slide from the truth of assertion (a) to that of assertion (b) was to have illuminated the (false but common) intuition that UN is necessary for severity. Instead we have been mired in Worrall's resistance to taking assertion (a) as true for use-constructed procedures – something I took to be a matter of mere definition, which just goes to show that one cannot always guess where the source of difficulties resides. Hopefully now no obstacles should remain to our agreement on this issue.

4 Concluding Comment on the Idea of a Single Account of Evidence (Remarks on Chapters 2, 3, and 4)

I do not claim that all of science involves collecting and drawing inferences from evidence, only that my account is focused on inference. As varied as are the claims that we may wish to infer, I do not see that we need more than one conception of what is required for evidence to warrant or corroborate a claim. Worrall berates me for holding a "one-size-fits-all" account of inference that always worries about how well a method has probed for the errors that threaten the inference. Similar sentiments are voiced by Chalmers and Musgrave. Granted data may be used in various ways, and we want hypotheses to be not just well tested, but also informative; however, if we are talking about the warrant to accord a given inference, then I stand guilty as charged.

I cannot really understand how anyone could be happy with their account of inference if it did not provide a unified requirement. In the context of this chapter, Worrall's introduction of "conditional evidence" was of no help in discriminating warranted from unwarranted use-constructions. The severity desideratum seems to be what matters. Similarly, Chalmers's "arguments from coincidence" and Musgrave's "inference to the best tested portion of an explanation" in Chapters 2 and 3, respectively, are all subsumed by the severity account. Different considerations arise in *applying* the severity definition, and different degrees of severity are demanded in different cases, but in all cases the underlying goal is the same. The whole point of the approach I take is to emphasize that what needs to have been probed are the threats of error in the case at hand. Even if one adds decision-theoretic criteria, which we will see leads Laudan to argue for different standards of evidence (Chapter 9), my point is that, *given the standards*, whether they are satisfied (by the data in question) does not change.

The deepest source of the disagreements raised by my critics, I see now, may be located in our attitudes toward solving classic problems of evidence, inference, and testing. The experimental account I favor was developed precisely in opposition to the philosophy of science that imagines all inferences to be paradigm-laden in the sense Kuhnians often espouse, wherein it is imagined scientists within paradigm T_1 circularly defend T_1 against anomaly and have trouble breaking out of their prisons. In this I am apparently on the side of Popper, whereas Musgrave, Chalmers, Worrall (and Laudan!) concede more to Lakatos and Kuhn. It is to be hoped that current-day Popperians move to a position that combines the best insights of Popper with the panoply of experimental tools and methods we now have available.

At the same time, let me emphasize, there are numerous gaps that need filling to build on the experimentalist approach associated with the error-statistical account. The example of use-novelty and double counting is an excellent case in point. Although in some cases, understanding the way formal error probabilities may be altered by double counting provides striking illumination for entirely informal examples, in other cases (unfortunately), it turns out that whether error probabilities are or should be altered, even in statistics, is unclear and requires philosophical–methodological insights into the goals of inference. This is typical of the "two-way street" we see throughout this volume. To help solve problems in practice, philosophers of science need to take seriously how they arise and are dealt with, and not be tempted to define them away. Conversely, in building the general experimentalist approach that I label the error-statistical philosophy of science, we may at least find a roomier framework for re-asking many philosophical problems about inductive inference, evidence and testing.

References

Giere, R.N. (1983), "Testing Theoretical Hypotheses," pp. 269–98 in J. Earman (ed.), *Testing Scientific Theories*, Minnesota Studies in the Philosophy of Science, vol. 10, University of Minnesota Press, Minneapolis.

Hitchcock, C., and Sober, E. (2004), "Prediction Versus Accommodation and the Risk of Overfitting," *British Journal for the Philosophy of Science*, 55: 1–34.

Mayo, D.G. (1991), "Novel Evidence and Severe Tests," *Philosophy of Science*, 58: 523–52.

Mayo, D.G. (1996), *Error and the Growth of Experimental Knowledge* (Chapters 8, 9, 10), University of Chicago Press, Chicago.

Mayo, D.G. (2008), "How to Discount Double Counting When It Counts," *British Journal for the Philosophy of Science*, 59: 857–79.

Musgrave, A. (1974), "Logical Versus Historical Theories of Confirmation," *British Journal for the Philosophy of Science*, 25: 1–23.

Musgrave, A. (1989), "Deductive Heuristics," pp. 15–32 in K. Gavroglu, Y. Goudaroulis, and P. Nicolacopoulos (eds.), *Imre Lakatos and Theories of Scientific Change*, Kluwer, Dordrecht.

Worrall, J. (1989), "Fresnel, Poisson, and the White Spot: The Role of Successful Prediction in the Acceptance of Scientific Theories," pp. 135–57 in D. Gooding, T. Pinch and S. Schaffer (eds.), The Uses of Experiment: Studies in the Natural Sciences, Cambridge University Press, Cambridge.

Related Exchanges

Musgrave, A.D. (2006), "Responses," pp. 301–4 in C. Cheyne and J. Worrall (eds.), *Rationality and Reality: Conversations with Alan Musgrave*, Kluwer Studies in the History and Philosophy of Science, Springer, Dordrecht, The Netherlands.

Worrall, J. (2002), "New Evidence for Old," in P. Gardenførs, J. Wolenski, and K. Kijania-Placek (eds.), *In the Scope of Logic, Methodology and Philosophy of Science* (vol. 1 of the 11th International Congress of Logic, Methodology, and Philosophy of Science, Cracow, August 1999), Kluwer, Dordrecht.

Worrall, J. (2006), "History and Theory-Confirmation," pp. 31–61 in J. Worrall and C. Cheyne (eds.), *Rationality and Reality: Conversations with Alan Musgrave*, Springer, Dordrecht, The Netherlands.

FIVE

Induction and Severe Testing

Mill's Sins or Mayo's Errors?

Peter Achinstein[1]

Although I have offered some criticisms of her views on evidence and testing (Achinstein, 2001, pp. 132–40), I very much admire Deborah Mayo's book (1996) and her other work on evidence. As she herself notes in the course of showing how misguided my criticism is, we actually agree on two important points. We agree that whether *e*, if true, is evidence that *h*, in the most important sense of "evidence," is an objective fact, not a subjective one of the sort many Bayesians have in mind. And we agree that it is an empirical fact, not an a priori one of the sort Carnap has in mind. Here I will take a broader and more historical approach than I have done previously and raise some general questions about her philosophy of evidence, while looking at a few simple examples in terms of which to raise those questions. It is my hope that, in addition to being of some historical interest, this chapter will help clarify differences between us.

1 Mill under Siege

One of Mayo's heroes is Charles Peirce. Chapter 12 of Mayo's major work, which we are honoring, is called "Error Statistics and Peircean Error Correction." She has some very convincing quotes from Peirce suggesting that he was a model error-statistical philosopher. Now I would not have the Popperian boldness to say that Mayo is mistaken about Peirce; that is not my aim here. Instead I want to look at the view of induction of a philosopher whom Peirce excoriates, Mayo doesn't much like either, and philosophers of science from Whewell to the present day have rejected. The philosopher

---

[1] I am very grateful to Linda S. Brown for helping me to philosophically reflect properly, and to Deborah Mayo for encouraging me to philosophically reflect properly on her views.

is John Stuart Mill. I say that Peirce excoriated Mill. Here is a quote from Peirce:

John Stuart Mill endeavored to explain the reasonings of science by the nominalistic metaphysics of his father. The superficial perspecuity of that kind of metaphysics rendered his logic extremely popular with those who think, but do not think profoundly; who know something of science, but more from the outside than the inside, and who for one reason or another delight in the simplest theories even if they fail to cover the facts. (Peirce, 1931–1935, 1.70)

What are Mill's sins – besides being an outsider to science, delighting in oversimplifications, and having the father he did? Well, there are many. In what follows I will focus on the sins Peirce notes that are either mentioned by Mayo or that are or seem inimical to her error-statistical philosophy. I do so because I want to give Mill a better run for his money – a fairer chance than Peirce gave him. Also, doing so may help to bring out strengths and weaknesses in error-statistical ideas and in Mill.

Here are four sins:

1. First and foremost, Mill's characterization of induction completely omits the idea of severe testing, which is central for Peirce and Mayo. For example, such inferences completely ignore the conditions under which the observed sample was taken, which is very important in severe testing.

2. Second, Mill's inductions conform to what later became known as the "straight rule." They are inductions by simple enumeration that license inferences of the form "All observed A's are B's; therefore, all A's are B's." Moreover, they are "puerile" inferences from simple, familiar properties directly observed in a sample to the existence of those properties in the general population. Such inferences cannot get us very far in science. Peirce, following in the footsteps of Whewell earlier in the nineteenth century, has considerable disdain for Mill's inductive account of Kepler's inference to the elliptical orbit of Mars (Peirce, 1931–1935, 7.419).

3. Third, Mill supposes that an induction is an argument in which the conclusion is inferred to be true and assigned a high probability. But according to Peirce (and Mayo), an induction or statistical inference "does not assign any probability to the inductive or hypothetic conclusion" (Peirce, 1931–1935, 2.748).

4. Finally, Mill assumes that all inductions presuppose a principle of the uniformity of nature, a presupposition that, besides being vague, is not warranted or needed.

In an attempt to give at least a partial defense of Mill, I will also invoke Isaac Newton, who proposed a view of induction that is similar in important respects to Mill's view. Newton, at least, cannot be accused by Peirce of not knowing real science. And although Mill offers an account of induction that is philosophically superior to Newton's, Newton shows more clearly than Mill how induction works in scientific practice. He explicitly demonstrates how his four Rules for the Study of Natural Philosophy, including rules 3 and 4 involving induction, were used by him in defending his law of gravity.

2 A Mill–Newton View of Induction: Sins 1 and 2

Mill offers the following classic definition of induction:

Induction, then, is that operation of the mind by which we infer that what we know to be true in a particular case or cases, will be true in all cases which resemble the former in certain assignable respects. In other words, Induction is the process by which we conclude that what is true of certain individuals of a class is true of the whole class, or that what is true at certain times will be true in similar circumstances at all times. (Mill, 1888, p. 188)

Mill declared inductive reasoning to be a necessary first stage in a process he called the "deductive method." He regarded the latter as the method for obtaining knowledge in the sciences dealing with complex phenomena. It establishes the laws necessary to analyze and explain such phenomena.[2]

Unlike Mill, Newton never explicitly defines "induction" in his works. But, like Mill, he considers it to be a necessary component of scientific reasoning to general propositions: "In . . . experimental philosophy, propositions are deduced from the phenomena and are made general by induction. The impenetrability, mobility, and impetus of bodies, and the laws of motion and the law of gravity have been found by this method."[3] The closest Newton comes to a definition of induction is his rule 3 at the beginning of Book III of the *Principia*:

Rule 3: Those qualities of bodies that cannot be intended or remitted and that belong to all bodies on which experiments can be made should be taken as qualities of all bodies universally.

Some commentators (including a former colleague of mine, Mandelbaum [1964]) take this rule to be a rule of "transdiction" rather than "induction."

---

[2] The second step, "ratiocination," involves deductive calculation. The third step, "verification," involves the comparison of the results of ratiocination with the "results of direct observation" (Mill, 1888, p. 303).

[3] Newton (1999, Book 3, *General Scholium*, p. 943).

By this is meant a rule for inferring what is true about all unobservables from what is true about all observables, whether or not they have been observed. In his discussion, however, Newton himself gives some examples involving inductions (from what has been observed to what may or may not be observable) and others involving transdictions (from all observables). And in his rule 4, we are told to regard propositions inferred by induction from phenomena as true, or very nearly true, until other phenomena are discovered that require us to modify our conclusion.

Let us focus first on Mill. It is quite correct to note in Sin 1 that Mill's characterization of induction, given earlier, completely omits the idea of severe testing. But Mill's response to this charge of sin will surely be this: "We must distinguish my *definition* of induction, or an inductive generalization, from the question of when an induction is justified." Recall one of Mill's definitions of an inductive inference as one in which "we conclude that what is true of certain individuals of a class is true of the whole class." But he does not say, and indeed he explicitly denies, that any induction so defined is justified. Indeed, on the next page following this definition he writes that whether inductive instances are "sufficient evidence to prove a general law" depends on "the number and nature of the instances." He goes on to explicitly reject the straight rule of enumerative induction as being universally valid. Some inductions with this form are valid; some are not.

Later in the book, in a section on the "evidence of the law of universal causation," he writes:

When a fact has been observed a certain number of times to be true, and is not in any instance known to be false; if we at once affirm that fact as an universal truth or law of nature, without either testing it by any of the four methods of induction [Mill's famous methods for establishing causal laws], or deducing it from other known laws, we shall in general err grossly. (Mill, 1888, p. 373)

In an even later section called "Fallacies of Generalization," Mill again emphasizes that inductive generalizations are not always valid but are subject to different sorts of "fallacies" or mistakes in reasoning, including failure to look for negative instances and making the generalization broader than the evidence allows. My point is simply that Mill defines inductive generalization in such a way that there are good and bad inductions.

Moreover, for Mill – and this point is quite relevant for, and similar to, Mayo's idea – whether a particular inductive inference from "all observed As are Bs" to "all As are Bs" is a good one is, by and large, an *empirical* issue, not an a priori one. Mill notes that the number of observed instances of a generalization that are needed to infer its truth depends on the kinds of instances and their properties. He writes (p. 205) that we may need only one observed instance of a chemical fact about a substance to validly generalize to all instances of that substance, whereas many observed instances of black crows are required to generalize about all crows. Presumably this is due to the empirical fact that instances of chemical properties of substances tend to be uniform, whereas bird coloration, even in the same species, tends not to be. Finally, in making an inductive generalization to a law of nature, Mill requires that we vary the instances and circumstances in which they are obtained in a manner described by his "four methods of experimental inquiry." Whether the instances and circumstances have been varied, and to what extent, is an empirical question and is not determined a priori by the fact that all the observed members of the class have some property.

In short, although Mill *defines* induction in a way that satisfies the formal idea of induction by simple enumeration, he explicitly denies that any induction that has this form is thereby valid. Whether it is valid depends on nonformal empirical facts regarding the sample, the sampling, and the properties being generalized. I would take Mill to be espousing at least some version of the idea of "severe testing."

Although Newton does not define "induction," the examples of the inductive generalizations he offers in the *Principia* conform to Mill's definition. And I think it is reasonable to say that Newton, like Mill, does not claim that any induction from "all observed As are Bs" to "all As are Bs" is valid. For one thing, he never makes this statement or anything like it. For another, his inductive rule 3 has restrictions, namely, to inductions about bodies and their qualities that "cannot be intended and remitted." In one interpretation offered by Newton commentators, this rule cannot be used to make inductive generalizations on qualities of things that are not bodies (e.g., on qualities of forces or waves) or on all qualities of things that are bodies (e.g., colors or temperatures of bodies). Moreover, Newton seems to defend his third rule by appeal to a principle of uniformity of nature, much like the one Mill invokes. In his discussion of rule 3, Newton writes, "We [should not] depart from the analogy of nature, since nature is always simple and ever consonant with itself." Later I will discuss Mill's more elaborated view of such a principle. Suffice it to say now that, if Newton takes such a principle to be empirical rather than a priori, then whether some particular inductive generalization is warranted can depend empirically on whether certain uniformities exist in nature regarding the bodies and sorts of properties in question that help to warrant the inference.

However, one of the most important reasons for citing Newton is to combat the idea – formulated as part of Sin 2 – that inductive generalizations of the sort advocated by Newton and Mill are puerile ones like those involving the color of crows; that they cite only familiar properties whose presence can be established simply by opening one's eyes and looking; and that they are ones that someone like Mill, knowing little science, can readily understand and appreciate.

From astronomical observations regarding the various planets and their satellites, Newton establishes (what he calls) six "Phenomena." For example, Phenomenon 1 states that the satellites of Jupiter by radii drawn to the center of Jupiter describe areas proportional to the times, and their periods are as 3/2 power of the distances from the center of Jupiter. From the six phenomena together with principles established in Book 1, Newton derives a series of theorems or "propositions" concerning the forces by which the satellites of the planets and the planets themselves are continually drawn away from rectilinear motion and maintained in their orbits. These forces are directed to a center (the planet or the sun), and they vary inversely as the square of the distance from that center. Explicitly invoking his rule 1, which allows him to infer that one force exists here, not many, and his inductive rule 3, Newton infers that this force governs all bodies in the universe.

If someone claims that such an induction is puerile, or that the property being generalized – namely, being subject to an inverse square force – is something familiar and directly observable in the manner of a black crow, I would say what the lawyers do: *res ipse loquitur.*

Nor does Mill restrict inductions to generalizations involving familiar properties whose presence is ascertainable simply by opening one's eyes and looking. His definition of induction as "the process by which we conclude that what is true of certain individuals of a class is true of the whole class" has no such requirement. Moreover, he cites examples of inductions in astronomy where the individual facts from which the inductions are made are not establishable in such a simple manner – for example, facts about the magnitude of particular planets in the solar system, their mutual distances, and the shape of the Earth and its rotation (Mill, 1888, p. 187).

3 Assigning Probability to an Inductive Conclusion: Sin 3

Let me turn to what Peirce, and I think Mayo, may regard as the worst sin of all – assigning a probability to the inductive conclusion or hypothesis. Peirce claims that "in the case of an analytic inference we know the probability of our conclusion (if the premises are true), but in the case of synthetic

inferences we only know the degree of trustworthiness of our proceeding"
(Peirce, 1931–1935, 2.693) Mayo (1996, p. 417) cites this passage from Peirce
very approvingly. An example of an analytic inference for Peirce is a mathematical demonstration. Inductions are among synthetic inferences. Another
quote Mayo gives from Peirce involves his claim that we can assign probability to a conclusion regarding a certain "arrangement of nature" only if

universes were as plenty as blackberries, if we could put a quantity of them in a bag,
shake them well up, draw out a sample, and examine them to see what proportion
of them had one arrangement and what proportion another. (Peirce, 1931–1935,
2.684)

Because this idea makes little sense, what Peirce claims we really want to
know in making a synthetic inductive inference is this: "[g]iven a synthetic
conclusion,... what proportion of all synthetic inferences relating to it
will be true within a given degree of approximation" (Peirce, 1931–1935,
2.686).

Mayo cites this passage, and says that "in more modern terminology,
what we want to know are the error probabilities associated with particular
methods of reaching conclusions" (Mayo, 1996, p. 416). The sin, then, is
assigning probability to an inductive conclusion and not recognizing that
what we really want is not this but error probabilities. So, first, do Newton
and/or Mill commit this sin? Second, is this really a sin?

Neither in their abstract formulations of inductive generalizations (Newton's rule 3; Mill's definition of "induction") nor in their examples of particular inductions to general conclusions of the form "all As are Bs" does the
term "probability" occur. Both write that from certain specific facts we can
conclude general ones – not that we can conclude general propositions with
probability, or that general propositions have a probability, or that they have
a probability conditional on the specific facts. From the inductive premises
we simply conclude that the generalization is true, or as Newton allows in
rule 4, "very nearly true," by which he appears to mean not "probably true"
but "approximately true" (as he does when he takes the orbits of the satellites
of Jupiter to be circles rather than ellipses). Nor, to make an induction to
the truth or approximate truth of the generalization, does Newton or Mill
explicitly require the assignment of a probability to that generalization.

However, before we conclude that no probabilistic sin has been committed, we ought to look a little more closely at Mill. Unlike Newton, Mill
offers views on probability. And there is some reason to suppose that he
would subscribe to the idea that one infers an inductive conclusion with
probability. In Chapter XVIII of Book 2 of *A System of Logic*, Mill advocates

what today we would call an *epistemic* concept of probability – a concept to be understood in terms of beliefs or expectations rather than in terms of belief-independent events or states of the world. He writes:

We must remember that the probability of an event is not a quality of the event itself, but a mere name for the degree of ground which we, or someone else, have for expecting it. The probability of an event to one person is a different thing from the probability of the same event to another, or to the same person after he has acquired additional evidence.... But its probability to us means the degree of expectation of its occurrence, which we are warranted in entertaining by our present evidence. (Mill, 1888, p. 351)

Mill offers examples from games of chance, but, unlike the classical Laplacian, he says that probabilities should be assigned on the basis of observed relative frequencies of outcomes. For example, he writes:

In the cast of a die, the probability of ace is one-sixth; not simply because there are six possible throws, of which ace is one, and because we do not know any reason why one should turn up rather than another, ... but because we do actually know, either by reasoning or by experience, that in a hundred or a million of throws, ace is thrown in about one-sixth of that number, or once in six times. (Mill, 1888, p. 355)

For Mill, this knowledge of relative frequencies of events can be inferred from specific experiments or from a "knowledge of the causes in operation which tend to produce" the event in question. Either way, he regards the determination of the probability of an event to be based on an induction. For example, from the fact that the tossing of a die – either this die or others – has yielded an ace approximately one-sixth of the time, we infer that this will continue to be the case. So, one-sixth represents the degree of expectation for ace that we are warranted in having by our present evidence.

From his discussion it sounds as if Mill is an objectivist rather than a subjectivist about epistemic probability (more in the mold of Carnap's *Logical Foundations of Probability* than that of subjective Bayesians). Moreover (like Carnap), he is an objectivist who relativizes probability to an epistemic situation of an actual or potential believer, so that the probability depends on the particular epistemic situation in question. (In *The Book of Evidence* I contrast this view with an objective epistemic view that does not relativize probability to particular epistemic situations; in what follows, however, I will focus on an interpretation of the former sort, since I take it to be Mill's.)

Is Mill restricting probability to events and, more specifically, to events such as those in games of chance, in which relative frequencies can be determined in principle? He certainly goes beyond games of chance. Indeed, he seems to want to assign probabilities to propositions or to allow conclusions to be drawn with a probability. In Chapter XXIII (pp. 386ff), "Of Approximate Generalizations and Probable Evidence," Mill considers propositions of the form "Most As are B" (and "X% of As are B"). He writes:

If the proposition, Most A are B, has been established by a sufficient induction, as an empirical law, we may conclude that any particular A is B *with a probability proportional to the preponderence of the number of affirmative instances over the number of exceptions.* (Mill, 1888, p. 390; emphasis added)

To be sure, we have a sort of relative frequency idea here, and the conclusion he speaks of is not general but about a particular A being B.[4] However, what I find more interesting is the idea of concluding a proposition "with a probability" and the idea that the proposition concluded can concern a specific A rather than a type – ideas congenial to epistemic views of probability.[5]

Now let us assume that someone, call her Deborah, has made an induction in Mill's sense (which will be understood broadly to include inferences from a subset of a class to a universal or statistical generalization about that class, or to a claim about an individual arbitrarily chosen within the class but not in the subset. And let us suppose that her induction is justified. My question is this: Can Mill claim that Deborah's induction is justified only if he assumes that, given her knowledge of the premises and given what else she knows, the degree to which someone in that situation is warranted in drawing the inductive conclusion is high? If, as I believe, this condition is necessary for Mill, and if, as I am suggesting, probability for Mill is to be understood as objective epistemic probability, and if such probability is applicable to inductive conclusions, given the premises, then objective epistemic probabilists are ready for action.[6] They need not assign a point probability, or an interval, or a threshold (as I personally do). It can be quite

---

[4] Mill (1888, p. 186) explicitly allows inferences to "individual facts" to count as inductions. (In the quotation above, I take it he is talking about the "odds," or, in terms of probability, the number of affirmative instances over the sum of affirmative instances and exceptions.) Mill goes on to say that we do so assuming that "we know nothing except that [the affirmative instances] fall within the class A" (p. 391).

[5] Mill gives various examples and arguments of this type, including ones in which a proposition about a specific individual is concluded with a numerical probability (1888, pp. 391–2). It is clear from his discussion that he is presupposing standard rules of probability.

[6] Here I include those (such as Mill and Carnap) who relativize probability to an epistemic situation, and anyone (such as myself) who claims that such relativizations can but need not be made.

vague – for example, "highly probable," or with Carnap, "probability > k" (whatever k is supposed to be, even if it varies from one case to another). But objective epistemic probabilists – again I include Mill – are committed to saying that the inference is justified only if the objective epistemic probability (the "posterior" probability) of the inductive conclusion, given the truth of the premises, is sufficiently high.

Is Mayo committed to rejecting this idea? First, she, like Peirce, is definitely committed to doing so if probability is construed only as relative frequency. Relative frequency theorists reject assigning probabilities to individual hypotheses (for the Peircean "blackberry" argument). Indeed, she considers Mill's inference from "most As are Bs" to a probabilistic conclusion about a particular A being B to be committing what she calls the "fallacy of probabilistic instantiation" (Mayo, 2005, p. 114). Second, and most important, she believes rather strongly that you can assess inductive generalizations and other inductive arguments without assigning any probability to the conclusion. The only probability needed here, and this is a relative frequency probability, is the probability of getting the sort of experimental result we did using the test we did, under the assumption that the hypothesis in the conclusion is false. What we want in our inductions, says Mayo, are hypotheses that are "highly probed," not ones that are "highly probable" – that is, that have a high posterior probability.

Is part of her reason for rejecting the requirement of high posterior probability for an inductive conclusion this: that the only usable notion of probability is the relative frequency one, so that assigning an objective epistemic probability to a hypothesis either makes little sense, or is not computable, or some such thing? That is, does she reject any concept of objective epistemic probability for hypotheses? Or is Mayo saying that such a concept is just not needed to determine the goodness of an inductive argument to an empirical hypothesis? I believe she holds both views. In what follows, however, I will consider only the second, and perhaps the more philosophically challenging, view that, even if hypotheses have posterior probabilities on the evidence, for a hypothesis to be justified by the experimental data this posterior probability – whether vague or precise – need not be high, and indeed may be very low. It is this claim that I want to question.

Using Mill's terminology for probability, but applying what he says to hypotheses (and not just events), the probability of a hypothesis is to be understood as the "degree of expectation [of its truth] which we are warranted in entertaining by our present evidence" (Mill, 1888, p. 351). Or, using terminology I have proposed elsewhere, the probability of a hypothesis *h*, given evidence *e*, is the degree of reasonableness of believing *h*, on the assumption of *e*. Now I ask, given that such a notion of probability makes sense, how could an inductive argument to a hypothesis from data be justified unless (to use my terminology) the degree of reasonableness of believing the conclusion given the data is high, or unless (to use Mill's terminology) the data warrant a high degree of expectation in the truth of the conclusion – that is, unless the posterior probability of the conclusion is high? (This I take to be a necessary condition, not a sufficient one; see Achinstein, 2001, ch. 6.)

I can see Mayo offering several responses. One, already mentioned, is to completely reject the idea of objective epistemic probability. A second, again mentioned earlier, is to say that Mill is committing the "fallacy of probabilistic instantiation." I do not think this is a fallacy when we consider epistemic probability. If 99% of As are B, then, given that a particular A was randomly selected, the degree of reasonableness of believing that this A is B is very high.[7]

A third response (which she may offer in addition to the first) is to say that, when we inductively infer a hypothesis, we infer not its truth or probability but its "reliability" in certain types of experimental situations. Although sometimes she says that from the fact that a hypothesis has passed a severe test (in her sense) we may infer the hypothesis (Achinstein, 2005, p. 99), and sometimes she says that we may infer that it is "correct,"[8] she also claims that passing a severe test shows that the hypothesis is "reliable." What does that mean? She writes: "Learning that hypothesis *H* is reliable, I propose, means learning that what *H* says about certain experimental results will often be close to the results actually produced – that *H* will or would often succeed in specified experimental applications" (Mayo, 1996, p. 10). Mayo wants to move beyond Karl Popper's view that passing a severe test simply means that, despite our efforts to falsify the hypothesis, the hypothesis remains unfalsified. Her idea of passing a severe test is such that, if a hypothesis does so, this tells us something important about how the hypothesis will perform in the future using this test. She writes: "Reliability deals with future performance, and corroboration, according to Popper, is only a 'report of past performance'" (Mayo, 1996, p. 9).

Let me formulate what I take to be Mayo's position concerning "passing a severe test" and "reliability." We have a hypothesis *h*, a test *T* for that

---

[7] See Mill (1888, pp. 390–1).
[8] "The evidence indicates the correctness of hypothesis H, when H passes a severe test" (Mayo, 1996, p. 64).

hypothesis, and some data *D* produced as a result of employing the test. For the hypothesis to pass a severe test *T* yielding data *D*, she requires (1) that the data "fit" the hypothesis in some fairly generous sense of "fit"[9]; and she requires (2) that the test *T* be a severe one in the sense that the probability is very high that *T* would produce a result that fits hypothesis *h* less well than data *D* if *h* were false (alternatively, the probability is very low that *T* would yield results that fit *h* as well as data *D* if *h* were false). In her view, the concept of probability here is relative frequency. If so, and taking a standard limiting relative frequency view, the probability she has in mind for claim 2 regarding a severe test can be expressed as follows:

Given that test *T* continues to be repeated, the relative frequency of outcomes fitting hypothesis *h* as well as data *D* will at some point in the testing sequence be and remain very low, under the assumption that *h* is false.

This is what is entailed by passing a severe test.

Now my question is this. Epistemically speaking, what, if anything, can one conclude from the fact that *h* has passed a severe test yielding data *D*?

Mill would want to conclude that "the degree of expectation . . . [in the truth of the hypothesis *h*] which we are warranted in entertaining by our evidence [i.e., by data *D*]" is high. More generally, I suggest, Mill requires that evidence should provide a good reason for believing a hypothesis. On my own view of evidence, the latter requires that the objective epistemic probability of the hypothesis *h*, given that data *D* were obtained as a result of test *T*, is very high. If Mayo were to accept this condition as necessary, then we would be in agreement that a posterior probability, however vaguely characterized, can be attributed to the hypothesis, and we would both agree that this probability is not a relative frequency.

The only problem I would anticipate here is that a posterior probability for *h* would presuppose the existence of a prior probability for *h* – a probability for *h* based on information other than data *D*. Suppose that such a prior probability for *h* is extremely low and that, despite the fact that *h* has passed a severe test, in Mayo's sense, with respect to data *D*, the posterior probability of *h* given data *D* is also very low. Then even though *h* has passed such a severe test with data *D*, I, and I believe Mill, would conclude that this fact does not warrant an inference to, or belief in, hypothesis *h*. If not, then "passing a severe test" in Mayo's sense would not be sufficient for inductive evidence in Mill's sense.

---

[9] Minimally she wants p(*D*;*h*) > p(*D*;*not-h*). (See Mayo, 2005, p. 124, fn. 3.)

To introduce a very simple example, suppose there is a disease S for which (while the patient is alive) there is one and only one test, T. This test can yield two different results: the test result turns red or it turns blue. Suppose that, of those who have disease S, 80% test red using T, so that P(red result with test T/disease S) = .8. And suppose that, of those who don't have disease S, 2 out of 100,000 test red, so that P(red result with test T/$-S$) = .00002. The probabilities here are to be construed as relative frequencies. Now let the hypothesis under consideration be that Irving has disease S. On the basis of the aforementioned probabilities, we can write the following:

   1. P(Irving's getting a red result with test T/ Irving has disease S) = .8

and

   2. P(Irving's getting a red result with test T/Irving does not have disease S) = .00002.

If so, then Irving's test result of red "fits" the hypothesis that Irving has S, in Mayo's sense, because probability 1 is much greater than probability 2. And because probability 2 is so low, Irving's getting the red result with test T should count as passing a severe test for the hypothesis that he has disease S.

Finally, suppose that only one person in ten million in the general population has disease S, so that P(S) = .0000001. Using Bayes' theorem, we compute that P(S/red result with test T) = .004; that is, four out of one thousand people who get the red result from test T (less than half of 1%) have disease S. I regard such a test for disease S as pretty lousy. But using these frequency probabilities as a basis for epistemic ones (represented with a small p), we obtain

$$p(\text{Irving has disease } S) = .0000001,$$

and

$$p(\text{Irving has disease } S/\text{Irving gets a red result with test } T) = .004.$$

If so, epistemically speaking, Irving's red test result gives very little reason to believe that he has disease S, despite the fact that Irving has passed a "severe test" for disease S in Mayo's sense. In such a case, if passing a severe test is supposed to give us a good reason to believe a hypothesis, it does not do the job.[10]

---

[10] Determining the posterior probability of having a certain disease (say), given a certain test result, without determining the prior probability of that disease (the "base rate") is an example of what is called the "base rate fallacy."

On the other hand, suppose that Mayo refuses to assign a posterior probability to *h*. Suppose she claims that we do not want, need, or have any such probability, whether or not this is epistemic probability. Then I, and I believe Mill, would have a problem understanding what passing a severe test has to do with something we regard as crucial in induction – namely, providing a good reason to believe *h*. We have data *D* obtained as a result of test *T*, and those data "fit" hypothesis *h* in Mayo's sense. Now we are told that if we were to continue indefinitely to test *h* by employing test *T*, and if *h* were false, the relative frequency of getting results that fit *h* as well as data *D* do would eventually become and stay low. Let us even suppose (as we did before) that *T* is the only test for disease *S* that can be made while the patient is living. We can say that, if the frequency probabilities are as previously reported, then giving the *T*-test to Irving and getting result "red" is as good as it gets for determining *S* – it is a better test for *S* than any other. If repeated and the hypothesis *h* is false, the frequency of outcomes "fitting" *h* as well as the present outcome does will remain low. In this sense, it is a "good test." But, given the additional information I have supplied, passing that test is not a good reason, or a good enough reason, to believe the hypothesis.

If we accept Mayo's definition of a hypothesis "passing a severe test," then, I think, we have to distinguish between

1. a hypothesis passing such a test with data results *D*, and
2. a hypothesis passing such a test with results *D* being a good reason to believe or infer that the hypothesis is true.

Establishing 1 is not sufficient for establishing 2.

The theory of evidence I have defended elsewhere distinguishes various concepts of evidence in use in the sciences. But the most basic of these (which I call "potential" evidence) in terms of which the others can be defined requires that some fact *e* is evidence that *h* only if *e* provides a good reason for believing *h*. Such a reason is provided, I claim, not simply when the (epistemic) posterior probability of *h*, given *e*, is high, but when the probability of an explanatory connection between *h* and *e*, given *e*, is high (which entails the former). I will not here defend the particulars of this view; I will say only that, where data from tests are concerned, the view of evidence I defend requires satisfying 2 above, not simply 1.

Finally, in this part of the discussion, let me note that Mayo's theory is being applied only to *experimental* evidence: data produced by experiments and observations as a result of what she calls a test. She is concerned with the question of whether the test results provide a severe test of the hypothesis and, hence, evidence for it. I agree that this question about evidence and inductive reasoning is important. But what happens if we take it a step higher and ask whether the facts described by propositions highly probed by the experimental and observational data can themselves be regarded as evidence for higher level propositions or theories. This is precisely what Newton does in the third book of the *Principia*.

On the basis of different observations of distances of the four known moons of Jupiter from Jupiter – astronomical observations made by Borelli, Townly, and Cassini – Newton infers Phenomenon 1: "that the satellites of Jupiter, by radii drawn to the center of Jupiter, describe areas proportional to the times, and their periodic times . . . are as 3/2 powers of the distances from that center." Similar so-called phenomena are inferred about the moons of Saturn, about our moon, and about the primary planets. Now appealing to his principles and theorems of mechanics in Book 1, Newton regards the six phenomena as providing very strong evidence for his universal law of gravity. (He speaks of "the argument from phenomena . . . for universal gravity" [Newton, 1999, p. 796].) More generally, how will this be understood in the error-statistical view of evidence? Should only the observational data such as those reported by Borelli, Townly, and Cassini be allowed to count as evidence for Newton's universal law of gravity? Or can the error-statistical theory be applied to Newton's phenomena themselves? Can we assign a probability to Phenomenon 1, given the assumption of falsity of Newton's law of gravity? If so, how can this probability be understood as a relative frequency? More generally, how are we to understand claims about evidence when the evidence consists in more or less theoretical facts and the hypothesis is even more theoretical? I will simply assert without argument that an account of evidence that appeals to objective epistemic posterior probabilities can make pretty good sense of this. Can Mayo?

4 The Principle of the Uniformity of Nature: Sin 4

Now, more briefly, we have the fourth and final sin – the claim that Mill appeals to an unnecessary, unwarranted, and vague principle of the uniformity of nature.

Mill's discussion here (1888, pp. 200–6) is admittedly somewhat confusing. First, he says, "there is an assumption involved in every case of induction," namely that "what happens once will, under a sufficient degree of similarity of circumstances, happen again, and not only again, but as often as the same circumstances recur." He calls this a "universal fact, which is our warrant for all inferences from experience." But second, he claims that this proposition, "that the course of nature is uniform," is itself "an instance of induction." It is not the first induction we make, but one of the last. It is founded on other inductions in which nature is concluded to be uniform in one respect or another. Third, he claims that, although the principle in question does not contribute to proving any more particular inductive conclusion, it is "a necessary condition of its being proved." Fourth, and perhaps most interesting, in an important footnote he makes this claim:

> But though it is a condition of the validity of every induction that there be uniformity in the course of nature, it is not a necessary condition that the uniformity should pervade all nature. It is enough that it pervades the particular class of phenomena to which the induction relates.... Neither would it be correct to say that every induction by which we infer any truth implies the general fact of uniformity as *foreknown*, even in reference to the kind of phenomena concerned. It implies *either* that this general fact is already known, *or* that we may now know it. (Mill, 1888, p. 203)

In this footnote Mill seems to abandon his earlier claim that the principle of uniformity of nature is "our warrant for all inferences from experience."

Possibly a more appealing way to view Mill's principle is as follows. Instead of asserting boldly but rather vaguely that nature is uniform, Mill is claiming somewhat less vaguely, and less boldly, that there are uniformities in nature – general laws governing various types of phenomena – which are inductively and validly inferred, perhaps using Mill's methods. That such laws exist is for Mill an empirical claim. Suppose that phenomena governed by such laws bear a similarity to others which by induction we infer are governed by some similar set of laws. Then, Mill may be saying, we can use the fact that one set of phenomena is so governed to strengthen the inference to laws regarding the second set.

So, for example, from the inductively inferred Phenomenon 1 regarding the motions of the moons of Jupiter, together with mechanical principles from Book 1, Newton infers Proposition 1: the force governing the motions of the moons of Jupiter is a central inverse-square force exerted by Jupiter on its moons. Similarly from the inductively inferred Phenomenon 2 regarding the motions of the moons of Saturn, Newton infers that a central inverse-square force is exerted by Saturn on its moons. The latter inference is strengthened, Mill may be saying, by the existence of the prior uniformity inferred in the case of Jupiter.

Finally, Mill's uniformity principle can be understood not just as a claim that uniformities exist in nature, and that these can, in appropriate circumstances, be used to strengthen claims about other uniformities, but also as a methodological injunction in science: look for such uniformities. So understood, what's all the fuss?

5 Mill: Saint or Sinner?

Shall we conclude from this discussion that Mill is a saint regarding induction? I do not know that I have proved that, but at the very least I hope I have cast some doubt on the view that he is an unworthy sinner. If it is a sin to assign a probability to an inductive conclusion, as Mayo, following Peirce, seems to believe, then at least we can understand the good motive behind it: When we want evidence sufficient to give us a good reason to believe a hypothesis, Mill would say, then, as good as Mayo's "severe-testing" is, we want more.

Appendix: The Stories of Isaac and "Stand and Deliver"

Mayo introduces an example in which there is a severe test, or battery of tests, T to determine whether high school students are ready for college (Achinstein, 2005, pp. 96–7, 115–17) The test is severe in the sense that passing it is very difficult to do if one is not ready for college. Mayo imagines a student Isaac who has taken the test and achieved a high score X, which is very rarely achieved by those who are not college-ready. She considers Isaac's test results X as strong evidence for the hypothesis

H : Isaac is college-ready.

Now suppose we take the probability that Isaac would get those test results, given that he is college-ready, to be extremely high, so that

$p(X/H)$ is practically 1,

whereas the probability that Isaac would get those test results, given that he is not college-ready, is very low, say

$$p(X/\sim H) = .05.$$

But, says Mayo's imagined critic, suppose that Isaac was randomly selected from a population in which college readiness is extremely rare, say one out of one thousand. The critic infers that

(A) $p(H) = .001.$

If so, then the posterior probability that Isaac is college-ready, given his high test results, would be very low; that is,

(B) $p(H/X)$ is very low,

even though the probability in B would have increased from that in A. The critic regards conclusion B as a counterexample to Mayo's claim that Isaac's test results X provide strong evidence that Isaac is college-ready.

Mayo's response is to say that to infer conclusions A and B is to commit the fallacy of probabilistic instantiation. Perhaps, even worse, it is to engage in discrimination, since the fact that we are assigning such a low prior probability to Isaac's being college-ready prevents the posterior probability of his being college-ready, given his test results, from being high; so we are holding poor disadvantaged Isaac to a higher standard than we do test-takers from an advantaged population.

My response to the probabilistic fallacy charge is to say that it would be true if the probabilities in question were construed as relative frequencies. However, as I stressed in the body of the chapter, I am concerned with epistemic probability. If all we know is that Isaac was chosen at random from a very disadvantaged population, very few of whose members are college ready, say one out of one thousand, then we would be justified in believing that it is very unlikely that Isaac is college-ready (i.e., conclusion A and, hence, B).

The reader may recall the 1988 movie *Stand and Deliver* (based on actual events) in which a high school math teacher in a poor Hispanic area of Los Angeles with lots of student dropouts teaches his students calculus with their goal being to pass the advanced placement calculus test. Miraculously all the students pass the test, many with flying colors. However, officials from the Educational Testing Service (ETS) are very suspicious and insist that the students retake the test under the watchful eyes of their own representatives. Of course, the students feel discriminated against and are very resentful. But the ending is a happy one: once again the group passes the test.

Now two questions may be raised here. First, given that all they knew were the test results and the background of the students, were the officials from the ETS justified in believing that, despite the high test results, it is unlikely that the students have a good knowledge of basic calculus? I think that this is a reasonable epistemic attitude. They assigned a very low prior epistemic probability to the hypothesis that these students know calculus and, hence, a low posterior probability to this hypothesis given the first test results. We may suppose that after the results of the second test, conducted under more stringent testing conditions, the posterior probability of the hypothesis was significantly high. To be sure, we the viewers of the film had much more information than the ETS officials: we saw the actual training sessions of the students. So the posterior epistemic probability we assigned was determined not only by the prior probability but by the test results and this important additional information.

The second question, which involves a potential charge of discrimination, is one about what action to take given the epistemic probabilities. After the first test results, what, if anything, should the ETS officials have done? This depends on more than epistemic probabilities. Despite low probabilities, it might have been judged better on grounds mentioned by Mayo – discrimination – or on other moral, political, or practical grounds not to repeat the test but to allow the test results to count as official scores to be passed on to the colleges. Or, given the importance of providing the colleges with the best information ETS officials could provide, the most appropriate course might well have been to do just what they did.

References

Achinstein, P. (2001), *The Book of Evidence*, Oxford University Press, New York.

Achinstein, P., ed. (2005), *Scientific Evidence: Philosophical Theories and Applications*, Johns Hopkins University Press, Baltimore, MD.

Mandelbaum, M. (1964), *Philosophy, Science, and Sense Perception*, Johns Hopkins University Press, Baltimore, MD.

Mayo, D.G. (1996), *Error and the Growth of Experimental Knowledge*, University of Chicago Press, Chicago.

Mayo, D.G. (2005), "Evidence as Passing Severe Tests: Highly Probed vs. Highly Proved," pp. 95–127 in P. Achinstein (ed.), *Scientific Evidence: Philosophical Theories and Applications*, Johns Hopkins University Press, Baltimore, MD.

Mill, J.S. (1888), *A System of Logic*, 8th edition, Harper and Bros., New York.

Newton, I. (1999), *Principia* (trans. I.B. Cohen and A. Whitman), University of California Press.

Peirce, C.S. (1931–1935), *Collected Papers*, C. Hartshorne and P. Weiss (eds.), Harvard University Press, Cambridge, MA.

Sins of the Epistemic Probabilist
Exchanges with Peter Achinstein

Deborah G. Mayo

1 Achinstein's Sins

As Achinstein notes, he and I agree on several key requirements for an adequate account of evidence: it should be objective and not subjective, it should include considerations of flaws in data, and it should be empirical and not a priori. Where we differ, at least at the moment, concerns the role of probability in inductive inference. He takes the position that probability is necessary for assigning degrees of objective warrant or "rational" belief to hypotheses, whereas, for the error statistician, probability arises to characterize how well probed or tested hypotheses are (highly probable vs. highly probed). Questions to be considered are the following:

1. *Experimental Reasoning*: How should probability enter into inductive inference – by assigning degrees of belief or by characterizing the reliability of test procedures?
2. *Objectivity and Rationality*: How should degrees of objective warrant be assigned to scientific hypotheses?
3. *Metaphilosophy*: What influences do philosophy-laden assumptions have on interpretations of historical episodes: Mill?

I allude to our disagreeing "at the moment" because I hope that the product of this latest installment in our lengthy exchange may convince him to shift his stance, at least slightly.

In making his case, Achinstein calls upon Mill (also Newton, but I largely focus on Mill). Achinstein wishes "to give Mill a better run for his money" by portraying him as an epistemic probabilist of the sort he endorses. His question is whether Mill's inductive account contains some features "that are or seem inimical to [the] error-statistical philosophy." A question for us is whether Achinstein commits some fundamental errors, or "sins" in his terminology, that are common to epistemic (Bayesian) probabilists in philosophy of science, and are or seem to be inimical to the spirit of an "objective" epistemic account. What are these sins?

1. *First*, there are *sins of "confirmation bias"* or "reading one's preferred view into the accounts of historical figures" or found in scientific episodes, even where this reconstruction is at odds with the apparent historical evidence.
2. *Second*, there are *sins of omission*, whereby we are neither told how to obtain objective epistemic probabilities nor given criteria to evaluate the success of proposed assignments.

These shortcomings have undermined the Bayesian philosophers' attempts to elucidate inductive inference; considering Achinstein's discussion from this perspective is thus of general relevance for the popular contemporary project of Bayesian epistemologists.

1.1 Mill's Innocence: Mill as Error Statistician

Before considering whether to absolve Mill from the sins Achinstein considers, we should note that Achinstein omits the one sin discussed in Mayo (1996) that arises most directly in association with Mill: Mill denies that novel predictions count more than nonnovel ones (pp. 252, 255). My reference to Mill comes from Musgrave's seminal paper on the issue of novel prediction:

According to modern logical empiricist orthodoxy, in deciding whether hypothesis *h* is confirmed by evidence *e*... we must consider only the statements *h* and *e*, and the logical relations between them. It is quite irrelevant whether *e* was known first and *h* proposed to explain it, or whether *e* resulted from testing predictions drawn from *h*. (Musgrave, 1974, p. 2)

"We find" some variant of the logicist approach, Musgrave tells us, "in Mill, who was amazed at Whewell's view" that successfully predicting novel facts gives a hypothesis special weight. "In Mill's view, 'such predictions and their fulfillment are... well calculated to impress the uninformed' " (Mill, 1888, Book 3, p. 356). This logicism puts Mill's straight rule account at odds with the goal of severity.

Nevertheless, Achinstein gives evidence that Mill's conception is sufficiently in sync with error statistics so that he is prepared to say, "I would take Mill to be espousing at least some version of the idea of 'severe testing'" (p. 174). Here is why: Although Mill regards the form of induction to be "induction by simple enumeration," whether any particular instantiation is valid also "depends on nonformal empirical facts regarding the sample, the sampling, and the properties being generalized" (Achinstein, p. 174) – for example, Mill would eschew inadequately varied data, failure to look for negative instances, and generalizing beyond what the evidence allows. These features could well set the stage for expiating the sins of the straight ruler: The "fit" requirement, the first clause of severity, is met because observed cases of As that are Bs "fit" the hypothesis that all As are Bs; and the additional considerations Achinstein lists would help to satisfy severity's second condition – that the ways such a fit could occur despite the generalization being false have been checked and found absent.

Achinstein finds additional intriguing evidence of Mill's error-statistical leanings. Neither Mill's inductions (nor those of Newton), Achinstein points out, are "guilty" of assigning probabilities to an inductive conclusion:

> Neither in their abstract formulations of inductive generalizations... nor in their examples of particular inductions... does the term "probability" occur.... From the inductive premises we simply conclude that the generalization is true, or as Newton allows in rule 4, "very nearly true," by which he appears to mean not "probably true" but "approximately true." (p. 176)

Neither exemplar, apparently, requires the assignment of a probability to inductively inferred generalizations. One might have expected Achinstein to take this as at least casting doubt on his insistence that such posterior probabilities are necessary. He does not.

1.2 Mill's Guilt: Mill as Epistemic Probabilist

Instead, Achinstein declares that their (apparent) avoidance of any "probabilistic sin" is actually a transgression to be expiated, because it conflicts with his own account! "However, before we conclude that no probabilistic sin has been committed, we ought to look a little more closely at Mill." In particular, Achinstein thinks we ought to look at contexts where Mill *does* talk about probabilities – namely in speaking about events – and substitute what he says there to imagine he is talking about probabilities of hypotheses. By assigning to Mill what is a probabilistic sin to error statisticians, Mill is restored to good graces with Achinstein's Bayesian probabilist.

Mill offers examples from games of chance, Achinstein notes (p. 177), where probabilities of general outcomes are assigned on the basis of their observed relative frequencies of occurrence:

> In the cast of a die, the probability of ace is one-sixth... because we do actually know, either by reasoning or by experience, that in a hundred or a million of throws, ace is thrown about one-sixth of that number, or once in six times. (Mill, 1888, p. 355)

In more modern parlance, Mill's claim is that we may accept or infer ("by reasoning or by experience") a statistical hypothesis H that assigns probabilities to outcomes such as "ace." For the error statistician, this inductive inference takes place by passing *H* severely or with appropriately low error probabilities. Although there is an inference to a probabilistic model of experiment, which in turn assigns probabilities to outcomes, there is no probabilistic assignment to *H* itself, nor does there seem to be in Mill. So such statements from Mill do not help Achinstein show that Mill intends to assign posterior probabilities to hypotheses.[1]

The frequentist statistician has no trouble agreeing with the conditional probabilities of events stipulated by Achinstein. For example, we can assert the probability of spades given the outcome "ace" calculated under hypothesis *H* that cards are randomly selected from a normal deck, which we may write as P(spade | ace; *H*) – noting the distinction between the use of "|" and ";". However, we would not say that an event severely passes, or that one event severely passes another event. A statistical hypothesis must assign probabilities to outcomes, whereas the event "being an ace" does not.

I claim no Mill expertise, yet from Achinstein's own presentation, Mill distinguishes the assignment of probabilities to events from assigning probabilities to hypotheses, much as the error statistician. What then is Achinstein's justification for forcing Mill into a position that Mill apparently rejects? Achinstein seems guilty of the sin of "reading our preferred view into the accounts of historical episodes and/or figures."

2 The Error-Statistical Critique

2.1 Some Sleight-of-Hand Sins

Having converted Mill so that he speaks like a Bayesian, Achinstein turns to the business of critically evaluating the error statistician: "[O]bjective epistemic probabilists – again I include Mill – are committed to saying that the inference is justified only if the objective epistemic probability (the 'posterior' probability) of the inductive conclusion" is sufficiently high (p. 179). At times during our exchanges, I thought Achinstein meant this assertion as a tautology, playing on the equivocal uses of "probability" in ordinary language. If high epistemic probability is just shorthand for high inductive

---

[1] On the other hand, they do raise the question of how Achinstein's epistemic probabilist can come to accept a statistical model (on which probabilistic assignments to events are based). Achinstein's formal examples start out by assuming that we have accepted a statistical model of experiment, *H*, usually random sampling from a binomial distribution. Is this acceptance of *H* itself to be a matter of assigning *H* a high posterior probability through a Bayesian calculation? The alternatives would have to include all the ways the model could fail. If not, then Achinstein is inconsistent in claiming that warranting *H* must take the form of a Bayesian probability computation.

warrant, then his assertion may be uncontroversial, if also uninformative. If hypothesis *H* passes a highly severe test, there is no problem in saying there is a high degree of warrant for *H*. But there is a problem if this is to be cashed out as a posterior probability, attained from Bayes's theorem and satisfying the probability axioms. Nevertheless, Achinstein assures me that this is how he intends to cash out epistemic probability. High epistemic probability is to be understood as high posterior probability (obtainable from Bayes's theorem). I construe it that way in what follows.

Achinstein presupposes that if we speak of data warranting an inference or belief in *H*, then we must be talking about an epistemic posterior probability in *H*, but this is false. Musgrave's critical rationalist would reject this as yet another variant on "justificationism" (see Chapter 3). Achinstein's own heroes (Mill and Newton) attest to its falsity, because as he has convincingly shown, they speak of warranting hypotheses without assigning them posterior probabilities.

In the realm of formal statistical accounts of induction, it is not only the error statistician who is prepared to warrant hypotheses while eschewing the Bayesian algorithm. The "likelihoodist," for instance, might hold that data **x** warrants *H* to the extent that *H* is more likely (given **x**) than rivals. High likelihood for *H* means $P(\mathbf{x};H)$ is high, but high likelihood for *H* does *not* mean high probability for *H*. To equate $P(\mathbf{x}|H)$ and $P(H|\mathbf{x})$ immediately leads to contradictions – often called the prosecutor's fallacy. For example, the likelihood of *H* and not-*H* do not sum to 1 (see Chapters 7 and 9, this volume).

If Achinstein were to accept this, then we would be in agreement that probabilistic concepts (including frequentist ones) may be used to qualify the evidential warrant for hypothesis H while denying that this is given by a probability assignment to H. Achinstein may grant these other uses of probability for induction yet hold that they are inferior to his idea of objective epistemic posterior probabilities. His position might be that, unless the warrant is a posterior probability, then it does not provide an adequate objective inductive account. This is how I construe his position.

2.2 Sins of Omission: How Do We Apply the Method? What Is So Good About It?

For such a sweeping claim to have any substance, it must be backed up with (1) some guidance as to how we are to arrive at objective epistemic posteriors and (2) an indication of (desirable) inferential criteria that objective epistemic probability accounts satisfy.

So how do we get to Achinstein's objective epistemic posterior probabilities? By and large this is not one of Achinstein's concerns; he considers informal examples where intuitively good evidence exists for a claim *H*, and then he captures this by assigning *H* a high degree of epistemic probability. If he remained at the informal level, the disagreements he finds between us would likely vanish, for then high probability could be seen as a shorthand for high inductive warrant, the latter attained by a severe test. When he does give a probabilist computation, he runs into trouble.

Unlike the standard Bayesian, Achinstein does not claim that high posterior probability is sufficient for warrant or evidence, but does "take [it] to be a necessary condition" (p. 180). It is not sufficient for him because he requires, in addition, what he describes as a non-Bayesian "explanatory connection" between data and hypotheses (p. 183). Ignoring the problem of how to determine the explanatory connection, and whether it demands its own account of inference (rendering his effort circular), let us ask: Are we bound to accept the necessity? My question, echoing Achinstein, is this: *Epistemically speaking, what, if anything, can one conclude about hypothesis H itself from the fact that Achinstein's method accords H a high objective epistemic probability?*

That is, we require some indication that *H*'s earning high (low) marks on Achinstein's objective probability scale corresponds to actually having strong (weak) evidence of the truth of *H*. If Achinstein could show this, his account would be superior to existing accounts of epistemic probability! It will not do to consider examples where strong evidence for *H* is intuitively sound and then to say we have high Achinstein posterior epistemic probability in *H*; he would have to demonstrate it with posteriors arrived at through Achinsteinian means.

3 Achinstein's Straight Rule for Attaining Epistemic Probabilities

To allow maximum latitude for Achinstein to make his case, we agree to consider the Achinsteinian example. The most clear-cut examples instantiate a version of a "straight rule," where we are to consider a "hypothesis" that consists of asserting that a sample possesses a characteristic such as "having a disease" or "being college-ready." He is led to this peculiar notion of a hypothesis because he needs to use techniques for probability assignments appropriate only for events. But let us grant all of the premises for Achinstein's examples.[2] We have (see p. 187):

---

[2] In an earlier exchange with Colin Howson, I discuss converting these inadmissible hypotheses into legitimate statistical ones to aid the criticism as much as possible, and I assume the same here (Mayo, 1997).

Achinstein's Straight Rule for Objective Epistemic Probabilities: If (we know only that) a$_i$ is randomly selected from a population where p% have property C, then the objective epistemic probability that a$_i$ has C equals p.

Next, Achinstein will follow a variant on a Bayesian gambit designed to show that data **x** from test *T* can pass hypothesis *H* severely, even though *H* is accorded a low posterior probability – namely by assuming the prior probability of *H* is sufficiently low. (Such examples were posed post-EGEK, first by Colin Howson (1997a).) I have argued that in any such example it is the Bayesian posterior, and not the severity assignment, that is indicted (Mayo 2005, 2006). This criticism and several like it have been discussed elsewhere, most relevantly in a collection discussing Achinstein's account of evidence (Mayo, 2005)! I focus mainly on the update to our earlier exchange, but first a quick review.

3.1 College Readiness

We may use Achinstein's summary of one of the canonical examples (pp. 186–7). We are to imagine a student Isaac who has taken the test and achieved a high score X, which is very rarely achieved by those who are not college-ready. Let

$H(I)$: Isaac is college-ready.

Let H' be the denial of H:

$H'(I)$ Isaac is not college-ready, i.e., he is deficient.

I use $H(I)$ to emphasize that the claim is about Isaac}.

Let S abbreviate: Isaac gets a high score (as high as X).

In the error statistical account, S is evidence against Isaac's deficiency, and for $H(I)$. (Although we would consider degrees of readiness, I allow his dichotomy for the sake of the example.) Now we are to suppose that the probability that Isaac would get those test results, given that he is college-ready, is extremely high, so that

$P(S|H(I))$ is practically 1,

Whereas the probability that Isaac would get those test results, given that he is not college-ready, is very low, say,

$P(S|H'(I)) = .05.$

Suppose that Isaac was randomly selected from a population – call it Fewready Town – in which college readiness is extremely rare, say one out of one thousand. The critic infers that

$$(*) \, P(H(I)) = .001.$$

If so, then the posterior probability that Isaac is college-ready, given his high test results, would be very low; that is,

$$P(H(I)|S) \text{ is very low,}$$

even though in this case the posterior probability has increased from the prior in (*).

The critic – for example, Achinstein – regards the conclusion as problematic for the severity account because, or so the critic assumes, the frequentist would also accept (*) $P(H) = .001$. Here is the critic's flaw. Although the probability of college readiness in a randomly selected student from high schoolers from Fewready Town is .001, it does not follow that Isaac, the one we happened to select, has a probability of .001 of being college-ready (Mayo, 1997a, 2005, p. 117). To suppose it does is to commit what may be called a fallacy of probabilistic instantiation.

3.2 Fallacy of Probabilistic Instantiation

We may abbreviate by $P(H(x))$ the probability that a randomly selected member of Fewready has the property C. Then the probabilistic instantiation argues from the first two premises,

$$P(H(x)) = .001.$$

The randomly selected student is I.

To the inference:

$$(*) \, P(H(I)) = .001.$$

This is fallacious. We need not preclude that $H(I)$ has a legitimate frequentist prior; the frequentist probability that Isaac is college-ready might refer to generic and environmental factors that determine the chance of his deficiency – although I do not have a clue how one might compute it. But this is not what the probability in (*) gives us. Now consider Achinstein's new update to our exchanges on Isaac.

3.3 Achinstein's Update

Achinstein now accepts that this assignment in (*) is a sin for a frequentist:

[T]he probabilistic fallacy charge would be true if the probabilities in question were construed as relative frequencies. However . . . I am concerned with epistemic probability. If all we know is that Isaac was chosen at random from a very disadvantaged population, very few of whose members are college ready, say one out of one thousand, then we would be justified in believing that it is very [improbable] that Isaac is college-ready. (Achinstein, p. 187)

Hence, (*) gives a legitimate objective epistemic frequentist prior.

Therefore, even confronted with Isaac's high test scores, Achinstein's probabilist is justified in denying that the scores are good evidence for $H(I)$. His high scores are instead grounds for believing $H'(I)$, that Isaac is not college-ready. It is given that the posterior for $H'(I)$ is high, and certainly an explanatory connection between the test score and readiness exists. Although the posterior probability of readiness has increased, thanks to his passing scores, for Achinstein this does not suffice to provide epistemic warrant for H (he rejects the common Bayesian distinction between increased support and confirmation). Unless the posterior reaches a threshold of a fairly high number, he claims, the evidence is "lousy." The example considers only two outcomes: reaching the high scores or not, i.e., S or $\sim S$. Clearly a lower grade gives even less evidence of readiness; that is, $P(H'(I)|\sim S) > P(H'(I)|S)$. Therefore, whether Isaac scored a high score or not, Achinstein's epistemic probabilist reports justified high belief that Isaac is not ready. The probability of Achinstein finding evidence of Isaac's readiness even if in fact he is ready (H is true) is low if not zero. Therefore, Achinstein's account violates what we have been calling the most minimal principle for evidence!

The *weak severity principle*: Data x fail to provide good evidence for the truth of H if the inferential procedure had very little chance of providing evidence against H, even if H is false.

Here the relevant H would be $H'(I)$ – Isaac is not ready.

If Achinstein allows that there is high objective epistemic probability for H even though the procedure used was practically guaranteed to arrive at such a high posterior despite H being false, then (to echo him again) the error statistician, and I presume most of us, would have a problem understanding what high epistemic probability has to do with something we regard as crucial in induction, namely ensuring that an inference to H is based on genuine evidence – on data that actually discriminate between the truth and falsity of H.

This fallacious argument also highlights the flaw in trying to glean reasons for epistemic belief by means of just any conception of "low frequency of error." If we declared "unready" for any member of Fewready, we would rarely be wrong, but in each case the "test" has failed to discriminate the particular student's readiness from his unreadiness. We can imagine a context where we are made to bet on the generic event – the next student randomly selected from the population has property *C*. But this is very different from having probed whether this student, Isaac, is ready or not – the job our test needs to perform for scientific inference.

I cannot resist turning Achinstein's needling of me back on him: Is Achinstein really prepared to claim there is high epistemic warrant for $H'(I)$ even though the procedure had little or no probability of producing evidence against $H'(I)$ and for $H(I)$ even if $H(I)$ is true? If he is, then the error statistician, and perhaps Mill, would have a hard time understanding what his concept has to do with giving an objective warrant for belief.

Let us take this example a bit further to explain my ironic allegation regarding "reverse discrimination." Suppose, after arriving at high belief in Issac's unreadiness, Achinstein receives a report of an error: in fact Isaac was selected randomly, not from Fewready Town, but from a population where college readiness is common, Fewdeficient Town. The same score now warrants Achinstein's assignment of a strong objective epistemic belief in Isaac's readiness (i.e., $H(I)$). A high school student from Fewready Town would need to have scored quite a bit higher on these same tests than one selected from Fewdeficient Town for his scores to be considered evidence of his readiness. So I find it surprising that Achinstein is content to allow this kind of instantiation to give an objective epistemic warrant.

3.4 The Case of General Hypotheses

When we move from hypotheses like "Isaac is college-ready" (which are really events) to generalizations – which Achinstein makes clear he regards as mandatory if an inductive account is not to be "puerile" – the difficulty for the epistemic probabilist becomes far worse if, like Achinstein, we are to obtain epistemic probabilities via his frequentist straight rule.

To take an example discussed earlier, we may infer with severity that the relativistic light deflection is within $\pm \varepsilon$ units from the GTR prediction, by finding that we fail to reject a null hypothesis with a powerful test. But how can a frequentist prior be assigned to such a null hypothesis to obtain the epistemic posterior?

The epistemic probabilist would seem right at home with a suggestion some Bayesians put forward – that we can apply a version of Achinstein's straight rule to hypotheses. We can imagine that the null hypothesis is

H_0: There are no increased risks (or benefits) associated with hormone replacement therapy (HRT) in women who have taken HRT for 10 years.

Suppose we are testing for discrepancies from zero in both positive and negative directions (Mayo, 2003). In particular, to construe the truth of a general hypothesis as a kind of "event," it is imagined that we sample randomly from a population of hypotheses, some proportion of which are assumed true. The proportion of these hypotheses that have been found to be true in the past serves as the prior epistemic probability for H_0.

Therefore, if H_0 has been randomly selected from a pool of null hypotheses, 50% of which are true, we have

$$(*) \; P(H_0) = .5.$$

Although (*) is fallacious for a frequentist, once again Achinstein condones it as licensing an objective epistemic probability. But which pool of hypotheses should we use? The percentages "initially true" will vary considerably, and each would license a distinct "objective epistemic" prior. Moreover, it is hard to see that we would ever know the proportion of true nulls rather than merely the proportion that have thus far not been rejected by other statistical tests!

The result is a kind of "innocence by association," wherein a given H_0, asserting no change in risk, gets the benefit of having been drawn from a pool of true or not-yet-rejected nulls, much as the member from Fewready Town is "deficient by association." Perhaps the tests have been insufficiently sensitive to detect risks of interest. Why should that be grounds for denying evidence of a genuine risk with respect to a treatment (e.g., HRT) that *does* show statistically significant risks?

To conclude, here is our answer to Achinstein's question:

Does Mayo allow that *H* may pass with high severity when the posterior probability of *H* is not high?

In no case would *H* pass severely when the grounds for warranting *H* are weak. But high posteriors need not correspond to high evidential warrant. Whether the priors come from frequencies or from "objective" Bayesian priors, there are claims that we would want to say had passed severely that do not get a high posterior (see Chapter 7, this volume). In fact, statistically significant results that we would regard as passing the nonnull hypothesis severely can show a decrease in probability from the prior (.5) to the posterior (see Mayo, 2003, 2005, 2006).

4 Some Futuristic Suggestions for Epistemic Probabilists

In the introductory chapter of this volume, we mentioned Achinstein's concession that "standard philosophical theories about evidence are (and ought to be) ignored by scientists" (2001, p. 3) because they view the question of whether data **x** provide evidence for *H* as a matter of purely logical computation, whereas whether data provide evidence for hypotheses is not an a priori but rather an empirical matter. He appears to take those failures to show that philosophers can best see their job as delineating the concepts of evidence that scientists seem to use, perhaps based on rational reconstructions of figures from the historical record.

Some may deny it is sinful that *epistemic probabilists* omit the task of how to obtain, interpret, and justify their objective epistemic probabilities, and claim I confuse the job of logic with that of methodology (Buldt 2000). Colin Howson, who had already denied that it was part of the job of Bayesian logic to supply the elements for his (subjective) Bayesian computation, declared in 1997 that he was moving away from philosophy of statistics to focus on an even purer brand of "inductive logic"; and he has clearly galvanized the large contemporary movement under the banner of "Bayesian episte-mology" (see Glymour, Chapter 9, this volume). Two points: First, it is far from clear that Bayesian logics provide normative guidance about "ratio-nal" inference[3]; after all, error statistical inference embodies its own logic, and it would be good to explore which provides a better tool for under-standing scientific reasoning and inductive evidence. Second, there is a host of new foundational problems (of logic and method) that have arisen in Bayesian statistical practice and in Bayesian-frequentist "unifications" in the past decade that are omitted in the Bayesian epistemological literature (see Chapter 7.2). It is to be hoped that days of atonement will soon be upon us.

References

Achinstein, P. (2001), *The Book of Evidence*, Oxford University Press, Oxford.

Buldt, B. (2000), "Inductive Logic and the Growth of Methodological Knowledge, Com-ment on Mayo," pp. 345–54 in M. Carrier, G. Massey, and L. Ruetsche (eds.), *Science at Century's End, Philosophical Questions on the Progress and Limits of Science*, University of Pittsburgh Press, Pittsburgh.

Howson, C. (1997a), "A Logic of Induction," *Philosophy of Science*, 64: 268–90.

Mayo, D.G. (1996), *Error and the Growth of Experimental Knowledge*, The University of Chicago Press, Chicago.

Mayo, D.G. (1997a), "Response to Howson and Laudan," *Philosophy of Science*, 64: 323–33.

Mayo, D.G. (2003), "Could Fisher, Jeffreys and Neyman Have Agreed on Testing? Com-mentary on J. Berger's Fisher Address," *Statistical Science*, 18: 19–24.

Mayo, D.G. (2005), "Evidence as Passing Severe Tests: Highly Probed vs. Highly Proved," pp. 95–127 in P. Achinstein (ed.), *Scientific Evidence*, Johns Hopkins University Press, Baltimore, MD.

---

[3] Even for the task of analytic epistemology, I suggest that philosophers investigate whether appealing to error-statistical logic offers a better tool for characterizing probabilistic knowl-edge than the Bayesian model. Given Achinstein's threshold view of evidential warrant, it is hard to see why he would object to using a severity assessment to provide the degree of epistemic warrant he seeks.

Mayo, D.G. (2006), "Philosophy of Statistics," pp. 802–15 in S. Sarkar and J. Pfeifer (eds.), *Philosophy of Science: An Encyclopedia*, Routledge, London.

Mill, J.S. (1888), *A System of Logic*, 8th edition, Harper and Bros., New York.

Musgrave, A. (1974), "Logical versus Historical Theories of Confirmation," *British Journal for the Philosophy of Science*, 25: 1–23.

Related Exchanges

Achinstein, P. (2000), "Why Philosophical Theories of Evidence Are (and Ought to Be) Ignored By Scientists, pp. S180–S192 in D. Howard (ed.), " *Philosophy of Science*, 67 (Symposia Proceedings).

Giere, R.N. (1997b), "Scientific Inference: Two Points of View," pp. S180–S184 in L. Darden (ed.), *Philosophy of Science*, 64 (PSA 1996: Symposia Proceedings).

Howson, C. (1997b), "Error Probabilities in Error," pp. S185–S194 in L. Darden (ed.), *Philosophy of Science*, 64 (PSA 1996: Symposia Proceedings).

Kelly, K., Schulte, O., Juhl, C., (1997), "Learning Theory and the Philosophy of Science," *Philosophy of Science*, 64: 245–67.

Laudan, L. (1997) "How about Bust? Factoring Explanatory Power Back into Theory Evaluation," *Philosophy of Science*, 64(2): 306–16.

Mayo, D.G. (1997a), "Duhem's Problem, the Bayesian Way, and Error Statistics, or "What's Belief Got to Do with It?" 64: 222–44.

Mayo, D.G. (1997b), "Error Statistics and Learning From Error: Making a Virtue of Necessity," pp. S195–S212 in L. Darden (ed.), *Philosophy of Science*, 64 (PSA 1996: Symposia Proceedings).

Mayo, D.G. (2000a), "Experimental Practice and an Error Statistical Account of Evidence," pp. S193–S207 in D. Howard (ed.), *Philosophy of Science*, 67 (Symposia Proceedings).

Mayo, D.G. (2000b), "Models of Error and the Limits of Experimental Testing," pp. 317–44 in M. Carrier, G. Massey and L. Ruetsche (eds.), *Science at Century's End, Philosophical Questions on the Progress and Limits of Science*, University of Pittsburgh Press, Pittsburgh.

Woodward, J. (1997), "Data, Phenomena, and Reliability," pp. S163–S179 in L. Darden (ed.), *Philosophy of Science*. 64 (PSA 1996: Symposia Proceedings).

SIX

Theory Testing in Economics and the Error-Statistical Perspective

Aris Spanos

1 Introduction

For a domain of inquiry to live up to standards of scientific objectivity it is generally required that its theories be tested against empirical data. The central philosophical and methodological problems of economics may be traced to the unique character of both economic theory and its non-experimental (observational) data. Alternative ways of dealing with these problems are reflected in rival methodologies of economics. My goal here is not to promote any one such methodology at the expense of its rivals so much as to set the stage for understanding and making progress on certain crucial conundrums in the methodology and philosophy of economics. This goal, I maintain, requires an understanding of the changing roles of theory and data in the development of economic thought alongside the shifting philosophies of science, which explicitly or implicitly find their way into economic theorizing and econometric practice. Given that this requires both economists and philosophers of science to stand outside their usual practice and reflect on their own assumptions, it is not surprising that this goal has been rather elusive.

1.1 The Preeminence of Theory in Economic Modeling

Historically, theory has generally held the preeminent role in economics, and data have been given the subordinate role of "quantifying theories" presumed to be true. In this conception, whether in the classical (nineteenth-century) or neoclassical (twentieth-century) historical period or even in contemporary textbook econometrics, data do *not* so much *test* as allow

* I am most grateful to my coeditor Deborah G. Mayo for numerous constructive criticisms and suggestions which improved the chapter considerably.

instantiating theories: sophisticated econometric methods enable elaborate ways "to beat data into line" (as Kuhn would say) to accord with an assumed theory. Because the theory has little chance to be falsified, such instantiations are highly *insevere* tests of the theory in question. Were the theories known to be (approximately) true at the outset, this might not be problematic, but in fact mainstream economic theories have been invariably unreliable predictors of economic phenomena, and rival theories could easily be made to fit the same data equally well, if not better.

1.2 Data-Driven Modeling

Attempts to redress the balance and give data a more substantial role in theory testing were frustrated by several inveterate methodological/ philosophical problems, including:

1. the huge gap between economic theories and the available observational data,
2. the difficulty of assessing when a fitted model "accounts for the regularities in the data," and
3. relating statistical inferences to substantive hypotheses and claims.

Faced with problems 1–3, endeavors to redress the balance tended to focus primarily on *data-driven models*, whose adequacy was invariably assessed in terms of goodness-of-fit measures. These efforts often gave rise to models with better short-term predictions, but they shed very little (if any) light on understanding the underlying economic phenomena. Indeed, data-driven correlation, linear regression, factor analysis, and principal component analysis, relying on goodness of fit, have been notoriously *unreliable* when applied to observational data, especially in the social sciences. Furthermore, the arbitrariness of goodness-of-fit measures created a strong impression that one can "forge" significant correlations (or regression coefficients) at will, if one was prepared to persevere long enough "mining" the data. The (mistaken) impression that *statistical spuriousness* is both inevitable and endemic, especially when analyzing observational data, is almost universal among social scientists and philosophers. This impression has led to widely held (but erroneous) belief that *substantive information* provides the only safeguard against statistical spuriousness. As a result, the advocates of the preeminence of theory perspective in economics have persistently charged any practices that strive for "objectivity" in data analysis by relying on the "regularities" contained in the data, with "measurement without theory," "data mining," and "hunting" for statistical significance. Perhaps

unsurprisingly, with Kuhnian relativisms and "theory-laden" data holding
sway in philosophy of science, it became easy to suppose that this is as good
as it gets in science in general, making it respectable to abandon or redefine
objectivity as opinions shared by a group (McCloskey, 1985) or formalizable
in terms of subjective Bayesian degrees of belief (Leamer, 1978).

1.3 The "Third Way"

Some econometricians, including Sargan, Hendry, Granger and Sims, wishing to avoid both extreme practices – theory-driven versus data-driven
modeling – set out a "third way" by adapting important advances in time-
series modeling in the 1970s. Aspiring to give data "a voice of its own,"
which could reliably constrain economic theorizing, these practitioners
were thrust into the role of inventing new methods while striving to find
or adapt a suitable foundation in one or another philosophy of science
(Kuhn, Popper, Lakatos); see Hendry (1980). Although loosely reflecting
the Popperian conception of critical appraisal and a Lakatosian demand for
"progressiveness," none of these classic philosophical approaches – at least
as they were formulated in philosophy of science – provided an appropriate
framework wherein one could address the numerous charges (legitimate
or baseless) leveled against any modeling practice that did not conform
to "quantifying theories" presumed true. Indeed, spurred by the general
impression that all statistical methods that rely on "regularities" in the data
are highly susceptible to the statistical spuriousness problem, the uphold-
ers of the preeminence-of-theory viewpoint could not see any difference
between solely data-driven methods and the third way, scorning the latter
as just a more sophisticated form of "data mining" (Spanos, 2000).

One of the things that I learned from the ERROR 06 conference, out of
which the exchanges in this volume arose, is that the shortcomings of the
standard philosophies of science growing out of the Popper–Lakatos and
rival Carnapian traditions are directly related to the difficulties one had
to face in attempting to develop a sound foundation for the theory–data
confrontation in economics. In coming to understand these shortcomings,
in effect, I am challenging philosophers of science to reevaluate the assumptions of their own standard models and hopefully take some steps toward
a framework that can steer us clear of some of the current confusion on
theory testing using empirical data.

The fact that the third way in empirical modeling in economics is still
struggling to achieve the twin goals of developing an adequate methodology and an underpinning philosophical foundation has contributed

significantly to the lack of a shared research agenda and a common language among the proponents (see the collection of papers in Granger, 1990). What they seem to share, and importantly so, is the idea that recognizing the flaws and fallacies of both extreme practices (theory-driven and data-driven modeling) provides a foothold for developing more adequate modeling. Minimally, in the third way, there seems to be a core endeavor to accomplish the goals afforded by sound practices of frequentist statistical methods in *learning from data*: to collect and model data to obtain reliable "statistical knowledge" that stands independently of theories that they may be called on to explain or be tested against with a view to shed light on the phenomenon of interest – a primary error-statistical objective.

It is interesting to note that experimental economists also found themselves facing the same quandary as the third way proponents in striving to find a suitable philosophical foundation, beyond Kuhn, Popper, and Lakatos, that would provide a sound methodological framework for their efforts to test economic theories using experiments (see Smith, 2007; Sugden, 2008).

1.4 The Error-Statistical Modeling Framework

In an attempt to provide more structure for the discussion that follows, the main philosophical/methodological problems and issues in the theory–data confrontation in economics, such as problems 1–3 mentioned earlier, may be viewed in the context of a broader framework that requires one to integrate adequately three interrelated modeling stages to bridge the theory–data gap:

A. **From theory to testable hypotheses.** Fashion an abstract and idealized theory $\mathscr{T}$ into hypotheses or claims h which are testable in terms of the available data $z_0 := (z_1, z_2, \ldots, z_n)$, often observational (nonexperimental).

B. **From raw data to reliable evidence.** Transform a finite and incomplete set of raw data z_0 – containing uncertainties, impurities, and noise – into reliable "evidence" e pertinent for theory appraisal.

C. **Confronting hypotheses with evidence.** Relate h and e, in a statistically reliable way, to assess whether e provides evidence for or against h.

Stage A proved to be particularly challenging in economics because a huge gap often exists between abstract and idealized economic theories and the *available* observational data. Stage B brings out the problem of establishing a reliable summary of the information contained in the data independently of the theory in question; this raises the issue of statistical spuriousness. Stage C raises issues that have to do with how statistical inferences pertain to substantive information – the last link that gives rise to learning from data.

The primary objective of this chapter is to make a case that the *error-statistical perspective* (see Mayo, 1996) offers a highly promising framework in which we may begin to clarify some of these crucial philosophical/methodological problems and make headway in their resolution (see Spanos, 2010). This framework enables one to foreground and work out solutions to the problems and issues raised by attempts to successfully deal with modeling stages A–C, using a refined version of the Fisher–Neyman–Pearson statistical inference framework (see Mayo and Spanos, 2009).

To be more specific, the error-statistical perspective proposes a sequence of interconnected models as a way to bridge the gap between theory and data and to deal with stage A. A post-data evaluation of inference, in the form of severe testing (Mayo, 1996), can be used to deal with stage C. Dealing with stage B, and giving data a substantial role in theory testing, requires an account where *statistical knowledge* (in direct analogy to Mayo's experimental knowledge) has "a life of its own," which stems from the *statistical adequacy* of the estimated model – the probabilistic assumptions constituting the statistical model are valid for the particular data z_0.

Statistical Knowledge. In the context of error statistics, the notion of *statistical knowledge*, stemming from a statistically adequate model, allows data "to have a voice of its own," in the spirit of the proponents of the "third way" (Hendry, 2000), separate from the theory in question, and it succeeds in securing the frequentist goal of objectivity in theory testing. Indeed, the notion of statistical adequacy replaces goodness of fit as *the* criterion for (1) accounting for the regularities in the data, (2) securing the predictive ability of the estimated models, and (3) dealing affectively with the problem of spurious statistical results. Statistically spurious results can be explained away as statistical misspecification without having to invoke any substantive information. Indeed, it is shown that goodness-of-fit measures are themselves untrustworthy when the fitted model is statistically misspecified (Spanos, 1999).

Revisiting the theory–data confrontation in economics using the error-statistical perspective puts us in a position to understand some of the problems raised by contemporary methodological discussions in economics/econometrics and sets the stage for making progress in an area where chronic problems have stumped any significant advancement for almost three centuries.

Outline of the Chapter. Section 2 outlines the error-statistical modeling framework as it relates to theory testing in economics. Section 3 elaborates the notion of how "statistical knowledge" has a life of its own and how it relates to learning from error at different levels of modeling. Section 4 traces the development of theory testing in economics since the early nineteenth century, highlighting the preeminence of theory thesis in the theory–data confrontation and the subordinate role attributed to data. Section 4 discusses how the preeminence of theory has affected the development of econometrics during the twentieth century, and Section 5 sets the stage for where we are at the beginning of the twenty-first century and how the error-statistical modeling framework can provide the foundations and overarching framework for making progress in theory testing in economics.

2 Theory–Data Confrontation in Economics and the Error-Statistical Account

2.1 Philosophy of Science and Theory Testing

Viewing the developments in philosophy of science since the 1930s from the viewpoint of the theory–data confrontation in economics, it becomes clear why the philosophical discourses have left the econometrician hanging. To be more specific, the logical empiricists' perspective of induction, as primarily a logical relationship $L(\mathbf{e}, h)$ between evidence $\mathbf{e}$ – taken as objectively given – and a hypothesis h, essentially *assumes away* the crucial modeling stages A–C facing a practicing econometrician. Moreover, the *inductive logics* of logical empiricists had little affinity for the ways that practicing econometricians understand the theory–data confrontation in the context of the Fisher-Neyman-Pearson (F-N-P) frequentist statistical inference (see Cox and Hinkley, 1974). The post–logical empiricist developments in philosophy of science growing from the recognition of *Duhemian ambiguities, underdetermination,* and the problem of *theory-laden* observation made the problem of theory–data confrontation in economics seem even more hopeless.

2.2 The Error-Statistical Account and Theory Testing

The first promising signs that discussions in philosophy of science could be potentially relevant to issues pertaining to the theory–data confrontation in economics emerged from the "new experimentalist" tradition, which began focusing attention on modeling the processes that generated the raw data $\mathbf{z}_0$ instead of taking the observational facts $\mathbf{e}$ as objectively given.

Using the piecemeal activities involved and the strategies used in successful experiments, Hacking (1983) argued persuasively against the theory-dominated view of experiment in science and made a strong case that in scientific research an experiment can have a "life of its own" that is largely independent of large-scale theory. From the econometric perspective, this move was one in the right direction, but there was still some distance to go to relate it to nonexperimental data. This link was provided by Mayo (1996), who broadened the notion of an experiment:

I understand "experiment," . . . far more broadly than those who take it to require literal control or manipulation. Any planned inquiry in which there is a deliberate and reliable argument from error may be said to be experimental. (p. 7)

In addition, she fleshed out three crucial insights arising from new experimentalism:

1. Understanding the role of experiment is the key to circumventing doubts about the objectivity of observation.
2. Experiment has a life of its own apart from high level theorizing (pointing to a local yet crucially important type of progress).
3. The cornerstone of experimental knowledge is its ability to discriminate backgrounds: signal from noise, real effect from artifact, and so on. (p. 63)

In her attempt to formalize these insights into a coherent epistemology of experiment, she proposed the *error-statistical account*, whose underlying reasoning is based on a refinement/extension of the F-N-P frequentist approach to statistical inference; the name stems from its focus on error probabilities (see Mayo and Spanos, 2009).

Contrary to the Popperian and *Growth of Knowledge* traditions' call for "going bigger" (from theories to paradigms, to scientific research programs and research traditions), to deal with such problems as theory-laden observation, underdetermination, and Duhemian ambiguities, Mayo argued that theory testing should be piecemeal and should "go smaller":

[I]n contrast to the thrust of holistic models, I take these very problems to show that we need to look to the force of low-level methods of experiment and inference. The fact that theory testing depends on intermediate theories of data, instruments, and experiment, and that the data are theory laden, inexact and "noisy," only underscores the necessity for numerous local experiments, shrewdly interconnected. (Mayo, 1996, p. 58)

The error-statistical account can deal with link A – *from theory to testable hypotheses* – by proposing a hierarchy of models – primary (theory), experimental (structural), and data (statistical) models – aiming to bridge the gap between theory and data in a piecemeal way that enables the modeler to "probe and learn from error" at each stage of the modeling (Mayo, 1996, p. 128). It can also deal with link B – *from raw data to reliable evidence* – by allowing data z_0 to have "a voice of its own" in the context of a *validated statistical model*; this provides the observational "facts" (**e**) pertinent for theory appraisal. Moreover, it constitutes a philosophical/methodological framework that has an inherent affinity for the ways practicing econometricians understand the problem in the context of the frequentist inductive inference. Indeed, it addresses link C – *confronting testable hypotheses with evidence* – by providing a coherent framework that deals effectively with the question: When do data z_0 provide evidence for or against *H*?

This question can be answered unambiguously only when the model-based statistical testing associated with F-N-P frequentist inference is supplemented with post-data assessment *severity evaluations* (see pp. 21–22, this volume). This solution provides an inferential construal of tests, based on their capacity to detect different discrepancies with data z_0, and can be used to address the *fallacies* of acceptance and rejection (see Mayo and Spanos, 2006).

The fundamental intuition underlying the error-statistical account of evidence is that if a hypothesis *H* "passes" a test *T* with data z_0, but *T* had very low capacity to detect departures from *H* when present, then data z_0 do not provide good evidence for the verity of *H*; its passing *T* with z_0 is not a good indication that *H* is true. Learning from error, according to Mayo, amounts to deliberate and reliable argument from error based on severe testing: "a testing procedure with an overwhelmingly good chance of revealing the presence of specific error, if it exists – but not otherwise" (1996, p. 7).

Mere fit is insufficient for z_0 to pass *H* severely; such a good fit must be something very difficult to achieve, and so highly improbable, were *H* to be in error. More important, the fact that data z_0 were used to arrive at a good fit with $H(z_0)$ does not preclude counting z_0 as good evidence for $H(z_0)$ – it all depends on whether the procedure for arriving at $H(z_0)$ would find evidence erroneously with very low probability.

As a framework for inductive inference, error statistics enhances the F-N-P framework in several different ways, the most crucial being the following (see Mayo and Spanos, 2009):

1. Emphasize the learning from data objective of empirical modeling.
2. Pay due attention to the validity of the premises of induction.
3. Emphasize the central role of error probabilities in assessing the reliability (capacity) of inference, both predata as well as post-data.

4. Supplement the original F-N-P framework with a post-data assessment of inference in the form of severity evaluations in order to provide an evidential construal of tests.
5. Bridge the gap between theory and data using a sequence of interconnected models: primary, experimental, and data models.
6. Actively encourage a thorough probing of the different ways an inductive inference might be in error by localizing error probing in the context of the different models.

2.3 Error Statistics and Model-Based Induction

The statistical underpinnings of the error-statistical approach are squarely within the frequentist inference framework pioneered by Fisher (1922). He initiated the recasting of statistical induction by turning the prevailing strategy of commencing with the data in search of a descriptive model on its head. He viewed the data $z_0 := (z_1, z_2, \ldots, z_n)$ as a realization of (1) a "random sample" from (2) a prespecified "hypothetical infinite population," formalizing both in purely probabilistic terms. He formalized the notion of a random sample as a set of *independent and identically distributed* (IID) random variables $Z := (Z_1, Z_2, \ldots, Z_n)$ giving rise to data z_0, and the "infinite hypothetical population" in the form of a distribution function $f(z; \theta)$, indexed by a set of unknown parameter(s) θ. He then combined the two to define the notion of a *parametric statistical model*. He defined the problem of *specification* as the initial choice of the statistical model that ensures that the data constitute a "typical realization": "The postulate of randomness thus resolves itself into the question, 'Of what population is this a random sample?'" (Fisher, 1922, p. 313), emphasizing the fact that "the adequacy of our choice may be tested *posteriori*" (p. 314).

The revolutionary nature of Fisher's recasting of induction is difficult to exaggerate because he rendered the vague notions of "uniformity of nature" and "representativeness of the sample" into testable probabilistic assumptions, and his concept of a statistical model provided the cornerstone for a new form of statistical induction that elucidated the ambiguities and weaknesses of the pre-Fisher perspective (see Spanos, 2010).

The frequentist approach to statistics was pioneered by Fisher (1921, 1922, 1925, 1935) and continued with Neyman and Pearson (1928, 1933); the Neyman–Pearson contributions in framing hypothesis testing in model-based inductive terms were particularly crucial for the development of the new paradigm (see also Neyman, 1950, 1977; Pearson, 1966). Although a detailed discussion of this F-N-P frequentist paradigm, spearheaded by *model-based statistical induction*, is beyond the scope of this chapter (see

Spanos, 2006c), it is important to state some of its key features for the discussion that follows.

1. The *chance regularity* (stochasticity, haphazardness) in statistical data is *inherent*.
2. It is a reflection of the probabilistic structure of the processes generating the data.
3. This chance regularity is captured by a *prespecified statistical model* chosen to render the data a truly typical realization of the generic stochastic process specified by this model.
4. Ascertainable error probabilities – based on the prespecified statistical model – provide the cornerstone for assessing the optimality and reliability of inference methods (see Neyman, 1977).

Unfortunately, the relationship between statistical and substantive information has not been adequately addressed by the F-N-P approach and remains an issue to this day (see Cox, 1990; Lehmann, 1990). In the next section it is argued that the sequence of interlinked models mentioned earlier (theory, structural, and statistical) can be used to shed ample light on this issue.

3 Error Statistics and Empirical Modeling in Economics

3.1 "Statistical Knowledge" with "a Life of Its Own"

Spanos (1986) proposed a modeling framework for econometrics, named *probabilistic reduction*, which largely overlaps with the error-statistical account in Mayo (1996) but is lacking the post-data severity component. The most surprising overlap is the sequence of interlinked models that aims to provide a bridge between theory and data as shown in Figure 6.1. Even though the sequence of models in Figure 6.1 (theory, structural, and statistical) were motivated primarily by the problems and issues relating to econometric modeling, a direct analogy exists between these and Mayo's primary, experimental, and data models, respectively. The main difference is one of emphasis, where Spanos (1986) provides a more formal and detailed account of what a statistical model is and how its validity vis-à-vis the data are secured using thorough *misspecification testing* and *respecification* in the context of the F-N-P statistical framework.

A validated statistical model gives rise to *statistical knowledge* with "a life of its own," in direct analogy to Mayo's (1996) *experimental knowledge* (see also Spanos, 1999).

Although a detailed discussion of the more formal aspects of statistical model specification, misspecification testing, and respecification

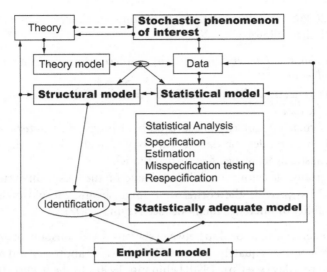

Figure 6.1. Sequence of interlinked models in empirical modeling.

(Figure 6.1), designed to secure a statistically adequate model, are beyond the scope of this chapter (see Spanos, 2006a, 2006b, 2006c), it is important to discuss briefly how the aforementioned interlinked models can be used to deal effectively with bridging over the three interrelated modeling stages A–C mentioned in the introduction.

Using the different models (theory, structural, and statistical) in Figure 6.1 one can deal with link A – *from theory to testable hypotheses* – by framing ways to bridge the gap between theory and data that enmesh the structural and statistical models.

From the theory side of the bridge, one constructs a *theory model* (a mathematical formulation of a theory), which might involve latent variables. In an attempt to connect the theory model to the available data, one needs to transform the theory model into an *estimable* (in light of the data) form, called a *structural model.* The construction of structural models must take into account the gap between the concepts envisaged by the theory model (intentions, plans) and what the available data actually measure (for further details see Spanos, 1995).

With respect to rendering a theory model estimable, the problem facing an economist is not very different from the situation confronting a physicist wishing to test the kinetic theory of gasses. Due to the unobservability of molecules, the physicist resorts to heating the gas to measure the expansion

Table 6.1. *Simple Normal Model*

Statistical GM: $X_k = \mu + u_k, \ k \in \mathbb{N}: = \{1, 2, \ldots, n, \ldots\}$
[1] Normality: $X_k \sim N(.,.)$
[2] Mean homogeneity: $E(X_k): = \mu$
[3] Variance homogeneity: $\text{Var}(X_k): = \sigma^2$
[4] Independence: $\{X_k, k \in \mathbb{N}\}$ is independent

of the gas under constant pressure in order to test an observable implication of the theory in question.

The structural model contains a theory's substantive subject matter information in light of the available data $\mathbf{z}_0$. How does a structural model connect to a statistical model?

Statistical Models. The statistical model $\mathcal{M}_\theta(\mathbf{z})$ is built exclusively using information contained in the data and is chosen in such a way to meet two interrelated aims:

1. to account for the chance regularities in data $\mathbf{z}_0$ by choosing a probabilistic structure for the stochastic process $\{Z_t, \ t \in \mathbb{N}\}$ underlying $\mathbf{z}_0$ to render $\mathbf{z}_0$ a typical realization thereof, and
2. to parameterize this probabilistic structure of $\{Z_t, \ t \in \mathbb{N}\}$ in the form of an adequate *statistical model* $\mathcal{M}_\theta(\mathbf{z})$ that would embed (nest) the structural model in its context.

Objective 1 is designed to deal with link B – *from raw data to reliable evidence* – using the notion of a *validated statistical model*: $\mathcal{M}_\theta(\mathbf{z}) = \{f(\mathbf{z}; \theta), \ \theta \in \Theta\}, \mathbf{z} \in \mathbb{R}^n$, where $f(\mathbf{z}; \theta)$ denotes the joint distribution of the sample $\mathbf{Z}:= (Z_1, Z_2, \ldots, Z_n)$. Formally, $\mathcal{M}_\theta(\mathbf{z})$ can be viewed as a reduction from a generic $f(\mathbf{z}; \theta)$ after imposing a certain probabilistic structure on $\{Z_t, \ t \in \mathbb{N}\}$ that reflects the chance regularity patterns in data $\mathbf{z}_0$ (see Spanos, 1989, 1999).

To give some idea of what this entails, consider the simple Normal model given in Table 6.1, which comprises a statistical generating mechanism, and the probabilistic assumptions 1 through 4 defining a Normal Independent and Identically Distributed (NIID) process. This model, specified in terms of $f(x_k; \theta)$, can be formally viewed as a reduction from the joint distribution as follows:

$$f(x_1, x_2, \ldots, x_n; \varphi) \overset{I}{=} \prod_{k=1}^{n} f_k(x_k; \theta_k) \overset{IID}{=} \prod_{k=1}^{n} f(x_k; \theta)$$
$$\overset{NIID}{=} \prod_{k=1}^{n} \left(\frac{1}{\sigma\sqrt{2\pi}} \exp\left\{ -\frac{(x_k - \mu)^2}{2\sigma^2} \right\} \right)$$

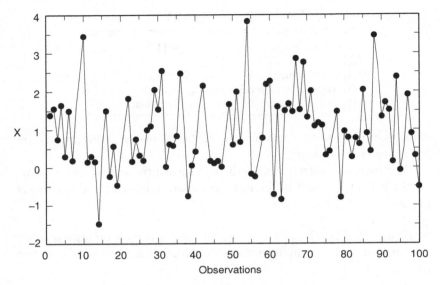

Figure 6.2. A realization of an NIID process.

This is a purely *probabilistic construct* which depends only on the joint distribution of the stochastic process $\{X_k, k \in \mathbb{N}\}$, say $f(\mathbf{x};\theta)$, where the unknown parameters are $\theta := (\mu, \sigma^2)$.

Using analogical reasoning one can infer that the data in Figure 6.2 could be realistically viewed as a realization of an NIID process $\{X_k, k \in \mathbb{N}\}$, but Figure 6.3 cannot because it exhibits cycles that indicate the presence of positive Markov dependence (see Spanos, 1999, ch. 5).

Spurious Correlation/Regression. When $\mathbf{X}_k$ is a vector with m components, say $\mathbf{X}_k := (X_{1k}, X_{2k}, \ldots, X_{mk})$, the statistical model in Table 6.1 becomes the simple *multivariate Normal model*, with assumptions [2]–[3] taking the form $E(\mathbf{X}_k) = \mu$, $\mathrm{Cov}(\mathbf{X}_k) = \Omega$. This model underlies several statistical procedures widely used in the social sciences, including correlation, linear regression, factor analysis, and principal component analysis. The spurious statistical inference results, considered endemic and inevitable when using observational data, can often be explained away as statistically meaningless when any of the model assumptions [1]–[4] are invalid for data $\mathbf{x}_k$, $k = 1, \ldots, n$.

Example: Trending Data. When assumption 2 is false, for example, the data exhibit trends ($E(X_k)$ changes with k) as in Figures 6.4 and 6.5, the sample correlation $\hat{\rho} = \dfrac{\sum_{k=1}^{n}(X_k - \overline{X})(Y_k - \overline{Y})}{\sqrt{\sum_{k=1}^{n}(X_k - \overline{X})^2 \sum_{k=1}^{n}(Y_k - \overline{Y})^2}}$ provides a "bad" (biased,

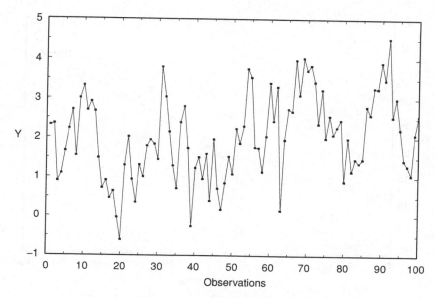

Figure 6.3. A realization of a Normal Markov stationary process.

inconsistent, etc.) estimator of the "population" correlation coefficient: $\rho = \frac{E[X_t - E(X_t)][Y_t - E(Y_t)]}{\sqrt{E[X_t - E(X_t)]^2 \cdot E[Y_t - E(Y_t)]^2}}$, rendering any inference concerning ρ *statistically meaningless.* This occurs because $\overline{X} = (1/n) \sum_{k=1}^{n} X_k$ and $\overline{Y} = (1/n) \sum_{k=1}^{n} Y_k$ are "bad" (inconsistent) estimators of $E(X_k) = \mu_x(k)$ and

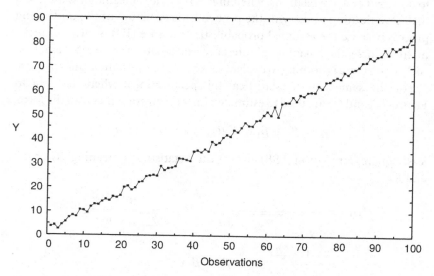

Figure 6.4. A realization of an NI, non-ID process.

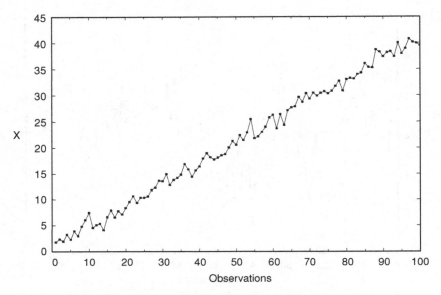

Figure 6.5. A realization of an NI, non-ID process.

$E(Y_k) = \mu_y(k)$, respectively. Intuitively, when the means $(E(X_k), E(Y_k))$ are changing with k as in Figures 6.4 and 6.5, the sample mean deviations $\sum_{k=1}^{n}(X_k - \overline{X})^2$, $\sum_{k=1}^{n}(Y_k - \overline{Y})^2$ (used in evaluating $\hat{\rho}$) are artificially overinflated because they are measured from fixed points $(\overline{X}, \overline{Y})$, giving rise to *apparent* high correlations, when in fact they yield meaningless numbers that have nothing to do with the probabilistic structure of the underlying process because the assumed probabilistic structure (ID) is false. To make matters worse, the reliability of inference exacerbates as n → ∞. Hence, in such a case there is nothing to explain (away) using substantive information.

For the same reason (statistical misspecification), when the data in Figures 6.4 and 6.5 are used to estimate a linear regression model of the form

$$y_k = \beta_0 + \beta_1 x_k + \mu_k, \quad k = 1, \ldots, n,$$

least-squares estimation (LSE) gives rise to a statistically meaningless t-ratio and R^2:

$$\tau(\beta_1) = \frac{\hat{\beta}_1}{s\sqrt{\sum_{k=1}^{n}(x_k - \overline{x})^2}}, \quad R^2 = \frac{(n-2)s^2}{\sum_{k=1}^{n}(y_k - \overline{y})^2},$$

$$s^2 = \frac{1}{n-2}\sum_{k=1}^{n}(y_k - \hat{\beta}_0 - \hat{\beta}_1 x_k)^2,$$

because both statistics $(\tau(\beta_1), R^2)$ involve deviations from constant sample means $(\bar{y}, \bar{x})$, which are inappropriate because their true means are changing with k (see Spanos, 1999).

The prespecified statistical model $\mathcal{M}_\theta(\mathbf{z})$ accounts for the statistical (chance) regularities in the data $\mathbf{z}_0$ only when its probabilistic assumptions are valid for $\mathbf{z}_0$. Validation of a statistical model takes the form of thorough Mis-Specification (M-S) testing of the assumptions comprising it (e.g., assumptions [1]-[4]; see Spanos, 1999, ch. 15). A statistically adequate (validated) model has a "life of its own" in the sense that it accounts for the statistical regularities in data $\mathbf{z}_0$ independently of any substantive information contained in the structural model in question.

Testing Substantive Information. Objective 2, associated with the specification of a statistical model $\mathcal{M}_\theta(\mathbf{z})$, has to do with the embedding of the structural model in its context, connecting the two sides of the theory–data gap and allowing one to assess the validity of the substantive information vis-à-vis data $\mathbf{z}_0$. The main objective of the embedding is to render the structural (estimable) model a special case (reparametrization/restriction) of the statistical model. Formally this process takes the form of structural parameter *identification*: the structural parameters φ are *uniquely* defined by the statistical parameters θ. Usually, more statistical parameters exist than structural parameters, which enables one to parameterize the substantive information in the form of *overidentifying restrictions*:

$$H_0: \mathbf{G}(\varphi, \theta) = 0 \text{ vs. } H_1: \mathbf{G}(\varphi, \theta) \neq 0,$$

which can be tested in the context of the statistical model in question, $\mathcal{M}_\theta(\mathbf{z})$ (Spanos, 1990).

How these restrictions can be properly appraised "evidentially" is the concern of modeling stage (C)– *confronting testable hypotheses with evidence*. When the substantive information in h passes a severe test in the context of a validated statistical model (e), it pertains to the substantive adequacy of the structural model vis-à-vis data $\mathbf{z}_0$. Then one can impose h to fuse the substantive and statistical information in the context of an empirical model (see Figure 6.1). The empirical model inherits its statistical and substantive meaningfulness from the statistical and structural models, respectively.

Statistical Knowledge. Of crucial importance is that *a statistically adequate model*

1. is independent (separate) from the substantive information, and, therefore,

2. can be used to provide the broader inductive premises for evaluating
   its adequacy.

The independence stems from the fact that the statistical model is erected
on purely probabilistic information by capturing the "chance regularities"
exhibited by data z_0. In Figures 6.1–6.5, the chance regularities have nothing
to do with any substantive subject matter information because data z_0 are
viewed as a realization of a *generic* stochastic process $\{Z_t, \ t \in \mathbb{N}\}$, irrespective of what substantive variable Z_t quantifies. In this sense, a statistically
adequate model provides a form of *statistical knowledge*, analogous to what
Mayo calls *experimental knowledge*, against which the substantive information is appraised. As remarked by Mayo (1996), the sequence of models
(theory, structural, statistical) performs double duty: "it provides a means
for spelling out the relationship between data and hypotheses – one that
organizes but does not oversimplify actual scientific episodes – and organizes the key tasks of a philosophy of experiment" (pp. 129–30). The notion
of "statistical knowledge," like experimental knowledge, is instrumental in
discriminating backgrounds: "signal from noise, real effect from artifact,
and so on" (p. 63). The notion of statistical knowledge formalizes the sense
in which data z_0 has a "voice of its own," separate from the one ideated
by a theory. This notion is crucial for theory testing because it ensures the
reliability and scientific objectivity of the theory–data confrontation.

3.2 Probing for Different "Local" Errors

The typology of different models, as exemplified in Figure 6.1, was designed
to delineate questions concerning different types of errors and to render their
probing much more effective than the traditional approach of lumping them
together in one overall error term. These models can be used to illustrate the
error-statistical emphasis on a thorough probing of the different ways an
inductive inference might be in error, by localizing the error probing in the
context of the different models (see Mayo, 1996). The reliability of evidence
is assessed at all levels of modeling by using error-statistical procedures
based on learning from error reasoning. When any of these potential errors
are present, the empirical evidence is rendered untrustworthy.

In the context of a *statistical model*, the primary sources of error are
inaccurate data and statistical misspecification. Let us briefly discuss the two.

1. *Inaccurate data:* data z_0 are marred by systematic errors imbued by
the collection and compilation process. The inaccuracy and inadequacies
of economic data as of the late 1950s has been documented by Morgenstern

(1950/1963). Since then, the accuracy of economic data has been improving steadily, but not enough attention has been paid to issues concerning how the collection, processing, and aggregation of data in economics might imbue systematic errors that can undermine their statistical analysis (see Abadir and Talmain, 2002).

2. *Statistical misspecification:* some of the probabilistic assumptions comprising the statistical model (premises of induction) are invalid for data z_0.

A misspecified statistical model gives rise to unreliable inferences because the actual error probabilities are very different from the nominal ones – the ones assumed to hold if the premises are true. Applying a .05 significance-level t-test, when the actual type I error is .95, renders the test highly unreliable (see Spanos, 2005).

In the context of a *structural (estimable) model*, the relevant source of error is:

3. *Incongruous measurement:* data z_0 do not measure the concepts ξ envisioned by the particular theory model (e.g., intentions vs. realizations; see Spanos, 1995).

This issue is fundamental because typically theory models are built on intentions on behalf of economic agents but the available data measure realizations of ongoing convoluted processes. Moreover, the overwhelming majority of economic data are collected by government agencies and private institutions for their own purposes and not by the modelers themselves.

In the context of the *theory* (and *empirical) model*, the primary source of error is:

4. *Substantive inadequacy:* the circumstances envisaged by the theory differ "systematically" from the actual Data-Generating Mechanism (DGM). This might involve investigating the validity of causal claims of *ceteris paribus* clauses, as well as searching for missing confounding factors (see Guala, 2005; Hoover, 2001a). Substantive adequacy concerns the extent to which the empirical model "captures" the aspects of the reality it purports to explain, shedding light on the phenomenon of interest (i.e., "learning from data"). It concerns the extent to which the inferences based on the empirical model are germane to the phenomenon of interest. This issue is particularly relevant when modeling with experimental data.

3.3 Experimental Economics and Theory Testing

The notion of experimental knowledge and the idea of an experiment having a "life of its own" have recently been introduced into experimental economics in an attempt to provide sound philosophical underpinnings to endeavors to test economic theories by designing and conducting experiments pertaining to economic behavior under laboratory conditions (see Guala, 2005). Some notable pioneers in experimental economics, like Vernon Smith (2007, ch. 13), came to appreciate that the error-statistical perspective provides a promising framework wherein many of the issues experimental economists have been grappling with can find a natural home and a number of systematic ways of being addressed. After declaring that "[e]xperimental knowledge drives the experimental method" (p. 305), he proclaims: "My personal experience as an experimental economist since 1956 resonates well with Mayo's (1996) critique of Lakatos" (Smith, 2007, p. 311). Similarly, Sugden (2008) acknowledged that "the account of scientific method that comes closest to what I have in mind is that of Deborah Mayo (1996)."

Viewed in the context of the error-statistical perspective discussed earlier, experimental economics attempts to bridge the theory–data gap by bringing the data closer to the theory (see Figure 6.1). Instead of using observational data, which are the result of a highly complicated ongoing actual DGM that involves numerous influencing factors, one generates experimental data under "controlled" conditions that aim to correspond closely to the conditions envisaged by economic theory. These data can then be used to evaluate theoretical predictions concerning economic behavior. In this sense, the structural (estimable) model – in view of the data – is often much closer to the theory model than in the case of observational data. Despite this *important difference*, the statistical modeling of experimental data in economics – with a view to learn about economic phenomena – raises the same issues and problems of reliability as for observational data, where the primary differences are in emphasis.

To be more specific, the same sequence of interconnecting models (theory, structural, statistical, and empirical; see Figure 6.1) are equally elucidating when experimental data is modeled, which raises very similar issues as pertains to probing for errors at different levels of modeling and the resulting reliability of inductive inferences. In particular, the various "canonical" errors (1–4), mentioned earlier, are no less applicable to experimental data, but with differences in emphasis. For example, (1) *inaccurate data* are easier to address directly in experimental economics because the experimenter, the data gatherer, and modeler are usually one and the same person (or group). Moreover, hypotheses of interest in experimental economics are usually rendered testable by embedding them into a *statistical model* (implicitly or explicitly), raising the problem of (2) *statistical misspecification*. The fact that the data are the result of a "controlled" experiment makes it *easier* in practice to ensure that the data can be realistically viewed as a realization of a random sample (IID), but that in itself does *not* suffice to secure the *statistical adequacy* of the estimated model and, thus, the statistical reliability of inference. Experimental design techniques, such as *randomization* and *blocking*, are applied in particular situations, but statistical adequacy – secured by thorough Mis-Specification (M-S) testing – provides the only way to establish the absence of systematic effects in a particular data. Similarly, the problem of (3) *incongruous measurement* is an issue that is easier to deal with in the context of generating one's own experimental data.

On the other hand, the problem of (4) *substantive inadequacy* is often more difficult to address when modeling experimental data because one needs to demonstrate that the inferences drawn from the experimental data (generated in a laboratory) are germane to, and shed light on, the phenomenon of interest. This problem is particularly critical in experimental economics, where it has unfolded in a variety of guises. Guala (2005) provides an illuminating discussion of these issues under the banner of "external validity": legitimately generalize inferences based on experimental data – generated in a laboratory's artificial setup – to the real-world phenomenon in question. This issue lies at the heart of the problem of theory testing in economics because in the absence of external validity it is not obvious how one would assess the ability of the theory in question to explain (or shed light on) the phenomenon of interest.

A particularly interesting methodological issue that has arisen in experimental economics concerns several "exhibits" that have been accumulated over the past decade or so and are considered "anomalies" for mainstream economic theory (see Sugden, 2005, 2008). These exhibits are essentially viewed as inductively established "replicable empirical effects" that, in the eyes of experimental economists, constitute "experimental knowledge." As argued by Mayo (1996, 2008), establishing experimental knowledge requires a thorough probing of all the different ways such claims can be in error at all levels of modeling. In particular, replicability by itself does not suffice in such a context because, for instance, all the experiments conducted may ignore the same confounding factor giving rise to the detected exhibit. In addition, statistical adequacy constitutes a necessary condition to eliminate the possibility that the "exhibit" in question is *not* a statistical artifact. Moreover, going from statistical significance to substantive significance raises crucial methodological issues (the fallacies mentioned earlier), which need to be addressed before such exhibits can qualify as experimental knowledge.

4 Theory–Data Confrontation before the Mid-Twentieth Century

4.1 Theory–Data Confrontation in Classical Economics

It is generally accepted that *economics* (originally called *political economy*) was founded as a separate scientific discipline by Adam Smith (1776), who envisioned the intended scope of political economy to be concerned with causes and explanations underlying economic phenomena. His method was an eclectic modification and adaptation of the "Newtonian method," as it was understood at that time, with a strong empiricist bend blended with the theoretical, institutional, and historical dimensions (see Redman, 1997). "With Ricardo (1817) economics took a major step toward abstract models, rigid and artificial definitions, syllogistic reasoning – and the direct application of the results to policy" (see Sowell, 2006, p. 80). In particular, the notion of a "theory" in economics was understood as a system of propositions deductively derived from certain premises defined in terms of particular postulates (principles). One of the earliest examples of *initial postulates* was given by Senior (1836), who claimed that political economy is ultimately based only on four general postulates: "(1) every man desires to maximize wealth with as little sacrifice as possible, (2) population is limited by the available resources, (3) capital enhances the productivity of labor, and (4) agriculture exhibits diminishing returns" (p. 26).

An early example of *economic theory propositions* was given by Ricardo (1817) and concerned the "long-run" equilibrium tendencies in key variables: (1) the price of corn would tend to increase, (2) the share of rent in the national income would tend to go up, (3) the rate of profit on capital would tend to fall, and (4) the real wages would tend to stay constant.

Methodological differences among political economists during the nineteenth century were mainly focused on the relationship between the initial postulates and the deductively derived propositions, on one side, and real-world data, on the other. On one side, Malthus (1836) argued in favor of anchoring the initial postulates on observation and then confronting the derived propositions with real-world data. On the other side, Ricardo (1817) considered the initial postulates as self-evident truths, not subject to empirical scrutiny, and viewed the relationship between theory and data as based on one's "impressions" concerning the broad agreement between the deductively derived propositions and "tendencies" in historical data. It is important to stress that economic data began to be collected systematically from the mid-seventeenth century (see Porter, 1836).

Political economy, from its very beginnings, was entangled in philosophical and methodological discussions concerning the nature and structure of its theories and methods. Methodological discussions concerning induction versus deduction, the logical structure of economic theories, the status of economic postulates and deductively inferred propositions, and the verification and testing of such postulates and propositions were considered an integral part of the broader discourse concerning "learning about economic phenomena of interest" (see Redman, 1997). Outstanding examples of such discussions during the nineteenth century include Malthus (1936), Senior (1836), McCulloch (1864), Mill (1844, 1871), Cairnes (1888), and Keynes (1890). Political economists who are equally well-known as philosophers include Jevons (1871) and Mill (1884).

During the first half of the nineteenth century, the idea of observation as a collection of "facts" was very unclear and fuzzy because real-world data usually come in the form of historical data series, which include numerous observations on several different variables without any obvious way to summarize them into a form that could be used for theory appraisal. The descriptive statistics tradition did not form a coherent body of knowledge until the end of the nineteenth century, and even then it did not provide a way to deal effectively with modeling stages B and C (see p. 205). Viewed retrospectively, the theory–data confrontation during the eighteenth and nineteenth centuries amounted to nothing more than pointing out that one's "impressions" concerning "tendencies" in historical data do *not* seem to contradict the deductively derived propositions in question.

This state of affairs was clearly highly unsatisfactory, rendering theory testing against data an unavailing exercise with very little credibility. Indeed, apparent empirical falsification did not seem to have any negative effect on the credibility of certain economic theories. As argued by Blaug (1958), when Ricardo's propositions (a)–(d) are appraised in terms of their observational implications, they constitute a dismal failure; none of them was borne out by the historical data up to the 1850s. Despite their empirical inadequacy, Ricardo's theories are viewed as a great success story by the subsequent economics literature to this day: "(1) As a theory and method, the Ricardian paradigm provided the basis for further advances in logical deduction, and (2) it also had the practical value of providing a basis for economic policy" (see Ekelund and Hebert, 1975, p. 110). The clarity and economy of the logical deduction from a few initial postulates, to the neglect of the role of the data, captured the imagination of the overwhelming majority of political economists during the second half of the nineteenth century.

Mill (1844, 1871) was instrumental in shaping a methodological framework for Ricardo's deductive turn, which justified the emphasis on *deductively derived propositions* and the subordinate role attributed to data. He argued that causal mechanisms underlying economic phenomena are too complicated – they involve too many contributing factors – to be disentangled using *observational data*. This concept is in contrast to physical phenomena whose underlying causal mechanisms are not as complicated – they involve only a few dominating factors – and the use of *experimental data* can help to untangle them by "controlling" the "disturbing" factors. Hence, economic theories can only establish *general tendencies* and not precise enough implications whose validity can be assessed using observational data. These tendencies are framed in terms of the primary causal contributing factors with the rest of the numerous (possible) disturbing factors relegated to *ceteris paribus* clauses whose appropriateness cannot, in general, be assessed using observational data: empirical evidence contrary to the implications of a theory can always be explained away as being caused by counteracting disturbing factors. As a result, Mill (1844) rendered the theory–data gap unavoidable and attributed to the data the auxiliary role of investigating the *ceteris paribus* clauses to shed light on the unaccounted by the theory disturbing factors that prevent the establishment of the tendencies predicted by the theory in question.

The subordinate role of data in theory appraisal was relegated even further by Cairnes (1888), who pronounced data more or less irrelevant for appraising the truth of deductively established economic theories. He turned the weaknesses acknowledged by Mill on their head and claimed that the focus on the deductive component of economic modeling, and the gap between theory and data, were in fact strengths not weaknesses. His argument was that given the "self-evident truth" of the initial postulates, and the deductive validity of the propositions that follow, the question of verification using data does not even arise because the truth of the premises ensures the truth of the conclusions! As a result, so the argument goes, the deductive nature of economic theories bestows upon them a superior status that even physical theories do not enjoy. The reason, allegedly, is that the premises of Newtonian mechanics are not "self-evident truths," as in economics – established introspectively via direct access to the ultimate causes of economic phenomena – but are mere inductive generalizations that must rely on experimentation and inductive inferences, which are known to be fallible (see Cairnes, 1888, pp. 72–94).

Keynes (1890) made an attempt to summarize the methodological discussions, and to synthesize the various views, that dominated political economy during the nineteenth century. His synthesis appears to be closer to Mill's position, with an added emphasis on the potential role of statistics in theory assessment. Despite these pronouncements, his discussion sounded more like lip service and did nothing to promote the use of statistics in theory appraisal vis-à-vis data.

4.2 Theory–Data Confrontation in Neoclassical Economics

The *marginalist revolution* in economics, which began in the 1870s, brought with it a mathematical turn in economic theorizing which gave rise to economic models (Bowley, 1924) based on optimization at the level of the individual agent (consumer or producer; see Backhouse, 2002). Marshall (1891), in a most influential textbook, retained Mill's methodological stance concerning the preeminence of theory over data in economic theorizing, and he demonstrated how neoclassical economics can be erected using marginal analysis based on calculus.

During the first part of the twentieth century, the majority of economists, led by Robbins (1932), reverted to Cairnes' extreme methodological position, which pronounced data more or less irrelevant for appraising the truth of deductively established propositions. Robbins was aware of the development of statistical techniques since the 1870s but dismissed their application to theory appraisal in economics on the basis of the argument that such techniques are only applicable to data that can be considered "random samples" from a particular population. Because no experimental data existed in economics that could potentially qualify as such a "random sample," the statistical analysis of economic data had no role to play in theory assessment. This argument stems from ignorance concerning the applicability and relevance of modern statistical methods, but unfortunately this ignorance lingers on to this day (Mirowski, 1994). Fortunately, Robbins lived long enough to regret his assertions against the use of statistics in theory assessment, describing them as exaggerated reactions to the claims of institutionalists and "crude" econometricians like Beveridge (see Robbins, 1971, p. 149).

These apparently extreme positions reflect the methodological/philosophical presuppositions thought to underlie and justify the mainstream economic theorizing and modeling of the period. Actual or perceived characteristics (and limitations) of theory and of data in economics are often used as the basis for erecting a conception of the goals of economic methodology that may be satisfied by a science with those features.

Despite several dissenting voices like Hutchison (1938), who used ideas from logical positivism and Karl Popper to restate Malthus's thesis that both the initial postulates and the deductively derived propositions should be subjected to empirical scrutiny, the prevailing view in economics during the first half of the twentieth century adopted the preeminence-of-theory viewpoint and gave data the subordinate role of availing the "quantifying of theoretical relationships." The problem, as it was seen at the time, was that economic theory models, derived on the basis of self-evident postulates, are necessarily valid *if only* one can verify them by bestowing "empirical content" onto them using the "right" data and the appropriate statistical techniques.

In the 1940s, methodological discussions in economics focused primarily on the realism of the initial postulates. An example was whether the assumption that firms "maximize profits" is realistic or not (see Machlup, 1963). Friedman (1953) settled that issue for economists by arguing that the realism of a theory's initial postulates is irrelevant, and the success of a theory should be solely assessed in terms of its empirical predictive ability. Its appeal among practicing economists can be explained by the fact that Friedman's basic argument justified what most economists were doing, including the preeminence of theory over data: "we cannot perceive 'facts' without a theory" (Friedman, 1953, p. 34). The paper generated a huge literature, both in the methodology of economics (Caldwell, 1984; Maki, 2009) and in the philosophy of science (Musgrave, 1981; Nagel, 1963).

5 Early Econometrics and Theory Testing

5.1 Early Pioneers in Econometrics

By the early twentieth century, the descriptive statistics tradition (see Mills, 1924) – with the development of new techniques such as regression, correlation, and related curve-fitting methods based on least-squares associated with Galton, Pearson, and Yule – offered economists more powerful tools, in addition to a statistical framework, for theory appraisal in economics.

It is not surprising that the first attempts at theory appraisal vis-à-vis real-world data focused primarily on the *demand/supply* model, which was popularized by Marshall's (1891) influential textbook and was seen as the cornerstone of the newly established neoclassical tradition. These early empirical studies of demand and supply (see Hendry and Morgan, 1995; Morgan, 1990), when viewed retrospectively from the error-statistical perspective, seem inexorably naïve. They seem to underappreciate the enormity of the gap between the theory (intentions to buy or sell corresponding to hypothetical prices) and the available data on quantities transacted and the corresponding prices (see also Spanos, 1995). Moreover, they appear to be unaware of the numerous possible errors, inaccurate data, statistical misspecification, incongruous measurement, and substantive inadequacy raised in Section 3.2 (see also Spanos, 2006a). As argued later in this chapter, these weaknesses can be explained by the fact that (1) these empirical studies were dominated by the preeminence-of-theory viewpoint, which considered theory appraisal as simply the "quantification of theoretical relationships" using data, and (2) the statistical framework associated with the descriptive statistics tradition (see Bowley, 1926; Mills, 1924) was inadequate for the task.

5.2 Ragnar Frisch and the Econometric Society

Notwithstanding the declarations of the founding statement of the Econometric Society in 1930 (Frisch, 1933, p. 106), the newly established econometric community viewed the theory–data confrontation as the "quantification of theoretical relationships" in economics: theory provides the structural relationships and data avail their quantification. This viewpoint went so far as to question the value of the newly established model-based approach to statistical induction associated with the Fisher-Neyman-Pearson (F-N-P) sampling theory methods because it was thought to be inextricably bound with *agricultural experimentation*. It was generally believed that these methods are relevant only for analyzing "random samples" from experimental data (see Frisch, 1934, p. 6). The prevailing viewpoint was that economic phenomena

1. are *not* amenable to the "experimental method,"
2. are influenced by numerous potential factors – hence, the *ceteris paribus* clause –
3. are intrinsically heterogeneous (spatial and temporal variability), and
4. involve data that are often vitiated by errors of measurement.

The intention was to develop a statistical framework that asserts the preeminence of theory perspective and could account for the aforementioned perceived features 1–4. This attitude led to an eclectic approach to statistical modeling and inference that was based on the pre-Fisher descriptive statistics paradigm but supplemented with *Quetelet's scheme* (1842). This scheme viewed data that exhibit randomness (chance regularity) as comprising two different components:

Q1 a *systematic (deterministic) component* (constant causes), determined by *substantive information,* and
Q2 a random part which represents the *non-systematic error* (accidental causes) *component* (see Desrosières, 1998).

This viewpoint is clearly at odds with the F-N-P perspective, which considers the stochastic nature of data to be inherent (see Section 2.3). However, for economists, the descriptive statistics framework in conjunction with the Quetelet scheme had an alluring appeal because (1) it preserves the preeminence of theory in the form of deterministic models, and (2) it offers a general way to use modern statistical inference techniques by attaching IID errors to structural models. Indeed, *tacking IID error terms* to structural models (3) renders the distinction between a statistical and a structural model virtually redundant. The prevailing view in the 1930s was that the "quantification of theoretical relationships" using data was unproblematic because the cogency of the theory would ensure statistical validity. Frisch (1934) was the first to implement explicitly the *error-tacking strategy* in econometrics. He argued that observable variables (X_{kt}, $k = 1, 2, \ldots, m$) in economics can be decomposed into a *latent systematic* (deterministic) component μ_{kt} and a *random white-noise* error ε_{kt}:

$$X_{kt} = \mu_{kt} + \varepsilon_{kt}, \quad k = 1, 2, \ldots, m, \tag{1a}$$

known as the *errors-in-variables* (or errors-of-measurement) scheme, where

$$[i]\ E(\varepsilon_{kt}) = 0, \quad [ii]\ E(\varepsilon_{kt}^2) = \sigma_k^2, \quad [iii]\ E(\varepsilon_{kt}\varepsilon_{js}) = 0,$$
$$k \neq j, \quad j, k = 1, \ldots, m, \quad t \neq s, \quad t, s = 1, 2, \ldots, n. \tag{1b}$$

Economic theory furnishes the deterministic relationships among the (latent) systematic components in the form of the system of m potential equations:

$$\alpha_{1j}\mu_{1t} + \alpha_{2j}\mu_{2t} + \cdots + \alpha_{mj}\mu_{mt} = 0, \ j = 1, 2, \ldots, m. \tag{2}$$

Combining (1) and (2) gives rise to the formulation:

$$\alpha_{1j}X_{1t} + \alpha_{2j}X_{2t} + \cdots + \alpha_{mj}X_{mt} = \varepsilon_t, \quad t = 1, 2, \ldots, n. \tag{3}$$

Although this can be written as a regression model, such a perspective is incorrect because it ignores the particular structure of the error term: $\varepsilon_t = \alpha_{1j}\varepsilon_{1t} + \alpha_{2j}\varepsilon_{2t} + \cdots + \alpha_{mj}\varepsilon_{mt}$. Hence, Frisch proposed a novel solution to the estimation problem in the form of "curve fitting" (which he called *confluence analysis*) as a problem in vector space geometry. The idea was to determine $r \leq m$, the number of (exact) theoretical relationships among m latent variables:

$$\alpha_{1j}\mu_{1t} + \alpha_{2j}\mu_{2t} + \cdots + \alpha_{mj}\mu_{mt} = 0, \quad j = 1, 2, \ldots, r, \tag{4}$$

using the corank of the data matrix $\mathbf{X} := [x_{kt}]_{k=1,\ldots,m}^{t=1,\ldots,n}$ (see Kalman, 1982).

Frisch's scheme (1)–(4) gave rise to a statistical formulation that is clearly different from traditional regression. The idea was that such models distinguished empirical research in economics from other fields, including mainstream statistics, to custom-tailor statistical inference and modeling for economic phenomena and nonexperimental data.

Viewed retrospectively from the error-statistical perspective of Section 3, the Frisch scheme had little chance to succeed as a general modeling strategy that would give rise to learning from the data about economic phenomena because:

1. it embraced the preeminence of theory over data perspective and focused almost exclusively on the "quantification of theoretical relationships";
2. the bridging of the theory–data gap using the error-tacking strategy was clearly much too simplistic to provide adequate answers to this crucial problem;
3. with exception of measurement error, it ignored the other possible errors, including incongruous measurement, statistical misspecification, and substantive inadequacy; and
4. its highly restrictive probabilistic perspective (linear/homoskedastic/static systems) and its statistical framework based on the Quetelet scheme were totally inadequate in dealing with the modeling stages A through C that were discussed in Section 1.

5.3 Trygve Haavelmo and Frequentist Inference

The first systematic attempt to introduce the F-N-P approach into econometrics was made by Haavelmo (1944). In this classic monograph he argued fervently in favor of adopting the new statistical inference and addressed the concern expressed by Robbins (1932) and Frisch (1934) that such methods were only applicable to cases where the data can be viewed as "random samples" from static populations. The development of the theory of stochastic processes – a sequence of dated random variables, say $\{X_t, \ t \in \mathbb{N}\}$ – by Kolmogorov and Khintchin in the early 1930s (see Doob, 1953) was both crucial and timely because it significantly extended the intended scope of the F-N-P approach beyond the original IID frame-up.

The Haavelmo (1944) monograph constitutes the best example of viewing the confrontation between theory and data in the context of bridging of the gap between theory and data, where both the theory and the data are accorded a "life of their own." As argued in Spanos (1989), it contains a wealth of methodological insights, which, unfortunately, had no impact on the subsequent developments of econometrics.

5.4 The Cowles Commission Modeling Perspective

The part of Haavelmo's monograph that had the greatest impact on the subsequent literature was his proposed technical "solution" to the *simultaneity problem*. This was considered a major issue in economics because the dominating theory – general equilibrium – gives rise to multiequation systems, known as the *Simultaneous Equations Model* (SEM); see Hood and Koopmans (1953).

To bring out the essence of the simultaneity problem, consider a simple but typical theory-based *structural model* that concerns the behavior of two endogenous variables, y_{1t} and y_{2t}:

$$y_{1t} = \alpha_{10} + \alpha_{11} y_{2t} + \alpha_{12} x_{1t} + \alpha_{13} x_{2t} + \alpha_{14} x_{3t} + \varepsilon_{1t} \qquad (5)$$

$$y_{2t} = \alpha_{20} + \alpha_{21} y_{1t} + \alpha_{22} x_{1t} + \alpha_{23} x_{4t} + \alpha_{24} x_{5t} + \varepsilon_{2t}, \qquad (6)$$

where y_{1t} is money in circulation, y_{2t} is interest rate, x_{1t} is price level, x_{2t} is aggregate income, x_{3t} is consumer credit outstanding, x_{4t} is government expenditure, and x_{5t} is balance of payments deficit. What renders the formulation in (5) and (6) so different from Frisch's scheme (1)–(3), as well as multivariate regression, is the *feedback* (co-determination) between the endogenous variables y_{1t} and y_{2t}: interest influences money and vice versa at each point in time; they appear on both sides of the equations. The structural errors are assumed to have the following probabilistic structure:

$$\begin{pmatrix} \varepsilon_{1t} \\ \varepsilon_{2t} \end{pmatrix} \sim \text{NIID} \left(\begin{pmatrix} 0 \\ 0 \end{pmatrix} \begin{pmatrix} \omega_{11} & \omega_{12} \\ \omega_{12} & \omega_{22} \end{pmatrix} \right), \qquad t = 1, 2, \ldots, n. \qquad (7)$$

The perspective underlying the specification of (5)–(7) constitutes a variant of Quetelet's scheme, (Q1) and (Q2) (deterministic component plus random error), but replaces Frisch's errors in variables (1) with errors in equations (5) and (6). The problem, as perceived by the Cowles group at the time, was simply technical and concerned how one can use the data $Z_0 := (Z_1, Z_2, \ldots, Z_n)$, where $Z_t = (y_{1t}, y_{2t}, x_{1t}, x_{2t}, \ldots, x_{5t})$ to avail the quantification of the structural parameters $\alpha = (\alpha_{ij}, i = 1, 2; j = 1, \ldots, 4)$, in the form of *consistent estimators*. It was known at the time that using least squares to estimate the structural parameters α would yield *biased* and *inconsistent estimators* due to the presence of simultaneity (co-determination); y_{1t} affects y_{2t} and vice versa.

Haavelmo (1943, 1944) proposed a way to construct "good" estimators for α using data Z_0. One way to understand his structural estimation method is to view it in conjunction with the so-called *reduced form* model, which arises by "solving" (5) and (6) for (y_{1t}, y_{2t}):

$$y_{1t} = \beta_{01} + \beta_{11}x_{1t} + \beta_{12}x_{2t} + \beta_{13}x_{3t} + \beta_{14}x_{4t} + \beta_{15}x_{5t} + u_{1t} \tag{8}$$

$$y_{2t} = \beta_{02} + \beta_{21}x_{1t} + \beta_{22}x_{2t} + \beta_{23}x_{3t} + \beta_{24}x_{4t} + \beta_{25}x_{5t} + u_{2t}, \tag{9}$$

where the reduced-form errors are assumed to have the following probabilistic structure:

$$\begin{pmatrix} u_{1t} \\ u_{2t} \end{pmatrix} \sim \text{NIID}\left(\begin{pmatrix} 0 \\ 0 \end{pmatrix} \begin{pmatrix} \sigma_{11} & \sigma_{12} \\ \sigma_{12} & \sigma_{22} \end{pmatrix} \right), \quad t = 1, 2, \ldots, n. \tag{10}$$

The "solving" itself gives rise to an implicit relationship between the structural and statistical parameters, α and $\beta = (\beta_{ij}, i = 1, 2; j = 1, \ldots, 5)$, known as *identification restrictions*: $G(\alpha, \beta) = 0$.

Because the model of equations (8)–(10) is just a *multivariate linear regression model*, the least-squares estimators $\hat{\beta}$ are both *unbiased* and *consistent* for β, and, thus, solving $G(\alpha, \hat{\beta}) = 0$ for α gives rise to consistent estimators. In the preceding example, the *identification restrictions* take the following form:

$$\beta_{11} = \beta_{21}\alpha_{11} + \alpha_{12}, \quad \beta_{12} = \beta_{22}\alpha_{11} + \alpha_{13}, \quad \beta_{13} = \beta_{23}\alpha_{11} + \alpha_{14},$$
$$\beta_{14} = \beta_{24}\alpha_{11}, \quad \beta_{15} = \beta_{25}\alpha_{11} \tag{11a}$$

$$\beta_{21} = \beta_{11}\alpha_{21} + \alpha_{22}, \quad \beta_{24} = \beta_{14}\alpha_{21} + \alpha_{23}, \quad \beta_{25} = \beta_{15}\alpha_{21} + \alpha_{24},$$
$$\beta_{22} = \beta_{12}\alpha_{21}, \quad \beta_{23} = \beta_{13}\alpha_{21}. \tag{11b}$$

Viewed retrospectively from the error-statistical perspective of Section 3, the preceding SEM had little chance to succeed as a way to learn from the data about economic phenomena for the same reasons, 1–4, as the Frisch scheme. In particular, when the SEM is viewed from the error-statistical perspective of Figure 6.1, the reduced-form model is the (implicit) statistical model in the context of which the structural model is embedded; it constitutes a reparameterization/restriction. This observation suggests that, when the former is statistically misspecified, any inference based on it, including the estimation of the structural parameters, is likely to be unreliable. Moreover, the adequacy of the structural model requires, in addition to the statistical adequacy of the reduced form, the appraisal of the *overidentifying restrictions*:

$$H_0: \mathbf{G}(\boldsymbol{\alpha}, \boldsymbol{\beta}) = 0 \text{ vs. } H_1: \mathbf{G}(\boldsymbol{\alpha}, \boldsymbol{\beta}) \neq 0.$$

These restrictions are over and above the ones needed for identifying $\boldsymbol{\alpha}$. In the preceding example, there are two such restrictions because there are ten statistical parameters in model (8)–(10) but only eight structural parameters in model (5)–(7) (see Spanos, 1986, 1990).

The agenda of the Cowles Commission (see Koopmans, 1950) held the promise to address the inference problems associated with the SEM. However, when its proposed inference tools were applied to economic data in the early 1950s, the results were very disappointing in terms of accounting for the regularities in the data and yielding accurate predictions and/or policy evaluations (see Christ, 1951; Klein, 1950). Partly as a result of these disappointing results, some of the protagonists, including Frisch, Haavelmo, and Koopmans, moved away from econometrics into other areas of economics (see Epstein, 1987).

6 Theory–Data Confrontation During the Second Half of the Twentieth Century

6.1 Textbook Econometrics and Its Problems

The prevailing view of the theory–data confrontation was dominated by the preeminence-of-theory perspective that led to the conception that econometrics is concerned with the "quantification of theoretical relationships" or, equivalently, "to give empirical content to a priori reasoning in economics" (see Klein, 1962, p. 1). In view of this, econometricians of that period embraced the descriptive statistics perspective in conjunction with the error-tacking strategy to statistical modeling. Hence, the textbook approach to econometrics, as formulated by Johnston (1963) and Goldberger (1964), can be seen as a continuation or modification of the Cowles Commission agenda. The modifications came primarily in the form of (1) less emphasis on simultaneity and (2) less rigidity about statistical modeling by allowing non-IID error terms.

The textbook approach to econometrics revolves around the linear regression model:

$$y_t = \beta_0 + \boldsymbol{\beta}_1^\top \mathbf{x}_t + u_t, \quad t = 1, 2, \ldots, n, \tag{12}$$

$$[i]\ E(u_t) = 0, \quad [ii]\ E(u_t^2) = \sigma^2, \quad [iii]\ E(u_t u_s) = 0,$$
$$t \neq s,\ t, s = 1, 2, \ldots, n. \tag{13}$$

This statistical model is viewed from the preeminence-of-theory perspective as based on a *theoretical relationship,* $y_t = \beta_0 + \boldsymbol{\beta}_1^\top \mathbf{x}_t$, where a tacked on white-noise error term u_t represents random causes such as disturbing factors in the *ceteris paribus* clause, approximation errors, measurement errors, and so on. The underlying statistical theory revolves around the Gauss–Markov theorem, which turns out to be totally inadequate for reliable and precise inference (see Spanos, 2006a).

Anemic Confirmation. Theory appraisal takes the form of *anemic confirmation* in the following sense. We would estimate (12) by least squares, which yields $y_t = \hat{\beta}_0 + \hat{\boldsymbol{\beta}}_1^\top \mathbf{x}_t + \hat{u}_t$, and use a goodness-of-fit measure like R^2, together with the sign, magnitude, and statistical significance of the coefficients, as the primary criteria for assessing the broad agreement between theory and data. The influence of the F-N-P statistical paradigm comes into this analysis only via the various techniques (estimation, testing, and prediction methods) for carrying out the anemic confirmation exercise: instantiating a theory assumed a priori to be true.

When viewed from the error-statistical perspective of Figure 6.1, it becomes apparent that anemic confirmation, as a theory-appraisal procedure, suffers from the same weaknesses as the Frisch scheme (1–4) and the Cowles Commission approach, and as a modeling strategy it holds some additional flaws. First, the criteria used are lacking unless the estimated model is statistically adequate ([i]–[iii] in (13) are valid). Otherwise, the t-statistics do not have the assumed error probabilities and R^2 is an inconsistent estimator of the underlying goodness-of-fit parameter. Second, the probabilistic assumptions [i]–[iii] in (13) provide an incomplete specification for the linear regression model (see Spanos, 2006a). Third, any departures from the error assumptions [i]–[iii] in (13) are not treated as indicating that some systematic statistical information is not accounted for by the original model, where respecification of the model is called for but is a nuisance to be "fixed" by using different estimation methods, such as Generalized Least Squares (GLS), Generalized Method of Moments (GMM), and Instrumental Variables (IV), to replace the original Ordinary Least Squares (OLS); see Greene (2003).

These textbook error-fixing strategies, however, are questionable on several grounds:

1. By focusing exclusively on the error term, the textbook perspective is geared toward "saving the theory" and overlooks the ways in which the systematic component may be in error. That is, it ignores alternative theories that might fit the same data equally well or even better. As a result, it (ab)uses the data in ways that "appear" to provide empirical (inductive) support for the theory in question, when in fact the inferences are unwarranted.

2. The "error-fixing" strategies make extensive (ab)use of the fallacies of acceptance and rejection (see Mayo and Spanos, 2004).

3. The error-fixing strategies do not distinguish between different sources of error, as mentioned in Section 3.2. A well-known example of this is the conflating of statistical with substantive inadequacy (see Spanos, 2006b).

In addition to these problems, the anemic confirmation procedure is open to considerable abuse in the sense that one could try numerous combinations of variables in x_t, and several estimation methods, to fabricate the "correct" signs, magnitudes, and significance levels for the estimated coefficients and only report the last part of the "fishing" expedition; this practice was branded "cookbook econometrics" by Leamer (1978). Indeed, the criticisms leveled at cookbook econometrics made practitioners unduly sensitive to accusations of "double-use of data" to such an extent that even looking at data plots, or testing the model assumptions, was considered unwarranted (see Kennedy, 2003).

Leamer's (1978) suggestion to deal with cookbook econometrics was to formalize these ad hoc procedures using informal Bayesian procedures driven by the modeler's degrees of belief. The presumption was that these strategies must be formalized to account for the data mining and double-use of data activities. The problem is that when statistical adequacy is recognized as a precondition for any reliable searching – to ensure the fidelity of the nominal error probabilities – these cookbook econometrics strategies are guided by unreliable procedures, and formalizing them achieves nothing of any inferential value; applying a .05 significance-level test when the actual type I error is .97 would lead inference results astray (see Spanos and McGuirk, 2001). Hence, statistical misspecification and other sources of error are considerably more worrisome than "fragility" as understood in the context of Leamer's extreme bounds analysis (see Leamer and Leonard, 1983). The latter results are vacuous when the underlying model is misspecified because one cannot address the misspecification of a statistical model by modifying the priors. Instead a guiding principle is needed that can be used to distinguish between legitimate and illegitimate modeling strategies, including double use of data, preliminary data analysis, misspecification testing, and so on. The *severity principle* provides a more appropriate guiding rationale (see Mayo, 1996; Spanos, 2000).

6.2 The Discontent with Empirical Modeling in Economics

The initial optimism associated by the promise of the new statistical methods of the Cowles Commission to significantly improve empirical modeling in economics became pessimism by the late 1960s. After two decades of laborious efforts to build large theory-based macroeconometric models, and millions of dollars spent by central banks and other government institutions, the results were more than a little disappointing. The combination of the Cowles Commission and the newly established textbook approach to econometrics did very little to allay the doubts created in the 1950s that empirical modeling in economics was *not* an effective tool in learning from data about economic phenomena of interest, nor was it useful for appraising different theories or for forecasting and policy decision purposes (see Epstein, 1987).

The increasing discontent with empirical analysis in economics reached a crescendo in the early 1970s with leading economists like Leontief (1971) lambasting both economic theorists and econometricians for the prevailing state of affairs. He was especially disparaging of deliberate attempts to enshroud the lack of substance under a veil of sophisticated mathematical formulations, both in economic theory and in econometrics. More specifically, he diagnosed a major imbalance between abstract theorizing and its empirical foundation and blamed the "indifferent results" in empirical applications primarily on the unreliability of empirical evidence arising from *nontestable* probabilistic assumptions concerning errors:

[T]he validity of these statistical tools depends itself on the acceptance of certain convenient assumptions pertaining to stochastic properties of the phenomena which the particular models are intended to explain; *assumptions that can be seldom verified.* In no other field of empirical inquiry has so massive and sophisticated a statistical machinery been used with such indifferent results. (Leontief, 1971, p. 3, emphasis added)

Leontief's presumption that probabilistic assumptions about error terms, such as [i]–[iii] in (13), cannot be assessed is clearly false, but the connection between such assumptions and the corresponding assumptions in terms of the observable process $\{(y_t \mid \mathbf{X}_t = \mathbf{x}_t), \ t \in \mathbb{N}\}$ was not clear at the time (see Mayo and Spanos, 2004).

Similar discontent with the state of scientific knowledge in economics was voiced by other well-known economists, including Boulding (1970), Brown (1972), and Worswick (1972). Leontief's discontent with the indifferent results of econometric modeling was acknowledged more generally and did not leave econometricians indifferent.

6.3 Data-Driven Modeling: The Beginnings of the "Third Way"

The first credible challenge for the Cowles Commission's preeminence-of-theory perspective came in the form of a persistently *inferior predictive performance* (see Cooper, 1972) of its multi-equation structural models when compared with the single-equation (theory-free) data-driven model – the autoregressive integrated moving-average [ARIMA(p,d,q)] model,

$$y_t = \alpha_0 + \sum_{k=1}^{p} \alpha_k y_{t-k} + \sum_{\ell=1}^{q} \beta_\ell \varepsilon_{t-\ell} + \varepsilon_t, \quad \varepsilon_t \sim \text{NIID}(0, \sigma^2), \ t \in \mathbb{N},$$

$$(14)$$

popularized by Box and Jenkins (B-J) (1970). The ARIMA model in (14) aims to capture the temporal correlation of a single data series y_t by regression on its own past $(y_{t-1}, y_{t-2}, \ldots, y_{t-p})$ (autoregression) and past errors $(\varepsilon_{t-1}, \varepsilon_{t-2}, \ldots, \varepsilon_{t-q})$ (moving average), where any heterogeneity in the original data $\{y_t^*, \ t = 1, 2, \ldots, n\}$ is eliminated by "differencing" $y_t := \Delta^d y_t^*$. Despite the obvious weakness that ARIMA models ignore all substantive information stemming from economic theory, they persistently outperformed the multi-equation (macroeconometric) structural models on prediction grounds. The question that naturally arose at the time was "Why did these large macroeconometric models perform so poorly on prediction grounds?" The answer was rather simple and grew out of the "third way" mentioned in the introduction but has not been fully appreciated in econometrics to this day: empirical models that do not account for the *statistical regularities* in the data are likely to give rise to untrustworthy empirical evidence and very poor predictive performance (see Granger and Newbold, 1986).

The primary problem with the Cowles Commission's structural macroeconometric models was that economic theory gave rise to *static* multi-equation equilibrium models, which largely ignored the temporal (time) dependence and heterogeneity information in the data.

The predictive success of ARIMA modeling encouraged several econometricians to challenge the then-dominant perspective of preeminence of theory over data and to call for greater reliance on the statistical regularities in the data and less reliance on substantive subject-matter information.

Sims argued that substantive information in macroeconomics is often "incredible" and "cannot be taken seriously" (1980, pp. 1–2). Indeed, the only such information needed for empirical modeling is some low-level theory of *which variables* might be involved in explaining a certain phenomenon of interest, say $Z_t := (Z_{1t}, \ldots, Z_{mt})$, and the modeling should focus on the statistical information contained in Z. He proposed the vector autoregressive [VAR(p)] model:

$$Z_t = a_0 + \sum_{k=1}^{m} A_k Z_{t-k} + \varepsilon_t, \quad \varepsilon_t \sim \text{NIID}(0, \Omega), \ t \in \mathbb{N}, \tag{15}$$

which captures the *joint* temporal dependence and heterogeneity contained in the data. Note that the VAR(p) model is simply a multiequation extension of the AR(p) model, a variation on the ARIMA formulation, relying on the *Normality, Markov,* and *stationarity* assumptions.

The LSE tradition (Hendry, 2000; Sargan, 1964) also called for more reliance on statistical regularities in the data and proposed the Autoregressive Distributed Lag [ADL(p,q)] model:

$$y_t = \alpha_0 + \beta_0' x_t + \sum_{k=1}^{p} \alpha_k y_{t-k} + \sum_{\ell=1}^{q} \beta_\ell' x_{t-\ell} + \varepsilon_t,$$

$$\varepsilon_t \sim \text{NIID}(0, \sigma^2), \quad t \in \mathbb{N}, \tag{16}$$

which is based on the same probabilistic assumptions as (15) but retains an indirect link to the theory via the long-run equilibrium solution of (16) (see Hendry, 1995).

Viewed in the context of the error-statistical perspective, the formulations in (14)–(16) constitute proper statistical models because they are parameterizations of generic stochastic processes, assumed to be Normal, Markov, and stationary, whose *statistical adequacy* vis-à-vis data $y_0 := (y_1, \ldots, y_n)$ could, in principle, be assessed using M-S testing. Historically, the Box and Jenkins (1970) "diagnostic checks" provided the initial impetus for M-S testing, which was then strongly encouraged by the LSE tradition.

What is particularly interesting about the Box-Jenkins (B-J), Sims, and LSE approaches is that all three (1) share a data-oriented objective to allow the data a "voice of its own," stemming from "accounting for the statistical regularities in the data"; (2) rely on statistical models in the context of a frequentist statistical framework; and (3) emphasize the use of error probabilities in statistical induction; all three features are crucial to the error-statistical perspective discussed in Section 2.

In addition, these approaches put forward a number of right-headed innovations that are designed to enhance the reliability of the "voice" of the data in empirical modeling. In particular, the B-J approach views statistical modeling as an iterative process that involves several stages: identification, estimation, diagnostic checking, and prediction. Moreover, graphical techniques and exploratory data analysis were rendered legitimate tools in choosing a statistical model and to ensure its adequacy by assessing whether it accounts for the regularities in the data.

The LSE approach rejected the preeminence-of-theory perspective and proposed a number of procedures for securing data-oriented models that are congruent with data before relating them to existing theories. A model congruent with data is reached using sound frequentist statistical procedures such as the "general-to-specific" procedure: "Model selection starts from the most general feasible model and arrives at a parsimonious model by imposing acceptable restrictions" (Campos et al., 2005, p. xxxvii). The primary motivation for following this procedure stems from maintaining the reliability of the ensuing inferences by ensuring the "optimality" of the testing procedure and keeping track of the relevant error probabilities. What distinguishes the LSE approach from other more data-oriented traditions was its persistent emphasis on justifying the methodological foundations of its procedures using the scientific credentials of frequentist inference, and its intention to ultimately relate the estimated models congruent with data to economic theory (see Hendry, 2000).

6.4 The Preeminence-of-Theory Perspective Reaffirming Itself

The preeminence-of-theory tradition is so deeply entrenched in economics that it was able to brush aside the "crises" of the 1970s and 1980s by mounting a counterattack on the B-J, Sims, and LSE approaches, scorning them as indulging in "data mining." The tradition reaffirmed itself by explaining away the predictive failure of the theory-dominated structural models as due to the inadequacies of the prevailing theories – an anomaly that can be accounted for by better theories rather than paying more attention to the regularities in the data. For example, Lucas (1976) and Lucas and Sargent (1981) argued for enhanced new classical macroeconomic theories based on (1) modeling expectations and rendering structural models dynamic to account for the temporal structure of economic time series and (2) constructing structural models anchored on rational *individual* behavior, which are invariant to policy interventions, thereby enabling improvement in their predictive performance. To some extent, the Lucas-Sargent call for improved dynamic structural macro models is a move in the right direction because the macro theories of the past could not account for the dynamic effects in the economic data; but if the enhancement of the theory is at the expense of ignoring the statistical information, it is an empty gesture. The all-important constraint of theory by the data was largely thrown overboard. One has no way of distinguishing between "better" and "worse" theories vis-à-vis the data (i.e., theories that account for the regularities in the data and those that do not).

Kydland and Prescott (1991) represent a recent, more extreme, return to the preeminence-of-theory perspective that has enjoyed some surprising appeal despite the fact that, on the face of it, it is essentially a call to ignore the data information almost entirely. Instead of using data to *estimate* the parameters of structural models by applying proper statistical procedures, one should use data to "calibrate" the parameters by using "informal" procedures; that is, use data in conjunction with ad hoc methods to assign numerical values to the unknown structural parameters in such a way that ensures that, when simulated, these models yield artificial data that tend to "mimic" the behavior of the actual data in very broad terms. Given that "calibration" purposefully forsakes error probabilities and provides no way to assess the reliability of inference, how does one assess the adequacy of the calibrated model?

The issue of how confident we are in the econometric answer is a subtle one which cannot be resolved by computing some measure of how well the model economy mimics historical data. The degree of confidence in the answer depends on the confidence that is placed in the economic theory being used. (Kydland and Prescott, 1991, p. 171)

Indeed, the theory being appraised should be the final arbiter: "The model economy which better fits the data is not the one used. Rather currently established theory dictates which one is used" (Kydland and Prescott, 1991, p. 174). The idea that it should suffice that a theory "*is not obscenely at variance with the data*" (Sargent, 1976, p. 233) is to disregard the work that statistical inference can perform in favor of some discretional subjective appraisal combined with certain eyeballing-based checks of "fit." If this description of what the current prevailing view advocates is correct, it hardly recommends itself as an empirical methodology that lives up to standards of scientific objectivity that requires its theories be thoroughly tested against data (see Hoover, 2001b, 2006; Spanos, 2009).

6.5 The Prevailing View Concerning the Theory versus Data Debates

The data-oriented approaches, including the "third way," because of their emphasis on "accounting for the regularities" in the data, have been under an incessant attack from the prevailing preeminence-of-theory tradition, having been charged with a litany of unscientific practices with catchy names like "measurement without theory," "data mining," "pre-test bias," "ignoring selection effects," "repeated testing," "hunting for statistical significance," "lack of proper identification," "double-use of data," and so forth (see Hendry, 2000, p. 469). Indeed, the mainstream critics act as if their own philosophical foundations are firmly secure because of the ostensible mathematical "rigor" of their models and the longevity of their tradition, going back to Ricardo's clarity of logical deduction and Mill's subordination of data to theory. Estimating a structural model presumed "true," using some seemingly sophisticated method (e.g., GMM), and quoting some fit statistics and a few "significant" coefficients confirming the theory meets their primary objective – never mind the endemic statistical misspecifications and other ways the inductive inferences might be in error, or the inability of these structural models to account for the regularities in the data, their pervasive predictive failures, and their unsoundness as tools for policy formulation.

A plausible way to explain the purposeful neglect of statistical misspecification might be that the recent version of the preeminence-of-theory perspective confuses this problem with the issue of the realisticness of structural models (see Spanos, 2009, 2010).

Unfortunately for economics, the end result of the methodological discussions on the theory–data confrontation of the past thirty years or so is that the practitioners of the third way found themselves constantly on the defensive. Its proponents are being required to justify their every procedure, which does not conform to the prevailing preeminence-of-theory practices. Frustrated by the lack of a coherent philosophical foundation that could be used to defend their methods and procedures from these incessant attacks, "third way" proponents like Hendry find themselves acting as reluctant philosophers without much help from philosophy of science itself. Indeed, the prevailing philosophical perspectives seem to suggest that those who have not given up on objective methods are somehow deceiving themselves, drowning out the error-statistical discourse by which, by controlling the reliability of inference through frequentist error probabilities, one can address inveterate classic philosophical problems, including "how we learn from data."

7 Conclusions

The main thesis of this chapter is that the *error-statistical perspective* can provide a coherent framework, wherein the "third way" can foster its twin goals of developing an adequate methodology and a corresponding philosophical foundation. It can help its practitioners attain the goal of objectively criticizing theories by giving data "a voice of its own" and furnishing a *natural home* and a *common language* for their methods and procedures.

The error-statistical perspective provides a methodological framework that can be used not only to defend frequentist procedures against unwarranted charges like many of the ones mentioned earlier, but also to mount a counterattack that exposes the major weaknesses in the foundations of the preeminence-of-theory perspective and its unscientific practices. The currently practiced *instantiation* (anemic confirmation) of theories, which "is not obscenely at variance with the data," has obviated any form of constructive dialogue between theory and data and, thus, has undermined the scientific credentials of economics as an empirical science. Instead, what is needed is a methodology of error inquiry that encourages detection and identification of the different ways an inductive inference could be in error by applying effective procedures that would detect such errors when present with very high probability (Mayo, 1996).

An important dimension of the error-statistical framework is its broader technical modeling perspective (Section 3), where the normal/linear/homoskedastic models relating to the ARIMA, VAR, and ADL formulations can be systematically extended to greatly enrich the family of potential statistical models that can be specified by changing the Normal, Markov, and stationarity probabilistic assumptions (see Spanos, 1989, 2006a, 2010).

Most important, the error-statistical perspective can provide a trenchant framework for the theory–data confrontation, wherein the sequence of interconnected models (theory, structural, statistical, and empirical) discussed in Sections 2 and 3 can be used to bring out, and deal adequately with, the three interrelated modeling links: (A) from theory to testable hypotheses, (B) from raw data to reliable evidence, and (C) confronting theories and hypotheses with evidence. The notion of "statistical knowledge" is crucial for theory testing because, by bestowing on data "a voice of its own" separate from the one ideated by the theory in question, it succeeds in securing the reliability and scientific objectivity of the theory–data confrontation.

Moreover, statistical knowledge (like experimental knowledge) is independent of high-level theories and – in fields like economics, which rely on observational data – the accumulation of statistically adequate models is of paramount importance because it can help to guide the challenging search for substantive explanations. It can provide a guiding constraint for new theories that aspire to explain the particular phenomenon of interest insofar as they need to account for this statistical knowledge.

References

Abadir, K., and Talmain, G. (2002), "Aggregation, Persistence and Volatility in a Macro Model," *Review of Economic Studies*, 69: 749–79.

Backhouse, R.E. (2002), *The Penguin History of Economics*, Penguin, London.

Blaug, M. (1958), *Ricardian Economics: A Historical Study*, Yale University Press, New Haven, CT.

Boulding, K. (1970), *Economics as a Science*, McGraw-Hill, New York.

Bowley, A.L. (1924), *The Mathematical Groundwork of Economics*, Reprints of Economic Classics, 1960, Augustus, M. Kelley, New York.

Bowley, A.L. (1926), *Elements of Statistics*, 5th ed. Staples Press, London.

Box, G.E.P., and Jenkins, G.M. (1970), *Time Series Analysis: Forecasting and Control*, Holden-Day, San Francisco, CA.

Brown, E.H.P. (1972), "The Underdevelopment of Economics," *Economic Journal*, 82: 1–10.

Cairnes, J.E. (1888), *The Character and Logical Method of Political Economy*, Reprints of Economic Classics, 1965, Augustus, M. Kelley, New York.

Caldwell, B. (1984), *Appraisal and Criticism in Economics: A Book of Readings*, Allen & Unwin, London.

Campos, J., Ericsson, N.R., and Hendry, D.F. (eds.) (2005), *General-to-Specific Modeling*, Vol. 1–2, Edward Elgar, Northampton, MA.

Christ, C.F. (1951), "A Test of an Econometric Model for the United States 1921–1947," in *NBER Conference on Business Cycles*, NBER, New York.

Cooper, R.L. (1972), "The Predictive Performance of Quarterly Econometric Models of the United States," in B.G. Hickman, *Econometric Models of Cyclical Behavior*, Columbia University Press, New York.

Cox, D.R., and Hinkley, D.V. (1974), *Theoretical Statistics*, Chapman & Hall, London.

Desrosières, A. (1998), *The Politics of Large Numbers: A History of Statistical Reasoning*, Harvard University Press, Cambridge, MA.

Doob, J.L. (1953), *Stochastic Processes*, Wiley, New York.

Ekelund, R.B. and Hebert, R.F. (1975), *A History of Economic Theory and Method*, McGraw-Hill, New York.

Epstein, R.J. (1987), *A History of Econometrics*, North Holland, Amsterdam.

Fisher, R.A. (1921), "On the 'Probable Error' of a Coefficient Deduced from a Small Sample," *Metron*, 1: 2–32.

Fisher, R.A. (1922), "On the Mathematical Foundations of Theoretical Statistics," *Philosophical Transactions of the Royal Society A*, 222: 309–68.

Fisher, R.A. (1925), *Statistical Methods for Research Workers*, Oliver & Boyd, Edinburgh.

Fisher, R.A. (1935), *The Design of Experiments*, Oliver & Boyd, Edinburgh.

Friedman, M. (1953), "The Methodology of Positive Economics," pp. 3–43 in *Essays in Positive Economics*, Chicago University Press, Chicago.

Frisch, R. (1933), "Editorial," *Econometrica*, 1: 1–4.

Frisch, R. (1934), *Statistical Confluence Analysis by Means of Complete Regression Schemes*, Universitetets Okonomiske Institutt, Oslo.

Goldberger, A.S. (1964), *Econometric Theory*, Wiley, New York.

Granger, C.W.J., ed. (1990), *Modelling Economic Series*, Clarendon Press, Oxford.

Granger, C.W.J., and Newbold, P. (1986), *Forecasting Economic Time Series*, 2nd ed., Academic Press, London.

Greene, W.H. (2003), *Econometric Analysis*, 5th ed., Prentice Hall, Upper Saddle River, NJ.

Guala, F. (2005), *The Methodology of Experimental Economics*, Cambridge University Press, Cambridge.

Haavelmo, T. (1943), "The Statistical Implications of a System of Simultaneous Equations," *Econometrica*, 11: 1–12.

Haavelmo, T. (1944), "The Probability Approach to Econometrics," *Econometrica*, 12(Suppl): 1–115.

Hacking, I. (1983), *Representing and Intervening: Introductory Topics in the Philosophy of Natural Science*, Cambridge University Press, Cambridge.

Hendry, D.F. (1980), "Econometrics – Alchemy or Science?" *Economica*, 47: 387–406.

Hendry, D.F. (1995), *Dynamic Econometrics*, Oxford University Press, Oxford.

Hendry, D.F. (2000), *Econometrics: Alchemy or Science?* Blackwell, Oxford.

Hendry, D.F., and Morgan, M.S. (eds.) (1995), *The Foundations of Econometric Analysis*, Cambridge University Press, Cambridge.

Hood, W.C., and Koopmans, T.C. (eds.) (1953), *Studies in Econometric Method*, Cowles Commission Monograph, No. 14, Wiley, New York.

Hoover, K.D. (2001a), *Causality in Macroeconomics*, Cambridge University Press, Cambridge.

Hoover, K.D. (2001b), *The Methodology of Empirical Macroeconomics*, Cambridge University Press, Cambridge.

Hoover, K.D. (2006), "The Methodology of Econometrics," pp. 61–87, in T.C. Mills and K. Patterson (eds.), *New Palgrave Handbook of Econometrics*, vol. 1, Macmillan, London.

Hutchison, T.W. (1938), *The Significance and Basic Postulates of Economic Theory*, Reprints of Economic Classics, 1965, Augustus, M. Kelley, New York.

Jevons, W.S. (1871), *The Theory of Political Economy*, 4th ed. 1911, Reprints of Economic Classics, 1965, Augustus M. Kelley, New York.

Johnston, J. (1963), *Econometric Methods*, McGraw-Hill, New York.

Kalman, R.E. (1982), "System Identification from Noisy Data," in A.R. Bednarek and L. Cesari (eds.), *Dynamic Systems II*, Academic Press, London.

Kennedy, P. (2003), *A Guide to Econometrics*, 5th ed., MIT Press, Cambridge.

Keynes, J.N. (1890), *The Scope and Method of Political Economy*, 2nd ed. 1917, Reprints of Economic Classics, 1986, Augustus M. Kelley, New York.

Klein, L.R. (1950), *Economic Fluctuations in the United States* 1921–1941. Cowles Commission for Research in Economics, Monograph No. 11, Wiley, New York.

Klein, L.R. (1962), *An Introduction to Econometrics*, Prentice Hall, Upper Saddle River, NJ.

Koopmans, T.C. (1950), *Statistical Inference in Dynamic Economic Models*, Cowles Commission Monograph, No. 10, Wiley, New York.

Kydland, F., and Prescott, P. (1991), "The Econometrics of the General Equilibrium Approach to Business Cycles," *Scandinavian Journal of Economics*, 93: 161–178.

Leamer, E.E. (1978), *Specification Searches: Ad Hoc Inference with Nonexperimental Data*, Wiley, New York.

Leamer, E.E., and Leonard, H.B. (1983), "Reporting the Fragility of Regression Estimates," *Review of Economics and Statistics*, 65: 306–317.

Lehmann, E.L. (1986), *Testing Statistical Hypotheses*, 2nd ed., Wiley, New York.

Leontief, W.W. (1971), "Theoretical Assumptions and Nonobserved Facts," *American Economic Review*, 61: 1–7.

Lucas, R.E. (1976), "Econometric Policy Evaluation: A Critique," pp. 19–46 in K. Brunner and A.M. Metzer (eds.), *The Phillips Curve and Labour Markets*, Carnegie-Rochester Conference on Public Policy, I, North-Holland, Amsterdam.

Lucas, R.E., and Sargent, T.J. (1981), *Rational Expectations and Econometric Practice*, George Allen & Unwin, London.

Machlup, F. (1963), *Essays in Economic Semantic*, Prentice Hall, Upper Saddle River, NJ.

Maki, U. (ed.) (2009), *The Methodology of Positive Economics: Reflections on the Milton Friedman Legacy*, Cambridge University Press, Cambridge.

Malthus, T. R. (1936), *Principles of Political Economy*, Reprints of Economic Classics, 1986, Augustus M. Kelley, New York.

Marshall, A. (1891/1920), *Principles of Economics*, 8th ed., Macmillan, London.

Mayo, D.G. (1996), *Error and the Growth of Experimental Knowledge*, University of Chicago Press, Chicago.

Mayo, D.G. (2008), "Some Methodological Issues in Experimental Economics," *Philosophy of Science*, 75: 633–45.

Mayo, D.G., and Spanos, A. (2004), "Methodology in Practice: Statistical Misspecification Testing," *Philosophy of Science*, 71: 1007–25.

Mayo, D.G., and Spanos, A. (2006), "Severe Testing as a Basic Concept in a Neyman–Pearson Philosophy of Induction," *British Journal for the Philosophy of Science*, 57: 323–57.

Mayo, D.G., and Spanos, A. (2009), "Error Statistics," forthcoming in D. Gabbay, P. Thagard, and J. Woods (eds.), *Philosophy of Statistics, Handbook of Philosophy of Science*, Elsevier.

McCloskey, D.N. (1985), *The Rhetoric of Economics*, University of Wisconsin, Madison.

McCulloch, J.R. (1864), *The Principles of Political Economy*, Reprints of Economic Classics, 1965, Augustus M. Kelley, New York.

Mill, J.S. (1844), *Essays on Some Unsettled Questions of Political Economy*, Reprints of Economic Classics, 1974, Augustus M. Kelley, New York.

Mill, J.S. (1871), *Principles of Political Economy*, Reprints of Economic Classics, 1976, Augustus M. Kelley, New York.

Mill, J.S. (1888), *A System of Logic, Ratiocinative and Inductive*, 8th ed., Harper & Bros., New York.

Mills, F.C. (1924), *Statistical Methods* (reprinted 1938), Henry Holt, New York.

Mirowski, P. (1994), "What Are the Questions?" pp. 50–74 in R.E. Backhouse (ed.), *New Directions in Economic Methodology*, Routledge, London.

Morgan, M.S. (1990), *The History of Econometric Ideas*, Cambridge University Press, Cambridge.

Morgenstern, O. (1950/1963), *On the Accuracy of Economic Observations*, 2nd ed., Princeton University Press, NJ.

Musgrave, A. (1981), "Unreal Assumptions in Economic Theory: The F-Twist Untwisted," *Kyklos*, 34: 377–87.

Nagel, E. (1963), "Assumptions in Economic Theory," *American Economic Review*, 53: 211–219.

Neyman, J. (1950), *First Course in Probability and Statistics*, Henry Holt, New York.

Neyman, J. (1977), "Frequentist Probability and Frequentist Statistics," *Synthèse*, 36: 97–131.

Neyman, J., and Pearson, E.S. (1928), "On the Use and Interpretation of Certain Test Criteria for the Purposes of Statistical Inference," Parts I and II, *Biometrika*, 20A: 175–240, 263–94.

Neyman, J., and Pearson, E.S. (1933), "On the Problem of the Most Efficient Tests of Statistical Hypotheses," *Philosophical Transactions of the Royal Society of London, A*, 231: 289–337.

Pearson, E.S. (1966), "The Neyman–Pearson Story: 1926–34," In F.N. David (ed.), *Research Papers in Statistics: Festschrift for J. Neyman*, Wiley, New York.

Porter, G.R. (1836/1912), *The Progress of a Nation*, 2nd ed., 1912, Reprints of Economic Classics, 1970, Augustus M. Kelley, New York.

Quetelet, A. (1942), *A Treatise on Man and the Development of His Faculties*, Chambers, Edinburgh.

Redman, D. A. (1991), *Economics and the Philosophy of Science*, Oxford University Press, Oxford.

Ricardo, D. (1817), *Principles of Political Economy and Taxation*, vol. 1 of *The Collected Works of David Ricardo*, P. Sraffa and M. Dobb (eds.), Cambridge University Press, Cambridge.

Robbins, L. (1932), *An Essay on the Nature and Significance of Economic Science*, 2nd ed. 1935, Macmillan, London.

Robbins, L. (1971), *Autobiography of an Economist*, Macmillan, London.

Sargan, J. D. (1964), "Wages and Prices in the United Kingdom: A Study in Econometric Methodology," in P.E. Hart, G. Mills, and J.K. Whitaker (eds.), *Econometric Analysis for Economic Planning*, Butterworths, London.

Sargent, T. (1976), "A Classical Macroeconomic Model for the United States," *Journal of Political Economy*, 84: 207–38.

Senior, N.W. (1836), *An Outline of the Science of Political Economy*, Reprints of Economic Classics, 1965, Augustus M. Kelley, New York.

Sims, C.A. (1980), "Macroeconomics and Reality," *Econometrica*, 48: 1–48.

Smith, A. (1776), *An Inquiry into the Nature and Causes of the Wealth of Nations*, R.H. Campell, A.S. Skinner, and W.B. Todd (eds.), Clarendon Press, Oxford.

Smith, V.L. (2007), *Rationality in Economics: Constructivist and Ecological Forms*, Cambridge University Press, Cambridge.

Sowell, T. (2006), *On Classical Economics*, Yale University Press, New Haven, CT.

Spanos, A. (1986), *Statistical Foundations of Econometric Modelling*, Cambridge University Press, Cambridge.

Spanos, A. (1989), "On Re-reading Haavelmo: A Retrospective View of Econometric Modeling," *Econometric Theory*, 5: 405–29.

Spanos, A. (1990), "The Simultaneous Equations Model Revisited: Statistical Adequacy and Identification," *Journal of Econometrics*, 44: 87–108.

Spanos, A. (1995), "On Theory Testing in Econometrics: Modeling with Nonexperimental Data," *Journal of Econometrics*, 67: 189–226.

Spanos, A. (1999), "Probability Theory and Statistical Inference: Econometric Modeling with Observational Data," Cambridge University Press, Cambridge.

Spanos, A. (2000), "Revisiting Data Mining: 'Hunting' with or without a License," *Journal of Economic Methodology*, 7: 231–64.

Spanos, A. (2005), "Misspecification, Robustness and the Reliability of Inference: the Simple T-Test in the Presence of Markov Dependence," Working Paper, Virginia Tech.

Spanos, A. (2006a), "Econometrics in Retrospect and Prospect," pp. 3–58 in T.C. Mills and K. Patterson (eds.), *New Palgrave Handbook of Econometrics*, vol. 1, Macmillan, London.

Spanos, A. (2006b), "Revisiting the Omitted Variables Argument: Substantive vs. Statistical Adequacy," *Journal of Economic Methodology*, 13: 179–218.

Spanos, A. (2006c), "Where Do Statistical Models Come From? Revisiting the Problem of Specification," in J. Rojo (ed.), *Optimality: The Second Erich L. Lehmann Symposium*, Lecture Notes-Monograph Series, vol. 49, Institute of Mathematical Statistics, Beachwood, OH.

Spanos, A. (2009), "The Pre-Eminence of Theory vs. the European CVAR Perspective in Macroeconometric Modeling", in *The Open-Access, Open-Assessment E-Journal*, 3, 2009-10. http://www.economics-ejournal.org/economics/journalarticles/2009-10.

Spanos, A. (2010), "Philosophy of Econometrics," forthcoming in D. Gabbay, P. Thagard, and J. Woods (eds.), *Philosophy of Economics, the Handbook of Philosophy of Science*, Elsevier, North-Holland, Amsterdam.

Spanos, A., and McGuirk, A. (2001), "The Model Specification Problem from a Probabilistic Reduction Perspective," *Journal of the American Agricultural Association*, 83: 1168–76.

Sugden, R. (2005), "Experiments as Exhibits and Experiments as Tests," *Journal of Economic Methodology*, 12: 291–302.

Sugden, R. (2008), "The Changing Relationship between Theory and Experiment in Economics," *Philosophy of Science*, 75: 621–32.

Worswick, G.D.N. (1972), "Is Progress in Economic Science Possible?" *Economic Journal*, 82: 73–86.

SEVEN

New Perspectives on (Some Old) Problems of Frequentist Statistics

I Frequentist Statistics as a Theory of Inductive Inference[1]

Deborah G. Mayo and David Cox

1 Statistics and Inductive Philosophy

1.1 What Is the Philosophy of Statistics?

The philosophical foundations of statistics may be regarded as the study of the epistemological, conceptual, and logical problems revolving around the use and interpretation of statistical methods, broadly conceived. As with other domains of philosophy of science, work in statistical science progresses largely without worrying about "philosophical foundations." Nevertheless, even in statistical practice, debates about the different approaches to statistical analysis may influence and be influenced by general issues of the nature of inductive-statistical inference, and thus are concerned with foundational or philosophical matters. Even those who are largely concerned with applications are often interested in identifying general principles that underlie and justify the procedures they have come to value on relatively pragmatic grounds. At one level of analysis at least, statisticians and philosophers of science ask many of the same questions.

- What should be observed and what may justifiably be inferred from the resulting data?
- How well do data confirm or fit a model?
- What is a good test?
- Does failure to reject a hypothesis H constitute evidence confirming H?

[1] This paper appeared in *The Second Erich L. Lehmann Symposium: Optimality*, 2006, Lecture Notes-Monograph Series, Volume 49, Institute of Mathematical Statistics, pp. 96–123.

- How can it be determined whether an apparent anomaly is genuine? How can blame for an anomaly be assigned correctly?
- Is it relevant to the relation between data and a hypothesis if looking at the data influences the hypothesis to be examined?
- How can spurious relationships be distinguished from genuine regularities?
- How can a causal explanation and hypothesis be justified and tested?
- How can the gap between available data and theoretical claims be bridged reliably?

That these very general questions are entwined with long-standing debates in philosophy of science helps explain why the field of statistics tends to cross over, either explicitly or implicitly, into philosophical territory. Some may even regard statistics as a kind of "applied philosophy of science" (Fisher, 1935a–b; Kempthorne, 1976), and statistical theory as a kind of "applied philosophy of inductive inference." As Lehmann (1995) has emphasized, Neyman regarded his work not only as a contribution to statistics but also to inductive philosophy. A core question that permeates "inductive philosophy" both in statistics and philosophy is: What is the nature and role of probabilistic concepts, methods, and models in making inferences in the face of limited data, uncertainty and error? We take as our springboard the recommendation of Neyman (1955, p. 17) that we view statistical theory as essentially a "Frequentist Theory of Inductive Inference." The question then arises as to what conception(s) of inductive inference would allow this. Whether or not this is the only or even the most satisfactory account of inductive inference, it is interesting to explore how much progress towards an account of inductive inference, as opposed to inductive behavior, one might get from frequentist statistics (with a focus on testing and associated methods). These methods are, after all, often used for inferential ends, to learn about aspects of the underlying data-generating mechanism, and much confusion and criticism (e.g., as to whether and why error rates are to be adjusted) could be avoided if there was greater clarity on the roles in inference of hypothetical error probabilities.

Taking as a backdrop remarks by Fisher (1935a), Lehmann (1995) on Neyman, and by Popper (1959) on induction, we consider the roles of significance tests in bridging inductive gaps in traditional hypothetical deductive inference. Our goal is to identify a key principle of evidence by which hypothetical error probabilities may be used for inductive inference from specific data, and to consider how it may direct and justify (a) different uses and interpretations of statistical significance levels in testing a variety of different types of null hypotheses, and (b) when and why "selection effects" need to be taken account of in data dependent statistical testing.

1.2 The Role of Probability in Frequentist Induction

The defining feature of an inductive inference is that the premises (evidence statements) can be true while the conclusion inferred may be false without a logical contradiction: the conclusion is "evidence transcending." Probability naturally arises in capturing such evidence transcending inferences, but there is more than one way this can occur. Two distinct philosophical traditions for using probability in inference are summed up by Pearson (1955 , p. 228).

"For one school, the degree of confidence in a proposition, a quantity varying with the nature and extent of the evidence, provides the basic notion to which the numerical scale should be adjusted." The other school notes the relevance in ordinary life and in many branches of science of a knowledge of the relative frequency of occurrence of a particular class of events in a series of repetitions, and suggests that "it is through its link with relative frequency that probability has the most direct meaning for the human mind."

Frequentist induction, whatever its form, employs probability in the second manner. For instance, significance testing appeals to probability to characterize the proportion of cases in which a null hypothesis H_0 would be rejected in a hypothetical long-run of repeated sampling, an error probability. This difference in the role of probability corresponds to a difference in the form of inference deemed appropriate: The former use of probability traditionally has been tied to the view that a probabilistic account of induction involves quantifying a degree of support or confirmation in claims or hypotheses.

Some followers of the frequentist approach agree, preferring the term "inductive behavior" to describe the role of probability in frequentist statistics. Here the inductive reasoner "decides to infer" the conclusion, and probability quantifies the associated risk of error. The idea that one role of probability arises in science to characterize the "riskiness" or probativeness or severity of the tests to which hypotheses are put is reminiscent of the philosophy of Karl Popper (1959). In particular, Lehmann (1995, p. 32) has noted the temporal and conceptual similarity of the ideas of Popper and Neyman on "finessing" the issue of induction by replacing inductive reasoning with a process of hypothesis testing.

It is true that Popper and Neyman have broadly analogous approaches based on the idea that we can speak of a hypothesis having been well-tested in some sense, quite different from its being accorded a degree of probability, belief, or confirmation; this is "finessing induction." Both also broadly shared the view that in order for data to "confirm" or "corroborate" a hypothesis H, that hypothesis would have to have been subjected to a test with high probability or power to have rejected it if false. But despite the close connection of the ideas, there appears to be no reference to Popper in the writings of Neyman (Lehmann, 1995, p. 31) and the references by Popper to Neyman are scant and scarcely relevant. Moreover, because Popper denied that any inductive claims were justifiable, his philosophy forced him to deny that even the method he espoused (conjecture and refutations) was reliable. Although H might be true, Popper made it clear that he regarded corroboration at most as a report of the past performance of H: it warranted no claims about its reliability in future applications. By contrast, a central feature of frequentist statistics is to actually assess and control the probability that a hypothesis would have rejected a hypothesis, if false. These probabilities come from formulating the data-generating process in terms of a statistical model.

Neyman throughout his work emphasizes the importance of a probabilistic model of the system under study and describes frequentist statistics as modelling the phenomenon of the stability of relative frequencies of results of repeated "trials," granting that there are other possibilities concerned with modelling psychological phenomena connected with intensities of belief, or with readiness to bet specified sums, etc., citing Carnap (1962), de Finetti (1974), and Savage (1964). In particular Neyman criticized the view of "frequentist" inference taken by Carnap for overlooking the key role of the stochastic model of the phenomenon studied. Statistical work related to the inductive philosophy of Carnap (1962) is that of Keynes (1921) and, with a more immediate impact on statistical applications, Jeffreys (1961).

1.3 Induction and Hypothetical-Deductive Inference

While "hypothetical-deductive inference" may be thought to "finesse" induction, in fact inductive inferences occur throughout empirical testing. Statistical testing ideas may be seen to fill these inductive gaps: If the hypothesis were deterministic we could find a relevant function of the data whose value (i) represents the relevant feature under test and (ii) can be predicted by the hypothesis. We calculate the function and then see whether the data agree or disagree with the prediction. If the data conflict with the prediction, then either the hypothesis is in error or some auxiliary or other background factor may be blamed for the anomaly (Duhem's problem).

Statistical considerations enter in two ways. If *H* is a statistical hypothesis, then usually no outcome strictly contradicts it. There are major problems involved in regarding data as inconsistent with *H* merely because they are highly improbable; all individual outcomes described in detail may have very small probabilities. Rather the issue, essentially following Popper (1959, pp. 86, 203), is whether the possibly anomalous outcome represents some systematic and reproducible effect.

The focus on falsification by Popper as the goal of tests, and falsification as the defining criterion for a scientific theory or hypothesis, clearly is strongly redolent of Fisher's thinking. While evidence of direct influence is virtually absent, the views of Popper agree with the statement by Fisher (1935a, p. 16) that every experiment may be said to exist only in order to give the facts the chance of disproving the null hypothesis. However, because Popper's position denies ever having grounds for inference about reliability, he denies that we can ever have grounds for inferring reproducible deviations.

The advantage in the modern statistical framework is that the probabilities arise from defining a probability model to represent the phenomenon of interest. Had Popper made use of the statistical testing ideas being developed at around the same time, he might have been able to substantiate his account of falsification.

The second issue concerns the problem of how to reason when the data "agree" with the prediction. The argument from *H* entails data *y*, and that *y* is observed, to the inference that *H* is correct is, of course, deductively invalid. A central problem for an inductive account is to be able nevertheless to warrant inferring *H* in some sense. However, the classical problem, even in deterministic cases, is that many rival hypotheses (some would say infinitely many) would also predict *y*, and thus would pass as well as *H*. In order for a test to be probative, one wants the prediction from *H* to be something that at the same time is in some sense very surprising and not easily accounted for were *H* false and important rivals to *H* correct. We now consider how the gaps in inductive testing may bridged by a specific kind of statistical procedure, the significance test.

2 Statistical Significance Tests

Although the statistical significance test has been encircled by controversies for over 50 years, and has been mired in misunderstandings in the literature, it illustrates in simple form a number of key features of the perspective on frequentist induction that we are considering. See for example Morrison and Henkel (1970) and Gibbons and Pratt (1975). So far as possible, we begin with the core elements of significance testing in a version very strongly related to but in some respects different from both Fisherian and Neyman-Pearson approaches, at least as usually formulated.

2.1 General Remarks and Definition

We suppose that we have empirical data denoted collectively by y and that we treat these as observed values of a random variable Y. We regard y as of interest only in so far as it provides information about the probability distribution of Y as defined by the relevant statistical model. This probability distribution is to be regarded as an often somewhat abstract and certainly idealized representation of the underlying data-generating process. Next we have a hypothesis about the probability distribution, sometimes called the hypothesis under test but more often conventionally called the null hypothesis and denoted by H_0. We shall later set out a number of quite different types of null hypotheses but for the moment we distinguish between those, sometimes called simple, that completely specify (in principle numerically) the distribution of Y and those, sometimes called composite, that completely specify certain aspects and which leave unspecified other aspects.

In many ways the most elementary, if somewhat hackneyed, example is that Y consists of n independent and identically distributed components normally distributed with unknown mean μ and possibly unknown standard deviation σ. A simple hypothesis is obtained if the value of σ is known, equal to σ_0, say, and the null hypothesis is that $\mu = \mu_0$, a given constant. A composite hypothesis in the same context would have σ unknown and again specify the value of μ.

Note that in this formulation it is required that some unknown aspect of the distribution, typically one or more unknown parameters, is precisely specified. The hypothesis that, for example, $\mu \leq \mu_0$ is not an acceptable formulation for a null hypothesis in a Fisherian test; while this more general form of null hypothesis is allowed in Neyman-Pearson formulations.

The immediate objective is to test the conformity of the particular data under analysis with H_0 in some respect to be specified. To do this we find a function $t = t(y)$ of the data, to be called the test statistic, such that:

- the larger the value of t the more inconsistent are the data with H_0;
- the corresponding random variable $T = t(Y)$ has a (numerically) known probability distribution when H_0 is true.

These two requirements parallel the corresponding deterministic ones. To assess whether there is a genuine discordancy (or reproducible deviation) from H_0 we define the so-called p-value corresponding to any t as:

$$p = p(t) = P(T > t; H_0)$$

regarded as a measure of concordance with H_0 in the respect tested. In at least the initial formulation alternative hypotheses lurk in the undergrowth but are not explicitly formulated probabilistically; also there is no question of setting in advance a preassigned threshold value and "rejecting" H_0 if and only if $p \leq \alpha$. Moreover, the justification for tests will not be limited to appeals to long-run behavior but will instead identify an inferential or evidential rationale. We now elaborate.

2.2 Inductive Behavior versus Inductive Inference

The reasoning may be regarded as a statistical version of the valid form of argument called in deductive logic *modus tollens*. This infers the denial of a hypothesis H from the combination that H entails E, together with the information that E is false. Because there was a high probability $(1-p)$ that a less significant result would have occurred were H_0 true, we may justify taking low p-values, properly computed, as evidence against H_0. Why? There are two main reasons:

Firstly such a rule provides low error rates (i.e., erroneous rejections) in the long run when H_0 is true, a behavioristic argument. In line with an error-assessment view of statistics we may give any particular value p, say, the following hypothetical interpretation: suppose that we were to treat the data as just decisive evidence against H_0. Then in hypothetical repetitions H_0 would be rejected in a long-run proportion p of the cases in which it is actually true. However, knowledge of these hypothetical error probabilities may be taken to underwrite a distinct justification.

This is that secondly such a rule provides a way to determine whether a specific data set is evidence of a discordancy from H_0.

In particular, a low p-value, so long as it is properly computed, provides evidence of a discrepancy from H_0 in the respect examined, while a p-value that is not small affords evidence of accordance or consistency with H_0 (where this is to be distinguished from positive evidence for H_0, as discussed below in Section 2.3). Interest in applications is typically in whether p is in some such range as $p > 0.1$ which can be regarded as reasonable accordance with H_0 in the respect tested, or whether p is near to such conventional numbers as 0.05, 0.01, 0.001. Typical practice in much applied work is to give the observed value of p in rather approximate form. A small value of p indicates that (i) H_0 is false (there is a discrepancy from H_0) or (ii) the basis of the statistical test is flawed, often that real errors have been underestimated, for example because of invalid independence assumptions, or (iii) the play of chance has been extreme.

It is part of the object of good study design and choice of method of analysis to avoid (ii) by ensuring that error assessments are relevant.

There is no suggestion whatever that the significance test would typically be the only analysis reported. In fact, a fundamental tenet of the conception of inductive learning most at home with the frequentist philosophy is that inductive inference requires building up incisive arguments and inferences by putting together several different piece-meal results. Although the complexity of the story makes it more difficult to set out neatly, as, for example, if a single algorithm is thought to capture the whole of inductive inference, the payoff is an account that approaches the kind of full-bodied arguments that scientists build up in order to obtain reliable knowledge and understanding of a field.

Amidst the complexity, significance test reasoning reflects a fairly straightforward conception of evaluating evidence anomalous for H_0 in a statistical context, the one Popper perhaps had in mind but lacked the tools to implement. The basic idea is that error probabilities may be used to evaluate the "riskiness" of the predictions H_0 is required to satisfy, by assessing the reliability with which the test discriminates whether (or not) the actual process giving rise to the data accords with that described in H_0. Knowledge of this probative capacity allows determining if there is strong evidence of discordancy. The reasoning is based on the following frequentist principle for identifying whether or not there is evidence against H_0:

FEV (i): y is (strong) evidence against H_0, i.e. (strong) evidence of discrepancy from H_0, if and only if, were H_0 a correct description of the mechanism generating **y**, then, with high probability this would have resulted in a less discordant result than is exemplified by **y**.

A corollary of **FEV** is that **y** is not (strong) evidence against H_0, if the probability of a more discordant result is not very low, even if H_0 is correct. That is, if there is a moderately high probability of a more discordant result, even were H_0 correct, then H_0 accords with **y** in the respect tested.

Somewhat more controversial is the interpretation of a failure to find a small p-value; but an adequate construal may be built on the above form of **FEV**.

2.3 Failure and Confirmation

The difficulty with regarding a modest value of p as evidence in favor of H_0 is that accordance between H_0 and y may occur even if rivals to H_0 seriously different from H_0 are true. This issue is particularly acute when the amount of data is limited. However, sometimes we can find evidence for H_0, understood as an assertion that a particular discrepancy, flaw, or error is absent, and we can do this by means of tests that, with high probability, would have reported a discrepancy had one been present. As much as Neyman is associated with automatic decision-like techniques, in practice at least, both he and E.S. Pearson regarded the appropriate choice of error probabilities as reflecting the specific context of interest (Neyman (1957); Pearson (1955)).

There are two different issues involved. One is whether a particular value of p is to be used as a threshold in each application. This is the procedure set out in most if not all formal accounts of Neyman-Pearson theory. The second issue is whether control of long-run error rates is a justification for frequentist tests or whether the ultimate justification of tests lies in their role in interpreting evidence in particular cases. In the account given here, the achieved value of p is reported, at least approximately, and the "accept/reject" account is purely hypothetical to give p an operational interpretation. E.S. Pearson (1955) is known to have disassociated himself from a narrow behaviorist interpretation (Mayo,1996). Neyman, at least in his discussion with Carnap (Neyman, 1955) seems also to hint at a distinction between behavioral and inferential interpretations.

In an attempt to clarify the nature of frequentist statistics, Neyman in this discussion was concerned with the term "degree of confirmation" used by Carnap. In the context of an example where an optimum test had failed to "reject" H_0, Neyman considered whether this "confirmed" H_0. He noted that this depends on the meaning of words such as "confirmation" and "confidence" and that in the context where H_0 had not been "rejected" it would be "dangerous" to regard this as confirmation of H_0 if the test in fact had little chance of detecting an important discrepancy from H_0 even if such a discrepancy were present. On the other hand if the test had appreciable power to detect the discrepancy the situation would be "radically different."

Neyman is highlighting an inductive fallacy associated with "negative results," namely that if data y yield a test result that is not statistically significantly different from H_0 (e.g., the null hypothesis of no effect), and yet the test has small probability of rejecting H_0, even when a serious discrepancy exists, then y is not good evidence for inferring that H_0 is confirmed by y. One may be confident in the absence of a discrepancy, according to this argument, only if the chance that the test would have correctly detected a discrepancy is high.

Neyman compares this situation with interpretations appropriate for inductive behavior. Here confirmation and confidence may be used to describe the choice of action, for example refraining from announcing a discovery or the decision to treat H_0 as satisfactory. The rationale is the pragmatic behavioristic one of controlling errors in the long-run. This distinction implies that even for Neyman evidence for deciding may require a distinct criterion than evidence for believing; but unfortunately Neyman did not set out the latter explicitly. We propose that the needed evidential principle is an adaptation of **FEV**(i) for the case of a p-value that is not small:

FEV(ii): A moderate p value is evidence of the absence of a discrepancy δ from H_0, only if there is a high probability the test would have given a worse fit with H_0 (i.e., smaller p value) were a discrepancy δ to exist. **FEV**(ii) especially arises in the context of embedded hypotheses (below).

What makes the kind of hypothetical reasoning relevant to the case at hand is not solely or primarily the long-run low error rates associated with using the tool (or test) in this manner; it is rather what those error rates reveal about the data-generating source or phenomenon. The error-based calculations provide reassurance that incorrect interpretations of the evidence are being avoided in the particular case. To distinguish between this "evidential" justification of the reasoning of significance tests, and the "behavioristic" one, it may help to consider a very informal example of applying this reasoning "to the specific case." Thus suppose that weight gain is measured by well-calibrated and stable methods, possibly using several measuring instruments and observers and the results show negligible change over a test period of interest. This may be regarded as grounds for inferring that the individual's weight gain is negligible within limits set by the sensitivity of the scales. Why?

While it is true that by following such a procedure in the long run one would rarely report weight gains erroneously, that is not the rationale for the particular inference. The justification is rather that the error probabilistic properties of the weighing procedure reflect what is actually the case in the specific instance. (This should be distinguished from the evidential interpretation of Neyman-Pearson theory suggested by Birnbaum (1977), which is not data-dependent.)

The significance test is a measuring device for accordance with a specified hypothesis calibrated, as with measuring devices in general, by its performance in repeated applications, in this case assessed typically theoretically or by simulation. Just as with the use of measuring instruments, applied to a specific case, we employ the performance features to make inferences about aspects of the particular thing that is measured, aspects that the measuring tool is appropriately capable of revealing.

Of course for this to hold the probabilistic long-run calculations must be as relevant as feasible to the case in hand. The implementation of this surfaces in statistical theory in discussions of conditional inference, the choice of appropriate distribution for the evaluation of p. Difficulties surrounding this seem more technical than conceptual and will not be dealt with here, except to note that the exercise of applying (or attempting to apply) **FEV** may help to guide the appropriate test specification.

3 Types of Null Hypothesis and Their Corresponding Inductive Inferences

In the statistical analysis of scientific and technological data, there is virtually always external information that should enter in reaching conclusions about what the data indicate with respect to the primary question of interest. Typically, these background considerations enter not by a probability assignment but by identifying the question to be asked, designing the study, interpreting the statistical results and relating those inferences to primary scientific ones and using them to extend and support underlying theory. Judgments about what is relevant and informative must be supplied for the tools to be used non-fallaciously and as intended. Nevertheless, there are a cluster of systematic uses that may be set out corresponding to types of test and types of null hypothesis.

3.1 Types of Null Hypothesis

We now describe a number of types of null hypothesis. The discussion amplifies that given by Cox (1958, 1977) and by Cox and Hinkley (1974). Our goal here is not to give a guide for the panoply of contexts a researcher might face, but rather to elucidate some of the different interpretations of test results and the associated p-values. In Section 4.3, we consider the deeper interpretation of the corresponding inductive inferences that, in our view, are (and are not) licensed by p-value reasoning.

1. Embedded Null Hypotheses. In these problems there is formulated, not only a probability model for the null hypothesis, but also models that represent other possibilities in which the null hypothesis is false and, usually, therefore represent possibilities we would wish to detect if present. Among the number of possible situations, in the most common there is a parametric family of distributions indexed by an unknown parameter θ partitioned into components $\theta = (\varphi, \lambda)$, such that the null hypothesis is that $\varphi = \varphi_0$, with λ an unknown nuisance parameter and, at least in the initial discussion with φ one-dimensional. Interest focuses on alternatives $\varphi > \varphi_0$.

This formulation has the technical advantage that it largely determines the appropriate test statistic $t(y)$ by the requirement of producing the most sensitive test possible with the data at hand.

There are two somewhat different versions of the above formulation. In one, the full family is a tentative formulation intended not so much as a possible base for ultimate interpretation but as a device for determining a suitable test statistic. An example is the use of a quadratic model to test adequacy of a linear relation; on the whole polynomial regressions are a poor base for final analysis but very convenient and interpretable for detecting small departures from a given form. In the second case the family is a solid base for interpretation. Confidence intervals for φ have a reasonable interpretation.

One other possibility, that arises very rarely, is that there is a simple null hypothesis and a single simple alternative, i.e., only two possible distributions are under consideration. If the two hypotheses are considered on an equal basis the analysis is typically better considered as one of hypothetical or actual discrimination, i.e., of determining which one of two (or more, generally a very limited number) of possibilities is appropriate, treating the possibilities on a conceptually equal basis.

There are two broad approaches in this case. One is to use the likelihood ratio as an index of relative fit, possibly in conjunction with an application of Bayes's theorem. The other, more in accord with the error probability approach, is to take each model in turn as a null hypothesis and the other as alternative leading to an assessment as to whether the data are in accord with both, one, or neither hypothesis. Essentially the same interpretation results by applying **FEV** to this case, when it is framed within a Neyman-Pearson framework.

We can call these three cases those of a formal family of alternatives, of a well-founded family of alternatives, and of a family of discrete possibilities.

2. Dividing Null Hypotheses. Quite often, especially but not only in technological applications, the focus of interest concerns a comparison of two or more conditions, processes or treatments with no particular reason for expecting the outcome to be exactly or nearly identical, e.g., compared with a standard a new drug may increase or may decrease survival rates.

One, in effect, combines two tests, the first to examine the possibility that $\mu > \mu_0$, say, the other for $\mu < \mu_0$. In this case, the two-sided test combines both one-sided tests, each with its own significance level. The significance level is twice the smaller p, because of a "selection effect" (Cox and Hinkley, 1974, p. 106). We return to this issue in Section 4. The null hypothesis of zero difference then divides the possible situations into two qualitatively different regions with respect to the feature tested, those in which one of the treatments is superior to the other and a second in which it is inferior.

3. Null Hypotheses of Absence of Structure. In quite a number of relatively empirically conceived investigations in fields without a very firm theory base, data are collected in the hope of finding structure, often in the form of dependencies between features beyond those already known. In epidemiology this takes the form of tests of potential risk factors for a disease of unknown etiology.

4. Null Hypotheses of Model Adequacy. Even in the fully embedded case where there is a full family of distributions under consideration, rich enough potentially to explain the data whether the null hypothesis is true or false, there is the possibility that there are important discrepancies with the model sufficient to justify extension, modification, or total replacement of the model used for interpretation. In many fields the initial models used for interpretation are quite tentative; in others, notably in some areas of physics, the models have a quite solid base in theory and extensive experimentation. But in all cases the possibility of model misspecification has to be faced even if only informally.

There is then an uneasy choice between a relatively focused test statistic designed to be sensitive against special kinds of model inadequacy (powerful against specific directions of departure), and so-called omnibus tests that make no strong choices about the nature of departures. Clearly the latter will tend to be insensitive, and often extremely insensitive, against specific alternatives. The two types broadly correspond to chi-squared tests with small and large numbers of degrees of freedom. For the focused test we may either choose a suitable test statistic or, almost equivalently, a notional family of alternatives. For example to examine agreement of n independent observations with a Poisson distribution we might in effect test the agreement of the sample variance with the sample mean by a chi-squared dispersion test (or its exact equivalent) or embed the Poisson distribution in, for example, a negative binomial family.

5. Substantively-Based Null Hypotheses. In certain special contexts, null results may indicate substantive evidence for scientific claims in contexts that merit a fifth category. Here, a theory T for which there is appreciable theoretical and/or empirical evidence predicts that H_0 is, at least to a very close approximation, the true situation.

(a) In one version, there may be results apparently anomalous for T, and a test is designed to have ample opportunity to reveal a discordancy with H_0 if the anomalous results are genuine.

(b) In a second version a rival theory T^* predicts a specified discrepancy from H_0, and the significance test is designed to discriminate between T and the rival theory T^* (in a thus far not tested domain).

For an example of (a) physical theory suggests that because the quantum of energy in nonionizing electro-magnetic fields, such as those from high voltage transmission lines, is much less than is required to break a molecular bond, there should be no carcinogenic effect from exposure to such fields. Thus in a randomized experiment in which two groups of mice are under identical conditions except that one group is exposed to such a field, the null hypothesis that the cancer incidence rates in the two groups are identical may well be exactly true and would be a prime focus of interest in analyzing the data. Of course the null hypothesis of this general kind does not have to be a model of zero effect; it might refer to agreement with previous well-established empirical findings or theory.

3.2 Some General Points

We have in the above described essentially one-sided tests. The extension to two-sided tests does involve some issues of definition but we shall not discuss these here.

Several of the types of null hypothesis involve an incomplete probability specification. That is, we may have only the null hypothesis clearly specified. It might be argued that a full probability formulation should always be attempted covering both null and feasible alternative possibilities. This may seem sensible in principle but as a strategy for direct use it is often not feasible; in any case models that would cover all reasonable possibilities would still be incomplete and would tend to make even simple problems complicated with substantial harmful side-effects.

Note, however, that in all the formulations used here some notion of explanations of the data alternative to the null hypothesis is involved by the choice of test statistic; the issue is when this choice is made via an explicit probabilistic formulation. The general principle of evidence **FEV** helps us to see that in specified contexts, the former suffices for carrying out an evidential appraisal (see Section 3.3).

It is, however, sometimes argued that the choice of test statistic can be based on the distribution of the data under the null hypothesis alone, in effect choosing minus the log probability as test statistic, thus summing probabilities over all sample points as or less probable than that observed. While this often leads to sensible results we shall not follow that route here.

3.3 Inductive Inferences Based on Outcomes of Tests

How does significance test reasoning underwrite inductive inferences or evidential evaluations in the various cases? The hypothetical operational interpretation of the p-value is clear, but what are the deeper implications either of a modest or of a small value of p? These depend strongly both on (i) the type of null hypothesis, and (ii) the nature of the departure or alternative being probed, as well as (iii) whether we are concerned with the interpretation of particular sets of data, as in most detailed statistical work, or whether we are considering a broad model for analysis and interpretation in a field of study. The latter is close to the traditional Neyman-Pearson formulation of fixing a critical level and accepting, in some sense, H_0 if $p > \alpha$ and rejecting H_0 otherwise. We consider some of the familiar shortcomings of a routine or mechanical use of p-values.

3.4 The Routine-Behavior Use of p-Values

Imagine one sets $\alpha = 0.05$ and that results lead to a publishable paper if and only for the relevant p, the data yield $p < 0.05$. The rationale is the behavioristic one outlined earlier. Now the great majority of statistical discussion, going back to Yates (1951) and earlier, deplores such an approach, both out of a concern that it encourages mechanical, automatic, and unthinking procedures, as well as a desire to emphasize estimation of relevant effects over testing of hypotheses. Indeed a few journals in some fields have in effect banned the use of p-values. In others, such as a number of areas of epidemiology, it is conventional to emphasize 95% confidence intervals, as indeed is in line with much mainstream statistical discussion. Of course, this does not free one from giving a proper frequentist account of the use and interpretation of confidence levels, which we do not do here (though see Section 3.6).

Nevertheless the relatively mechanical use of p-values, while open to parody, is not far from practice in some fields; it does serve as a screening device, recognizing the possibility of error, and decreasing the possibility of the publication of misleading results. A somewhat similar role of tests arises in the work of regulatory agents, in particular the Food and Drug Administration (FDA). While requiring studies to show p less than some preassigned level by a preordained test may be inflexible, and the choice of critical level arbitrary, nevertheless such procedures have virtues of impartiality and relative independence from unreasonable manipulation. While adhering to a fixed p-value may have the disadvantage of biasing the literature towards positive conclusions, it offers an appealing assurance of some known and desirable long-run properties. They will be seen to be particularly appropriate for Example 3 of Section 4.2.

3.5 The Inductive-Evidence Use of p-Values

We now turn to the use of significance tests which, while more common, is at the same time more controversial; namely as one tool to aid the analysis of specific sets of data, and/or base inductive inferences on data. The discussion presupposes that the probability distribution used to assess the p-value is as appropriate as possible to the specific data under analysis.

The general frequentist principle for inductive reasoning, **FEV**, or something like it, provides a guide for the appropriate statement about evidence or inference regarding each type of null hypothesis. Much as one makes inferences about changes in body mass based on performance characteristics of various scales, one may make inferences from significance test results by using error rate properties of tests. They indicate the capacity of the particular test to have revealed inconsistencies and discrepancies in the respects probed, and this in turn allows relating p-values to hypotheses about the process as statistically modeled. It follows that an adequate frequentist account of inference should strive to supply the information to implement **FEV**.

Embedded Nulls. In the case of embedded null hypotheses, it is straightforward to use small p-values as evidence of discrepancy from the null in the direction of the alternative. Suppose, however, that the data are found to accord with the null hypothesis (p not small). One may, if it is of interest, regard this as evidence that any discrepancy from the null is less than δ, using the same logic in significance testing. In such cases concordance with the null may provide evidence of the absence of a discrepancy from the null of various sizes, as stipulated in **FEV**(ii).

To infer the absence of a discrepancy from H_0 as large as δ we may examine the probability $\beta(\delta)$ of observing a worse fit with H_0 if $\mu = \mu_0 + \delta$. If that probability is near one then, following **FEV**(ii), the data are good evidence that $\mu < \mu_0 + \delta$. Thus $\beta(\delta)$ may be regarded as the stringency or severity with which the test has probed the discrepancy δ; equivalently one might say that $\mu < \mu_0 + \delta$ has passed a severe test (Mayo, 1996).

This avoids unwarranted interpretations of consistency with H_0 with insensitive tests. Such an assessment is more relevant to specific data than is the notion of power, which is calculated relative to a predesignated critical value beyond which the test "rejects" the null. That is, power appertains to a prespecified rejection region, not to the specific data under analysis.

Although oversensitivity is usually less likely to be a problem, if a test is so sensitive that a p-value as or even smaller than the one observed is probable even when $\mu < \mu_0 + \delta$, then a small value of p is not evidence of departure from H_0 in excess of δ.

If there is an explicit family of alternatives, it will be possible to give a set of confidence intervals for the unknown parameter defining H_0, and this would give a more extended basis for conclusions about the defining parameter.

Dividing and Absence of Structure Nulls. In the case of dividing nulls, discordancy with the null (using the two-sided value of p) indicates direction of departure (e.g., which of two treatments is superior); accordance with H_0 indicates that these data do not provide adequate evidence of any difference, even of the direction. One often hears criticisms that it is pointless to test a null hypothesis known to be false, but even if we do not expect two means, say, to be equal, the test is informative in order to divide the departures into qualitatively different types. The interpretation is analogous when the null hypothesis is one of absence of structure: a modest value of p indicates that the data are insufficiently sensitive to detect structure. If the data are limited this may be no more than a warning against over interpretation rather than evidence for thinking that indeed there is no structure present. That is because the test may have had little capacity to have detected any structure present. A small value of p, however, indicates evidence of a genuine effect; that to look for a substantive interpretation of such an effect would not be an intrinsically error-prone procedure.

Analogous reasoning applies when assessments about the probativeness or sensitivity of tests are informal. If the data are so extensive that accordance with the null hypothesis implies the absence of an effect of practical importance, and a reasonably high p-value is achieved, then it may be taken as evidence of the absence of an effect of practical importance. Likewise, if the data are of such a limited extent that it can be assumed that data in accord with the null hypothesis are consistent also with departures of scientific importance, then a high p-value does not warrant inferring the absence of scientifically important departures from the null hypothesis.

Nulls of Model Adequacy. When null hypotheses are assertions of model adequacy, the interpretation of test results will depend on whether one has a relatively focused test statistic designed to be sensitive against special kinds of model inadequacy, or so called omnibus tests. Concordance with the null in the former case gives evidence of absence of the type of departure that the test is sensitive in detecting, whereas, with the omnibus test, it is less informative. In both types of tests, small p-value is evidence of some departure, but so long as various alternative models could account for the observed violation (i.e., so long as this test had little ability to discriminate between them), these data by themselves may only provide provisional suggestions of alternative models to try.

Substantive Nulls. In the preceding cases, accordance with a null could at most provide evidence to rule out discrepancies of specified amounts or types, according to the ability of the test to have revealed the discrepancy. More can be said in the case of substantive nulls. If the null hypothesis represents a prediction from some theory being contemplated for general applicability, consistency with the null hypothesis may be regarded as some additional evidence for the theory, especially if the test and data are sufficiently sensitive to exclude major departures from the theory. As encapsulated in Fisher's aphorism (Cochran, 1965) that to help make observational studies more nearly bear a causal interpretation, one should make ones' theories elaborate, by which he meant one should plan a variety of tests of different consequences of a theory, to obtain a comprehensive check of its implications. The limited result that one set of data accords with the theory adds one piece to the evidence whose weight stems from accumulating an ability to refute alternative explanations.

In the first type of example under this rubric, there may be apparently anomalous results for a theory or hypothesis T, where T has successfully passed appreciable theoretical and/or empirical scrutiny. Were the apparently anomalous results for T genuine, it is expected that H_0 will be rejected, so that when it is not, the results are positive evidence against the reality of the anomaly. In a second type of case, one again has a well-tested theory T, and a rival theory T^* is determined to conflict with T in a thus far untested domain, with respect to an effect. By identifying the null with the prediction from T, any discrepancies in the direction of T^* are given a very good chance to be detected, such that, if no significant departure is found, this constitutes evidence for *T* in the respect tested.

Although the general theory of relativity, GTR, was not facing anomalies in the 1960s, rivals to the GTR predicted a breakdown of the weak equivalence principle (WEP) for massive self-gravitating bodies, e.g., the earth-moon system: this effect, called the Nordvedt effect would be 0 for GTR (identified with the null hypothesis) and non-0 for rivals. Measurements of the round trip travel times between the earth and moon (between 1969 and 1975) enabled the existence of such an anomaly for GTR to be probed. Finding no evidence against the null hypothesis set upper bounds to the possible violation of the WEP, and because the tests were sufficiently sensitive, these measurements provided good evidence that the Nordvedt effect is absent, and thus evidence for the null hypothesis (Will, 1993). Note that such a negative result does not provide evidence for all of GTR (in all its areas of prediction), but it does provide evidence for its correctness with respect to this effect. The logic is this: theory *T* predicts H_0 is at least a very close approximation to the true situation; rival theory T^* predicts a specified discrepancy from H_0, and the test has high probability of detecting such a discrepancy from *T* were T^* correct. Detecting no discrepancy is thus evidence for its absence.

3.6 Confidence Intervals

As noted above in many problems the provision of confidence intervals, in principle at a range of probability levels, gives the most productive frequentist analysis. If so, then confidence interval analysis should also fall under our general frequentist principle. It does. In one sided testing of $\mu = \mu_0$ against $\mu > \mu_0$, a small *p*-value corresponds to μ_0 being (just) excluded from the corresponding $(1-2p)$ (two-sided) confidence interval (or $1-p$ for the one-sided interval). Were $\mu = \mu_L$, the lower confidence bound, then a less discordant result would occur with high probability $(1-p)$. Thus **FEV** licenses taking this as evidence of inconsistency with $\mu = \mu_L$ (in the positive direction). Moreover, this reasoning shows the advantage of considering several confidence intervals at a range of levels, rather than just reporting whether or not a given parameter value is within the interval at a fixed confidence level.

Neyman developed the theory of confidence intervals *ab initio*, i.e., relying only implicitly rather than explicitly on his earlier work with E.S. Pearson on the theory of tests. It is to some extent a matter of presentation whether one regards interval estimation as so different in principle from testing hypotheses that it is best developed separately to preserve the conceptual distinction. On the other hand there are considerable advantages to regarding a confidence limit, interval, or region as the set of parameter values consistent with the data at some specified level, as assessed by testing each possible value in turn by some mutually concordant procedures. In particular this approach deals painlessly with confidence intervals that are null or which consist of all possible parameter values, at some specified significance level. Such null or infinite regions simply record that the data are inconsistent with all possible parameter values, or are consistent with all possible values. It is easy to construct examples where these seem entirely appropriate conclusions.

4 Some Complications: Selection Effects

The idealized formulation involved in the initial definition of a significance test in principle starts with a hypothesis and a test statistic, then obtains data, then applies the test and looks at the outcome. The hypothetical procedure involved in the definition of the test then matches reasonably closely what was done; the possible outcomes are the different possible values of the specified test statistic. This permits features of the distribution of the test statistic to be relevant for learning about corresponding features of the mechanism generating the data. There are various reasons why the procedure actually followed may be different and we now consider one broad aspect of that.

It often happens that either the null hypothesis or the test statistic are influenced by preliminary inspection of the data, so that the actual procedure generating the final test result is altered. This, in turn may alter the capabilities of the test to detect discrepancies from the null hypotheses reliably, calling for adjustments in its error probabilities.

To the extent that p is viewed as an aspect of the logical or mathematical relation between the data and the probability model such preliminary choices are irrelevant. This will not suffice in order to ensure that the p-values serve their intended purpose for frequentist inference, whether in behavioral or evidential contexts. To the extent that one wants the error-based calculations that give the test its meaning to be applicable to the tasks of frequentist statistics, the preliminary analysis and choice may be highly relevant.

The general point involved has been discussed extensively in both philosophical and statistical literatures, in the former under such headings as requiring novelty or avoiding *ad hoc* hypotheses, under the latter, as rules against peeking at the data or shopping for significance, and thus requiring selection effects to be taken into account. The general issue is whether the evidential bearing of data y on an inference or hypothesis H_0 is altered when H_0 has been either constructed or selected for testing in such a way as to result in a specific observed relation between H_0 and y, whether that is agreement or disagreement. Those who favor logical approaches to confirmation say no (e.g., Mill 1888, Keynes, 1921), whereas those closer to an error-statistical conception say yes (Whewell (1847), Pierce (1931–5)). Following the latter philosophy, Popper required that scientists set out in advance what outcomes they would regard as falsifying H_0, a requirement that even he came to reject; the entire issue in philosophy remains unresolved (Mayo, 1996).

Error-statistical considerations allow going further by providing criteria for when various data dependent selections matter and how to take account of their influence on error probabilities. In particular, if the null hypothesis is chosen for testing because the test statistic is large, the probability of finding some such discordance or other may be high even under the null. Thus, following **FEV**(i), we would not have genuine evidence of discordance with the null, and unless the p-value is modified appropriately, the inference would be misleading. To the extent that one wants the error-based calculations that give the test its meaning to supply reassurance that apparent inconsistency in the particular case is genuine and not merely due to chance, adjusting the p-value is called for.

Such adjustments often arise in cases involving data-dependent selections either in model selection or construction; often the question of adjusting p arises in cases involving multiple hypotheses testing, but it is important not to run cases together simply because there is data dependence or multiple hypothesis testing. We now outline some special cases to bring out the key points in different scenarios. Then we consider whether allowance for selection is called for in each case.

4.1 Examples

Example 1. An investigator has, say, 20 independent sets of data, each reporting on different but closely related effects. The investigator does all 20 tests and reports only the smallest p, which in fact is about 0.05, and its corresponding null hypothesis. The key points are the independence of the tests and the failure to report the results from insignificant tests.

Example 2. A highly idealized version of testing for a DNA match with a given specimen, perhaps of a criminal, is that a search through a data-base of possible matches is done one at a time, checking whether the hypothesis of agreement with the specimen is rejected. Suppose that sensitivity and specificity are both very high. That is, the probabilities of false negatives and false positives are both very small. The first individual, if any, from the data-base for which the hypothesis is rejected is declared to be the true match and the procedure stops there.

Example 3. A microarray study examines several thousand genes for potential expression of say a difference between Type 1 and Type 2 disease status. There are thus several thousand hypotheses under investigation in one step, each with its associated null hypothesis.

Example 4. To study the dependence of a response or outcome variable y on an explanatory variable x, it is intended to use a linear regression analysis of y on x. Inspection of the data suggests that it would be better to use the regression of log y on log x, for example because the relation is more nearly linear or because secondary assumptions, such as constancy of error variance, are more nearly satisfied.

Example 5. To study the dependence of a response or outcome variable y on a considerable number of potential explanatory variables x, a data-dependent procedure of variable selection is used to obtain a representation which is then fitted by standard methods and relevant hypotheses tested.

Example 6. Suppose that preliminary inspection of data suggests some totally unexpected effect or regularity not contemplated at the initial stages. By a formal test the effect is very "highly significant." What is it reasonable to conclude?

4.2 Need for Adjustments for Selection

There is not space to discuss all these examples in depth. A key issue concerns which of these situations need an adjustment for multiple testing or data dependent selection and what that adjustment should be. How does the general conception of the character of a frequentist theory of analysis and interpretation help to guide the answers?

We propose that it does so in the following manner: Firstly it must be considered whether the context is one where the key concern is the control of error rates in a series of applications (behavioristic goal), or whether it is a context of making a specific inductive inference or evaluating specific evidence (inferential goal). The relevant error probabilities may be altered for the former context and not for the latter. Secondly, the relevant sequence of repetitions on which to base frequencies needs to be identified. The general requirement is that we do not report discordance with a null hypothesis by means of a procedure that would report discordancies fairly frequently even though the null hypothesis is true. Ascertainment of the relevant hypothetical series on which this error frequency is to be calculated demands consideration of the nature of the problem or inference. More specifically, one must identify the particular obstacles that need to be avoided for a reliable inference in the particular case, and the capacity of the test, as a measuring instrument, to have revealed the presence of the obstacle.

When the goal is appraising specific evidence, our main interest, **FEV** gives some guidance. More specifically the problem arises when data are used to select a hypothesis to test or alter the specification of an underlying model in such a way that **FEV** is either violated or it cannot be determined whether **FEV** is satisfied (Mayo and Kruse, 2001).

Example 1. (*Hunting for Statistical Significance.*) The test procedure is very different from the case in which the single null found statistically significant was preset as the hypothesis to test; perhaps it is H_0, 13, the 13th null hypothesis out of the 20. In Example 1, the possible results are the possible statistically significant factors that might be found to show a "calculated" statistical significant departure from the null. Hence the type 1 error probability is the probability of finding at least one such significant difference out of 20, even though the global null is true (i.e., all twenty observed differences are due to chance). The probability that this procedure yields erroneous rejection differs from, and will be much greater than, 0.05 (and is approximately 0.64). There are different, and indeed many more, ways one can err in this example than when one null is prespecified, and this is reflected in the adjusted p-value.

This much is well known, but should this influence the interpretation of the result in a context of inductive inference? According to **FEV** it should. However the concern is not the avoidance of often announcing genuine effects erroneously in a series, the concern is that this test performs poorly as a tool for discriminating genuine from chance effects in this particular case. Because at least one such impressive departure, we know, is common even if all are due to chance, the test has scarcely reassured us that it has done a good job of avoiding such a mistake in this case. Even if there are other grounds for believing the genuineness of the one effect that is found, we deny that this test alone has supplied such evidence.

Frequentist calculations serve to examine the particular case, we have been saying, by characterizing the capability of tests to have uncovered mistakes in inference, and on those grounds, the "hunting procedure" has low capacity to have alerted us to, in effect, temper our enthusiasm, even where such tempering is warranted. If, on the other hand, one adjusts the p-value to reflect the overall error rate, the test again becomes a tool that serves this purpose.

Example 1 may be contrasted to a standard factorial experiment set up to investigate the effects of several explanatory variables simultaneously. Here there are a number of distinct questions, each with its associated hypothesis and each with its associated p-value. That we address the questions via the same set of data rather than via separate sets of data is in a sense a technical accident. Each p is correctly interpreted in the context of its own question. Difficulties arise for particular inferences only if we in effect throw away many of the questions and concentrate only on one, or more generally a small number, chosen just because they have the smallest p. For then we have altered the capacity of the test to have alerted us, by means of a correctly computed p-value, whether we have evidence for the inference of interest.

Example 2. (*Explaining a Known Effect by Eliminative Induction.*) Example 2 is superficially similar to Example 1, finding a DNA match being somewhat akin to finding a statistically significant departure from a null hypothesis: one searches through data and concentrates on the one case where a "match" with the criminal's DNA is found, ignoring the non-matches. If one adjusts for "hunting" in Example 1, shouldn't one do so in broadly the same way in Example 2? No.

In Example 1 the concern is that of inferring a genuine, "reproducible" effect, when in fact no such effect exists; in Example 2, there is a known effect or specific event, the criminal's DNA, and reliable procedures are used to track down the specific cause or source (as conveyed by the low "erroneous match" rate). The probability is high that we would not obtain a match with person i, if i were not the criminal; so, by **FEV**, finding the match is, at a qualitative level, good evidence that i is the criminal. Moreover, each non-match found, by the stipulations of the example, virtually excludes that person; thus, the more such negative results the stronger is the evidence when a match is finally found. The more negative results found, the more the inferred "match" is fortified; whereas in Example 1 this is not so.

Because at most one null hypothesis of innocence is false, evidence of innocence on one individual increases, even if only slightly, the chance of guilt of another. An assessment of error rates is certainly possible once the sampling procedure for testing is specified. Details will not be given here.

A broadly analogous situation concerns the anomaly of the orbit of Mercury: the numerous failed attempts to provide a Newtonian interpretation made it all the more impressive when Einstein's theory was found to predict the anomalous results precisely and without any *ad hoc* adjustments.

Example 3. (*Micro-Array Data.*) In the analysis of micro-array data, a reasonable starting assumption is that a very large number of null hypotheses are being tested and that some fairly small proportion of them are (strictly) false, a global null hypothesis of no real effects at all often being implausible. The problem is then one of selecting the sites where an effect can be regarded as established. Here, the need for an adjustment for multiple testing is warranted mainly by a pragmatic concern to avoid "too much noise in the network." The main interest is in how best to adjust error rates to indicate most effectively the gene hypotheses worth following up. An error-based analysis of the issues is then via the false-discovery rate, i.e., essentially the long-run proportion of sites selected as positive in which no effect is present. An alternative formulation is via an empirical Bayes model and the conclusions from this can be linked to the false discovery rate. The latter method may be preferable because an error rate specific to each selected gene may be found; the evidence in some cases is likely to be much stronger than in others and this distinction is blurred in an overall false-discovery rate. See Shaffer (2006) for a systematic review.

Example 4. (*Redefining the Test.*) If tests are run with different specifications, and the one giving the more extreme statistical significance is chosen, then adjustment for selection is required, although it may be difficult to ascertain the precise adjustment. By allowing the result to influence the choice of specification, one is altering the procedure giving rise to the p-value, and this may be unacceptable. While the substantive issue and hypothesis remain unchanged the precise specification of the probability model has been guided by preliminary analysis of the data in such a way as to alter the stochastic mechanism actually responsible for the test outcome.

An analogy might be testing a sharpshooter's ability by having him shoot and then drawing a bull's-eye around his results so as to yield the highest number of bull's-eyes, the so-called principle of the Texas marksman. The skill that one is allegedly testing and making inferences about is his ability to shoot when the target is given and fixed, while that is not the skill actually responsible for the resulting high score.

By contrast, if the choice of specification is guided not by considerations of the statistical significance of departure from the null hypothesis, but rather because the data indicate the need to allow for changes to achieve linearity or constancy of error variance, no allowance for selection seems needed. Quite the contrary: choosing the more empirically adequate specification gives reassurance that the calculated p-value is relevant for interpreting the evidence reliably (Mayo and Spanos, 2006). This might be justified more formally by regarding the specification choice as an informal maximum likelihood analysis, maximizing over a parameter orthogonal to those specifying the null hypothesis of interest.

Example 5. (*Data Mining.*) This example is analogous to Example 1, although how to make the adjustment for selection may not be clear because the procedure used in variable selection may be tortuous. Here too, the difficulties of selective reporting are bypassed by specifying all those reasonably simple models that are consistent with the data rather than by choosing only one model (Cox and Snell, 1974). The difficulties of implementing such a strategy are partly computational rather than conceptual. Examples of this sort are important in much relatively elaborate statistical analysis in that series of very informally specified choices may be made about the model formulation best for analysis and interpretation (Spanos, 2000).

Example 6. (*The Totally Unexpected Effect.*) This raises major problems. In laboratory sciences with data obtainable reasonably rapidly, an attempt to obtain independent replication of the conclusions would be virtually obligatory. In other contexts a search for other data bearing on the issue would be needed. High statistical significance on its own would be very difficult to interpret, essentially because selection has taken place and it is typically hard or impossible to specify with any realism the set over which selection has occurred. The considerations discussed in Examples 1–5, however, may give guidance. If, for example, the situation is as in Example 2 (explaining a known effect) the source may be reliably identified in a procedure that fortifies, rather than detracts from, the evidence. In a case akin to Example 1, there is a selection effect, but it is reasonably clear what is the set of possibilities over which this selection has taken place, allowing correction of the p-value. In other examples, there is a selection effect, but it may not be clear how to make the correction. In short, it would be very unwise to dismiss the possibility of learning from data something new in a totally unanticipated direction, but one must discriminate the contexts in order to gain guidance for what further analysis, if any, might be required.

5 Concluding Remarks

We have argued that error probabilities in frequentist tests may be used to evaluate the reliability or capacity with which the test discriminates whether or not the actual process giving rise to data is in accordance with that described in H_0. Knowledge of this probative capacity allows determination of whether there is strong evidence against H_0 based on the frequentist principle we set out in **FEV**. What makes the kind of hypothetical reasoning relevant to the case at hand is not the long-run low error rates associated with using the tool (or test) in this manner; it is rather what those error rates reveal about the data-generating source or phenomenon. We have not attempted to address the relation between the frequentist and Bayesian analyses of what may appear to be very similar issues. A fundamental tenet of the conception of inductive learning most at home with the frequentist philosophy is that inductive inference requires building up incisive arguments and inferences by putting together several different piece-meal results; we have set out considerations to guide these pieces. Although the complexity of the issues makes it more difficult to set out neatly, as, for example, one could by imagining that a single algorithm encompasses the whole of inductive inference, the payoff is an account that approaches the kind of arguments that scientists build up in order to obtain reliable knowledge and understanding of a field.

References

Birnbaum, A. (1977), "The Neyman-Pearson Theory as Decision Theory, and as Inference Theory; with a Criticism of the Lindley-Savage Argument for Bayesian Theory," *Synthese*, 36: 19–49.

Carnap, R. (1962), *Logical Foundations of Probability*, University of Chicago Press, Chicago.

Cochran, W.G. (1965), "The Planning of Observational Studies in Human Populations," (with discussion). *Journal of the Royal Statistical Society A*, 128: 234–65.

Cox, D.R. (1958), "Some Problems Connected with Statistical Inference," *Annals of Mathematical Statistics*, 29: 357–72.

Cox, D.R. (1977), "The Role of Significance Tests (with Discussion)," *Scandinavian Journal of Statistics*, 4: 49–70.

Cox, D.R., and Hinkley, D.V. (1974), *Theoretical Statistics*, Chapman and Hall, London.

Cox, D.R., and Snell, E.J. (1974), "The Choice of Variables in Observational Studies," *Journal of the Royal Statistical Society C*, 23: 51–9.

de Finetti, B. (1974), *Theory of Probability*, 2 Vols. English translation from Italian, Wiley, New York.

Fisher, R.A. (1935a), *Design of Experiments*, Oliver and Boyd, Edinburgh.

Fisher, R.A. (1935b), "The Logic of Inductive Inference," *Journal of the Royal Statistical Society*, 98: 39–54.

Gibbons, J.D., and Pratt, J.W. (1975), "P-Values: Interpretation and Methodology," *American Statistician*, 29: 20–5.

Jeffreys, H. (1961), *Theory of Probability*, 3rd ed., Oxford University Press, Clarendon, Oxford.

Kempthorne, O. (1976), "Statistics and the Philosophers," pp. 273–314 in Harper, W.L. and Hooker, C.A. (eds.), *Foundations of Probability Theory, Statistical Inference, and Statistical Theories of Science, vol. II: Foundations and Philosophy of Statistical Inference*, D. Reidel, Dordrecht, Holland.

Keynes, J.M. (1921), *A Treatise on Probability* (reprinted 1952), St. Martin's Press, New York.

Lehmann, E.L. (1993), "The Fisher and Neyman-Pearson Theories of Testing Hypotheses: One Theory or Two?" *Journal of the American Statistical Association*, 88: 1242–9.

Lehmann, E.L. (1995), "Neyman's Statistical Philosophy," *Probability and Mathematical Statistics*, 15: 29–36.

Mayo, D.G. (1996), *Error and the Growth of Experimental Knowledge*, University of Chicago Press, Chicago.

Mayo, D.G., and Kruse, M. (2001), "Principles of Inference and Their Consequences," pp. 381–403 in D. Cornfield and J. Williamson (eds.), *Foundations of Bayesianism*, Kluwer Academic, Dordrecht, The Netherlands.

Mayo, D.G., and Spanos, A. (2006), "Severe Testing as a Basic Concept in a Neyman-Pearson Philosophy of Induction," *British Journal of Philosophy of Science*, 57: 323–357.

Mill, J.S. (1888), *A System of Logic*, 8th ed., Harper and Bros., New York.

Morrison, D., and Henkel, R. (eds.) (1970), *The Significance Test Controversy*, Aldine, Chicago.

Neyman, J. (1955), "The Problem of Inductive Inference," *Communications in Pure and Applied Mathematics*, 8: 13–46.

Neyman, J. (1957), "Inductive Behavior as a Basic Concept of Philosophy of Science," *International Statistical Review*, 25: 7–22.

Pearson, E.S. (1955), "Statistical Concepts in Their Relation to Reality," *Journal of the Royal Statistical Society B*, 17: 204–7.

Pierce, C.S. (1931–1935). *Collected Papers*, vols. 1–6, Hartshorne and P. Weiss (eds.), Harvard University Press, Cambridge.

Popper, K. (1959), *The Logic of Scientific Discovery*, Basic Books, New York.

Savage, L.J. (1964), "The Foundations of Statistics Reconsidered," pp. 173–188 in H.E. Kyburg and H.E. Smokler (eds.), *Studies in Subjective Probability*, Wiley, New York.

Shaffer, J.P. (2006), "Recent Developments Towards Optimality in Multiple Hypothesis Testing," pp. 16–32 in J. Rojo (ed.), *Optimality: The Second Erich L. Lehmann Symposium*, Lecture Notes-Monograph Series, vol. 49, Institute of Mathematical Statistics, Beachwood, OH.

Spanos, A. (2000), "Revisiting Data Mining: 'Hunting' with or without a License," *Journal of Economic Methodology*, 7: 231–64.

Whewell, W. (1847), *The Philosophy of the Inductive Sciences. Founded Upon Their History*. 2nd ed., vols. 1 and 2 (reprinted 1967). Johnson Reprint, London.

Will, C. (1993), *Theory and Experiment in Gravitational Physics*, Cambridge University Press, Cambridge.

Yates, F. (1951), "The Influence of Statistical Methods for Research Workers on the Development of the Science of Statistics," *Journal of the American Statistical Association*, 46: 19–34.

II Objectivity and Conditionality in Frequentist Inference

David Cox and Deborah G. Mayo

1 Preliminaries

Statistical methods are used to some extent in virtually all areas of science, technology, public affairs, and private enterprise. The variety of applications makes any single unifying discussion difficult if not impossible. We concentrate on the role of statistics in research in the natural and social sciences and the associated technologies. Our aim is to give a relatively non-technical discussion of some of the conceptual issues involved and to bring out some connections with general epistemological problems of statistical inference in science. In the first part of this chapter (7(I)), we considered how frequentist statistics may serve as an account of inductive inference, but because this depends on being able to apply its methods to appropriately circumscribed contexts, we need to address some of the problems in obtaining the methods with the properties we wish them to have. Given the variety of judgments and background information this requires, it may be questioned whether any account of inductive learning can succeed in being "objective." However, statistical methods do, we think, promote the aim of achieving enhanced understanding of the real world, in some broad sense, and in this some notion of objectivity is crucial. We begin by briefly discussing this concept as it arises in statistical inference in science.

2 Objectivity

Objectivity in statistics, as in science more generally, is a matter of both aims and methods. Objective science, in our view, aims to find out what is the case as regards aspects of the world, independently of our beliefs, biases, and interests; thus objective methods aim for the critical control of inferences and hypotheses, constraining them by evidence and checks of error.

The statistician is sometimes regarded as the gatekeeper of objectivity when the aim is to learn about those aspects of the world that exhibit haphazard variability, especially where methods take into account the uncertainties and errors by using probabilistic ideas in one way or another. In one form, probability arises to quantify the relative frequencies of errors in a hypothetical long run; in a second context, probability purports to quantify the "rational" degree of belief, confirmation, or credibility in hypotheses. In the "frequentist" approach, the aim of objective learning about the world is framed within a statistical model of the process postulated to have generated data.

Frequentist methods achieve an objective connection to hypotheses about the data-generating process by being constrained and calibrated by the method's error probabilities in relation to these models: the probabilities derived from the modeled phenomena are equal or close to the actual relative frequencies of results in applying the method. In the second, degree of belief construal by contrast, objectivity is bought by attempting to identify ideally rational degrees of belief controlled by inner coherency. What are often called "objective" Bayesian methods fall under this second banner, and many, although of course not all, current Bayesian approaches appear to favor the use of special prior probabilities, representing in some sense an indifferent or neutral attitude (Berger, 2004). This is both because of the difficulty of eliciting subjective priors and because of the reluctance among scientists to allow subjective beliefs to be conflated with the information provided by data. However, since it is acknowledged that strictly noninformative priors do not exist, the "objective" (or default) priors are regarded largely as conventionally stipulated reference points to serve as weights in a Bayesian computation. We return to this issue in Section 11.

Given our view of what is required to achieve an objective connection to underlying data-generating processes, questions immediately arise as to how statistical methods can successfully accomplish this aim. We begin by considering the nature and role of statistical analysis in its relations to a very general conception of learning from data.

3 Roles of Statistics

Statistical methods, broadly conceived, are directed to the numerous gaps and uncertainties scientists face in learning about the world with limited and fallible data. Any account of scientific method that begins its work only once well-defined evidence claims and unambiguous hypotheses and theories are available forfeits the ability to be relevant to understanding the actual processes behind the success of science. Because the contexts in which statistical methods are most needed are ones that compel us to be most aware of errors and threats to reliability, considering the nature of statistical methods in the collection, modeling, and analysis of data is a good way to obtain a more realistic account of science. Statistical methods are called on at a variety of stages of inquiry even in explorations where only a vague research question is contemplated. A major chapter in statistical theory addresses the design of experiments and observational studies aiming to achieve unambiguous conclusions of as high a precision as is required. Preliminary checks of data quality and simple graphical and tabular displays of the data are made; sometimes, especially with very skillful design, little additional analysis may be needed. We focus, however, on cases where more formal analysis is required, both to extract as much information as possible from the data about the research questions of concern and to assess the security of any interpretation reached.

A central goal behind the cluster of ideas we may call frequentist methods is to extract what can be *learned from data* that can also be vouched for. Essential to this school is the recognition that it is typically necessary to *communicate* to others what has been learned and its associated uncertainty. A fundamental requirement that it sets for itself is to provide means to address legitimate critical questions and to give information about which conclusions are likely to stand up to further probing and where weak spots remain. The whole idea of Fisherian theory implicitly and of Neyman–Pearson more explicitly is that formal methods of statistical inference become relevant primarily when the probing one can otherwise accomplish is of relatively borderline effectiveness, so that the effects are neither totally swamped by noise nor so clear-cut that formal assessment of errors is relatively unimportant. The roles played by statistical methods in these equivocal cases are what make them especially relevant for the epistemological question of how reliable inferences are possible despite uncertainty and error. Where the recognition that data are always fallible presents a challenge to traditional empiricist foundations, the cornerstone of statistical induction is the ability to move from less accurate to more accurate data. Fisher put it thus:

It should never be true, though it is still often *said*, that the conclusions are no more accurate than the data on which they are based. Statistical data are always erroneous, in greater or less degree. The study of inductive reasoning is the study of the embryology of knowledge, of the processes by means of which truth is extracted from its native ore in which it is fused with much error. (Fisher, 1935, p. 39)

4 Formal Statistical Analysis

The basic inferential goal shared by formal statistical theories of inference is to pose and answer questions about aspects of statistical models in light of the data. To this end, empirical data, denoted by y, are viewed as the observed value of a vector random variable Y. The question of interest may be posed in terms of a probability distribution of Y as defined by the relevant statistical model. A model (or family of models) gives the probability distribution (or density) of Y, $f_Y(y; \theta)$, which may be regarded as an abstract and idealized representation of the underlying data-generating process. Statistical inferences are usually couched in terms of the unknown parameter θ.

Example 1. *Bernoulli trials.* Consider n independent trials $(Y_1, Y_2, \ldots, Y_n)$, each with a binary outcome – success or failure – where the probability of success at each trial is an unknown constant θ with a value between 0 and 1. This model is a standard probability model which is often used to represent "coin-tossing" trials, where the hypothesis of a "fair" coin is $\theta = .5$. It serves as a standard in modeling aspects of many cases that are appropriately analogous.

Crucial conceptual issues concern the nature of the probability model and in particular the role of probability in it. An important concept that arises in all model-based statistical inference is that of *likelihood*. If y is a realized data set from $f(y; \theta)$, the likelihood is a function of θ with y fixed: $\text{Lik}(\theta; y) = f(y; \theta)$.

In the case of binomial trials, the data $y = (y_1, y_2, \ldots, y_n)$ forms a sequence of r "successes" and $n - r$ "failures," with the $\text{Lik}(\theta; y) = \theta^r (1 - \theta)^{n-r}$.

Likelihoods do not obey the probability laws: for example, the sum of the likelihoods of a hypothesis and its denial is not 1.

4.1 Three Different Approaches

Three broad categories represent different approaches that are taken regarding model-based statistical inference. Each has given rise to philosophical and methodological controversies that have rumbled on for anywhere from fifty to two hundred years, which we do not plan to review here.[1] The following discussion is merely an outline of a few of the issues involved, which sets the stage for elucidating the unifying principle that enables frequentist

---

[1] See references in 7(IV).

methods, when properly formulated, to obtain objective information about underlying data-generating processes.

Likelihood Methods. The first approach rests on a comparative appraisal of rival statistical hypotheses H_0 and H_1 according to the ratio of their likelihoods. The basic premise is what Hacking (1965) called the *law of likelihood*: that the hypothesis with the higher likelihood has the higher evidential "support," is the more "plausible," or does a better job of "explaining" the data. The formal analysis is based on looking at ratios of likelihoods of two different parameter values $f_S(s; \theta_1)/f_S(s; \theta_0)$; the ratios depend only on statistic s, and are regarded as comparative summaries of "what the data convey" about θ.

Bayesian Methods. In Bayesian approaches, the parameter θ is modeled as a realized value of a random variable Θ with a probability distribution $f_\Theta(\theta)$, called the prior distribution. Having observed y, inference proceeds by computing the conditional distribution of Θ, given $Y = y$, the posterior distribution.

Under this broad category, two notable subgroups reflect contrasting uses of probability: (1) to represent personalistic degree of belief and (2) to represent impersonal or rational degree of belief or some broadly equivalent notion. The central point is that the focus of interest, ψ, is typically an unknown constant, usually a component of θ, and if we were to aim at talking about a probability distribution for ψ, an extended notion of probability, beyond immediately frequentist concepts, would usually then be unavoidable.

With a generalized notion of probability as degree of belief, it is possible, formally at least, to assign a probability distribution to ψ given the data. Many purport to use this notion to assess the strength of evidence or degree of credibility that some hypothesis about ψ is in some sense true. This is done by what used to be called inverse probability and is nowadays referred to as a Bayesian argument. Once the relevant probabilities are agreed on, the calculation uses Bayes's theorem, an entirely uncontroversial result in probability theory that stems immediately from the definition of conditional probability. We obtain the posterior distribution of the full parameter by multiplying the likelihood by the prior density and then multiplying by a suitable constant to make the total posterior probability equal to 1.

Frequentist (Sampling) Methods. The third paradigm makes use of the frequentist view of probability to characterize methods of analysis by means of their performance characteristics in a hypothetical sequence of repetitions. Two main formulations are used in the frequentist approach to the summarization of evidence about the parameter of interest, ψ. For simplicity, we suppose from now on that for each research question of interest ψ is one-dimensional.

The first is the provision of sets or intervals within which ψ is in some sense likely to lie (confidence intervals) and the other is the assessment of concordance and discordance with a specified value ψ_0 (significance tests). Although we concentrate on the latter, the estimation of ψ via sets of intervals at various confidence levels is the preferred method of analysis in many contexts. The two are closely connected, as our interpretation of tests makes plain: confidence intervals consist of parameter values that are not inconsistent with the data at specified levels. Parameter values outside a $(1-c)$ confidence interval are those that contradict the data at significance level c.

4.2 Our Focus

We concentrate here on the frequentist approach. The personalistic approach, whatever merits it may have as a representation of personal belief and personal decision making, is in our view inappropriate for the public communication of information that is the core of scientific research, and of other areas, too. To some extent, the objectivist Bayesian view claims to address the same issues as the frequentist approach; some of the reasons for preferring the frequentist approach are sketched in Section 12. The likelihood approach also shares the goal to learn "what can be said" about parameters of interest; however, except for very simple problems, the pure likelihood account is inadequate to address the complications common in applications and we do not specifically discuss it here.

The key difference between the frequentist approach and the other paradigms is its focus on the sampling distribution of the test (or other) statistic (i.e., its distribution in hypothetical repetition). In our view, the sampling distribution, when properly used and interpreted, is at the heart of the objectivity of frequentist methods. We will discuss the formulation and implementation of these methods in order to address central questions about the relevance and appropriate interpretation of its core notions of hypothetical repetitions and sampling distributions. In the first part of this chapter (Mayo and Cox), we considered the reasoning based on p-values; our considerations now pertain to constructing tests that (1) permit p-values to be calculated under a variety of null hypotheses and (2) ensure the relevance of the hypothetical long run that is used in particular inferences.

5 Embarking on Formal Frequentist Analysis

In a fairly wide variety of contexts, the formal analysis may be seen to proceed broadly as follows. First, we divide the features to be analyzed into two parts and denote their full set of values collectively by y and by x, which are typically multidimensional. A probability model is formulated according to which y is the observed value of a vector random variable Y whose distribution depends on x, regarded as fixed. Note that especially in observational studies it may be that x could have been regarded as random but, given the question of interest, we chose not do so. This choice leads to the relevant sampling distribution.

Example 2. *Conditioning by model formulation.* For a random sample of men, we measure systolic blood pressure, weight, height, and age. The research question may concern the relation between systolic blood pressure and weight, allowing for height and age, and if so, specifically for that question, one would condition on the last three variables and represent by a model the conditional distribution of Y, systolic blood pressure, given x, the other three variables: $f(y \mid x_1, x_2, x_3; \theta)$. The linear regression

$$Y_i = \beta_0 + \beta_1 x_{1i} + \beta_2 x_{2i} + \beta_3 x_{3i} + u_i, \quad i = 1, 2, \ldots, n,$$

is an example of a statistical model based on such a conditional distribution. Here the unknown parameters $(\beta_0, \beta_1, \beta_2, \beta_3)$ represent the effect of changing one explanatory variable while the others are held fixed and are typically the focus of interest. The u_i $(i = 1, 2, \ldots, n)$ are not directly observed random terms of zero expectation, representing the haphazard component of the variation.

One would condition on the explanatory variables even if, say, one knew the distribution of age in the population. We may call this *conditioning by model formulation*. This serves to constrain what is allowed to vary conceptually in determining the sampling distribution for inference.

Of course, for different purposes the additional information of how x varies *would* be relevant and the appropriate statistical model would reflect this. For example, if one were interested in the correlation between systolic blood pressure and age, the relevant distribution would be the joint distribution of Y and X_3, say $f(y, x_3; \psi)$. In that case, the question is how the two variables covary with each other and, thus, the sample space should include all their possible values and the associated probabilities.

Here we consider parametric models in which the probability density of Y is in the form $f_Y(y; \theta)$, where θ is typically a vector of parameters $\theta = (\psi, \lambda)$. Dependence on x is not shown explicitly in this notation. The *parameter of interest*, ψ, addresses the research question of concern; additional parameters are typically needed to complete the probability specification. Because these additions may get in the way of the primary focus of research, they are dubbed *nuisance parameters*. In contrasting different statistical approaches, we draw some comparisons based on how each handles such nuisance parameters.

Virtually all such models are to some extent provisional, which is precisely what is expected in the building up of knowledge. Probability models range from purely empirical representations of the pattern of observed haphazard variability to representations that include substantial elements of the underlying science base. The former get their importance from providing a framework for broad families of statistical methods – for example, some form of regression analysis – that find fruitful application across many fields of study. The latter provide a stronger link to interpretation. An intermediate class of models of increasing importance in observational studies, especially in the social sciences, represents a potential data-generating process and, hence, may point toward a causal interpretation. Parameters – especially the parameters of interest, ψ – are intended to encapsulate important aspects of the data-generating process separated off from the accidents of the specific data under analysis. Probability is to be regarded as directly or indirectly based on the empirical stability of frequencies under real or hypothetical repetition. This allows for a wide latitude of imaginative applications of probability, not limited to phenomena exhibiting actual repetitions.

Example 3. Cox and Brandwood (1959) put in order, possibly of time, the works of Plato, taking *Laws* and *Republic* as reference points. The data were the stresses on the last five syllables of each sentence and these were assumed to have in each book a probability distribution over the thirty-two possibilities. What does probability mean in such a case?

This is a good example of how historical questions, lacking literal repetitions, may be tackled statistically: an attribute such as the relative frequencies of the thirty-two ending types, together with deliberately chosen reference standards, may be used to discriminate patterns. By taking two large known works of Plato, between which the works being dated were written, as giving probability distributions for the thirty-two possible endings, one can assign a (stylometric) score to the relative frequencies of ending types observed in the works whose relative dates are unknown. This allows one to determine what would be expected *statistically* were the two works identical with respect to the particular stylometric score, and thereby to probe null hypotheses of form "these two works are identical (with respect to the time written)." The other works may thus be ordered according to their differences from, or affinity to, the two reference standards (given assumptions about changes in literary style).

Two general challenges now arise. How do we use the data as effectively as possible to learn about θ? How can we check on the appropriateness of the model?

6 Reducing the Data by Sufficiency

Suppose then that data $y = (y_1, y_2, \ldots, y_n)$ are modeled as a realization of random variable Y, and that a family F of possible distributions is specified; we seek a way to reduce the data so that what we wish to learn about the unknown θ may be "extracted from its native ore." We seek a function of the data, a *statistic* $S(Y)$, such that knowing its value $s(y)$ would suffice to encompass the statistical information in the n-dimensional data as regards the parameter of interest. This is called a *sufficient* statistic. We aim to choose a sufficient statistic that minimizes the dimensionality of $s = s(y)$ such that the distribution of the sample factorizes as

$$f(y; \theta) = f_S(s; \theta) f_{Y|S}(y|s),$$

where $f_{Y|S}(y \mid s)$, the conditional distribution of Y given the value of S, does not depend on the unknown parameter θ. In other words, knowing the distribution of the sufficient statistic S suffices to compute the probability of any given y. The process of reducing the data in this way may be called *reduction by sufficiency*.

Example 4. *Binomial model.* Consider n independent trials $(Y_1, Y_2, \ldots, Y_n)$, each with a binary outcome (success or failure), where the probability of success is an unknown constant θ. These are called Bernouilli trials. The sufficient statistic in this case is $S = \sum_{k=1}^n Y_k$, the number of successes, and has a binomial sampling distribution determined by the constants n and θ. It can be shown that the distribution of the sample reduces as earlier, where $f_S(s; \theta)$ is a binomial, and $f_{Y|S}(y|s)$ is a discrete uniform, distribution; all permutations of the sequence of successes and failures are equally likely. We now argue as follows. The experiment would be equivalent to having been given the data $y = (y_1, y_2, \ldots, y_n)$ in two stages:

First, we are told the value of $s(y)$ (e.g., $S = s$ successes out of n Bernouli trials). Then some inference can be drawn about θ using the sampling distribution of S, $f_S(s; \theta)$, in some way.

Second, we learn the value of the remaining parts of the data (e.g., the first k trials were all successes, the rest failures). Now, if the model is appropriate, then the second phase is equivalent to a random draw from a totally known distribution and could just as well be the outcome of a random number generator. Therefore, all the information about θ is, so long as the model is appropriate, locked in s and in the dependence on θ of the distribution of the random variable S.

This second stage is essentially to observe a realization of the conditional distribution of Y given $S = s$, generated by observing y in the distribution $f_{Y|S}(y \mid s)$. Because this conditional distribution is totally known, it can be used to assess the validity of the assumed model. Insofar as the remaining parts of the data show discordancy, in some relevant respect, with being from the known distribution, doubt is thrown on the model. In Example 4, for instance, the fact that any permutation of the r successes in n trials has known probability, assuming the correctness of the model, gives us a standard to check if the model is violated. It is crucial that any account of statistical inference provides a conceptual framework for this process of model criticism, even if in practice the criticism is often done relatively informally. The ability of the frequentist paradigm to offer a battery of simple significance tests for model checking and possible improvement is an important part of its ability to supply objective tools for learning.

The appropriate reduction to a sufficient statistic s is usually best found by considering the likelihood function, which is the probability of the data considered as a function of the unknown parameter θ. The aim is to factor this as a function of s times a function of y not involving θ. That is, we wish to write the following:

$$f_Y(y; \theta) = m_1(y)m_2(s; \theta),$$

say, taking the minimal s for which this factorization holds. One important aspect is that, for any given y from distribution $f_Y(y; \theta)$, in the *same experiment*, the ratio of likelihoods at two different values of θ depends on the data only through s.

7 Some Confusion over the Role of Sufficiency

Sufficiency as such is not specifically a frequentist concept. Unfortunately some confusion has appeared in the literature over the role of sufficiency in frequentist statistics. We can address this by considering two contrasting experimental procedures in relation to Example 4.

Example 5. *Binomial versus negative binomial.* Consider independent Bernoulli trials, where the probability of success is an unknown constant θ, but imagine two different experimental procedures by which they may be produced. Suppose first that a preassigned number of trials, n, is observed. As noted earlier, the sufficient statistic is r, the number of successes, and this is the observed value of a random variable R having a *binomial* distribution with parameters n and θ. Suppose now instead that trials continue until a preassigned number of successes, r has occurred after n trials. Such an observational process is often called inverse sampling. In this second case the sufficient statistic is n, the observed value of a random variable N having a *negative binomial distribution* determined by the constants r and θ. We may denote the two experiments by E_N and E_R, respectively.

Now it has been argued that, because r and n determine the likelihood in the same form proportional to $\theta^n(1 - \theta)^{n-r}$, whether arising from E_N or E_R, that in both cases the same inference should be drawn. It is clear, however, from the present perspective that the roles of n and r are quite different in the two situations and there is no necessary reason to draw the same conclusions. Experiments E_N and E_R have different sample spaces, and because the sampling distributions of the respective sufficient statistics differ, the same string of successes and failures would result in a difference in p-values (or confidence-level) assessments, depending on whether it arose from E_N or E_R, although the difference is typically minor. Perhaps the confusion stems in part because the various inference schools accept the broad, but not the detailed, implications of sufficiency: the difference emanates from holding different notions of inference. We now explain this.

7.1 Sufficiency Principle (General)

If random variable Y, in a given experiment E, has probability density $f_y(y; \theta)$ and S is minimal sufficient for θ, then as long as the model for E is adequate, identical inferences about θ should be drawn from data y' and y'' whenever $\mathbf{y}'$ and $\mathbf{y}''$ yield the same value of s.

We may abbreviate this as follows:

If s is minimal sufficient for θ in experiment E, and $s(y') = s(y'')$, then the inference from y' and y'' about θ should be identical; that is, $\mathrm{Infr}_E(y') = \mathrm{Infr}_E(y'')$.

However, when proposing to apply the sufficiency principle to a particular inference account, the relevant method for inference must be taken into account. That is, Infr_E is relative to the inference account.

7.2 Sufficiency in Sampling Theory

If a random variable Y, in a given experiment E, arises from $f(y; \theta)$, and the assumptions of the model are valid, then all the information about θ contained in the data is obtained from consideration of its minimal sufficient statistic S and its *sampling distribution* $f_S(s; \theta)$.

An inference in sampling theory, therefore, needs to include the relevant sampling distribution, whether it was for testing or estimation. Thus, in using the abbreviation $\mathrm{Infr}_E(y)$ to refer to an inference from y in a sampling theory experiment E, we assume for simplicity that E includes a statement of the probability model, parameters, and sampling distribution corresponding to the inference in question. This abbreviation emphasizes that the inference that is licensed is relative to the particular experiment, the type of inference, and the overall statistical approach being discussed.[2]

In the case of frequentist sampling theory, features of the experiment that alter the sampling distribution must be taken account of in determining what inferences about θ are warranted, and when the same inferences from given experiments may be drawn. Even if y' and y'' have proportional likelihoods but are associated with different relevant sampling distributions, corresponding to E' and E'', y' and y'' each provides different relevant information for inference. It is thus incorrect to suppose, within the sampling paradigm, that it is appropriate to equate $\mathrm{Infr}_{E'}(\mathbf{y}')$ and $\mathrm{Infr}_{E''}(\mathbf{y}'')$.

These points show that sampling theory violates what is called the *strong likelihood principle*.

---

[2] This abbreviation, like Birnbaum's Ev(E, x), may be used to discuss general claims about principles of evidence. Birnbaum's Ev(E, x), "the evidence about the parameter arising from experiment E and result x," is, for Birnbaum, the inference, conclusion or report, and thus is in sync with our notion (Birnbaum, 1962).

We prefer it because it helps avoid assuming a single measure of "the" evidence associated with an experimental outcome. By referring to the inference licensed by the result, it underscores the need to consider the associated methodology and context.

7.3 The Strong Likelihood Principle (SLP)

Suppose that we have *two* experiments, E' and E'', with different probability models $f'_{Y'}(y'; \theta)$ and $f''_{Y''}(y''; \theta)$, respectively, with the same unknown parameter θ. If y'^* and y''^* are observed data from E' and E'', respectively, where the likelihoods of y'^* and y''^* are proportional, *then* y'^* and y''^* have the identical evidential import for any inference about θ.

Here proportionality means that, for all θ, $f''_Y(y''; \theta) / f'_Y(y'; \theta)$ is equal to a constant that does not depend on θ. A sample of, say, six successes in twenty trials would, according to the SLP, have the identical evidential import whether it came from a binomial experiment, with sample size fixed at twenty, or from a negative binomial experiment where it took twenty trials to obtain six successes.

By contrast, suppose a frequentist is interested in making an inference about θ on the basis of data y' consisting of r successes in n trials in a binomial experiment E'. Relevant information would be lost if the report were reduced to the following: there were r successes in n Bernoulli trials, generated from *either* a binomial experiment with n fixed, y'^*, or a negative binomial experiment with r fixed, y''^* – concealing which was actually the source of the data. Information is lost because $\text{Infr}_{E'}(y'^*)$ is *not* equal to $\text{Infr}_{E''}(y''^*)$ due to the difference in the associated sampling distributions. Equivalences that hold with respect to a single experiment, as is the case with sufficiency, cannot be assumed to hold in comparing data from different experiments.

8 Sufficient Statistics and Test Statistics

How then are we to extract answers to the research question out of $f_S(s; \theta)$; all that the reduction to s has done is to reduce the dimensionality of the data. To establish a significance test, we need to choose an appropriate test statistic $T(Y)$ and find a distribution for assessing its concordancy with H_0. To warrant the interpretations of the various significance tests that we delineated in the first part of this chapter (Mayo and Cox), we need to consider how to identify test statistics to construct appropriate tests.

To interpret t, the observed value of T, we compare it with its predicted value under the null hypothesis by finding, for any observed value t, $p = P(T \geq t; H_0)$. That is, we examine how extreme t is in its probability distribution under H_0. Thus, we need both to choose an appropriate test statistic $T(Y)$ and also to compute its distribution in order to compare t with what is expected under H_0. To this end we find a suitable feature t of the data, in light of the previous discussion, a function of sufficient statistic s, such that

[1] The larger the value of t, the greater the discrepancy with the null hypothesis in the respect of concern.

[2] The probability distribution of the random variable T is exactly known when the null hypothesis is true, so that in particular the distribution does not depend on nuisance parameters.

We must then collect data y to compute $t(y)$, ensuring the data satisfy adequately the assumptions of the relevant probability model. Note that the p-value can itself be regarded as a random variable P; and the probability that P takes different values under alternatives to the null hypothesis may be calculated.

To satisfy condition [1], the larger the value of t, the smaller the corresponding p-value must be. Satisfying condition [2] is the frequentists' way of ensuring as far as possible that observed discordancies are attributable to discrepancies between the null hypothesis and the actual phenomena giving rise to the data. This is key to avoiding ambiguities in pinpointing the source of observed anomalies (Duhem's problem).

The choice of test statistic depends on the type of null hypothesis involved (see the delineation in 7(I), p. 257). We deal first with an important situation where an essentially unique answer is possible. Suppose there is a full model covering both null and alternative possibilities. With such "embedded" nulls, there is formulated not only a probability model for the null hypothesis but also models that represent other possibilities in which the null hypothesis is false and usually, therefore, represent possibilities whose presence we would wish to detect.

Among the number of possible situations, a common one involves a parametric family of distributions indexed by an unknown parameter θ. Suppose that θ is one-dimensional. We reduce by sufficiency to $S(Y)$. If $S(Y)$ itself is one-dimensional, the test statistic must be a function of $S(Y)$ and we can almost always arrange that $S(Y)$ itself can be taken as the test statistic and its distribution thus found.

The null hypothesis is typically not logically contradicted however far t is from what is expected under the null hypothesis except in those rare cases where certain values of t are logically impossible under the null hypothesis; however, the p-value indicates the level at which the data contradict the null hypothesis. In selecting tests or, in the embedded case, corresponding confidence limits, two perspectives are possible. One focuses on being able to give objective guarantees of low long-run error rates and optimality properties that hold regardless of unknown nuisance parameters. A second focuses on being able to objectively determine how consistent data are from various values of the parameter of interest. The former relates to the behavioristic perspective traditionally associated with Neyman–Pearson theory, the latter with the inductive inference perspective that we advance here.

Consider generating a statistic for the $1 - \alpha$ upper confidence bound, $\text{CI}^U(Y; \alpha)$ for estimating a normal mean μ. This statistic is directly related to a test of $\mu = \mu_0$ against $\mu < \mu_0$. In particular, Y is statistically significantly smaller than those values of μ in excess of $\text{CI}^U(Y; \alpha)$ at level α. Mathematically, the same intervals emerge from following the Neyman–Pearson or Fisherian perspective. Both aim to guarantee the sensitivity of the analysis by ensuring $P(\mu' < \text{CI}^U(Y; \alpha))$ is minimal for $\mu' > \mu$, subject to the requirement that, with high probability $(1 - \alpha)$, $\text{CI}^U(Y; \alpha)$ exceeds the true value of μ. That is,

$$P(\mu < \text{CI}^U(Y; \alpha)) = 1 - \alpha.$$

To contrast the differences in interpretation and justification, consider forming $\text{CI}^U(Y; \alpha)$ for a normal mean where the variance σ^2 is known. The observed upper limit is $\overline{y}_0 + k(\alpha)\sigma_y$, where $\overline{y}_0 = \sum_{k=1}^{n} -y_k$, $k(\alpha)$ is the upper α-point of the standard normal distribution, and $\sigma_y = \sigma/\sqrt{n}$. Consider the inference $\mu < \overline{y}_0 + k(\alpha)\sigma_y$. One rationale that may be given to warrant this inference is that it instantiates an inference rule that yields true claims with high probability $(1 - \alpha)$ since

$$P(\mu < \overline{Y} + k(\alpha)\sigma_y) = 1 - \alpha.$$

The procedure, it is often said, has high long-run "coverage probabilities." A somewhat different justification, based on the same probabilistic facts, is to view $\mu \leq \overline{y}_0 + k(\alpha)\sigma_y$ as an inference based on a type of *reductio ad absurdum* argument: suppose in fact that this inference is false and the true mean is μ', where $\mu' > \overline{y}_0 + k(\alpha)\sigma_y$. Then it is very probable that we would have observed a larger sample mean since

$$P(\overline{Y} > \overline{y}_0; \mu') > 1 - \alpha.$$

Therefore, one can reason, $\overline{y}_0$ is inconsistent at level $(1 - \alpha)$, with having been generated from a population with μ in excess of the upper confidence limit. This reasoning is captured in the frequentist principle of evidence FEV that we set out in 7(I), p. 254.

The Neyman–Pearson formulation arrives at essentially the same test or confidence interval but proceeds in a sense in the opposite direction. Rather than beginning with sufficient statistic S, optimality criteria are set up for arriving at the most sensitive analysis possible with the data. Solving the optimality problem, one arrives at a procedure in which the data enter via sufficient statistic *S*. In the Neyman–Pearson theory, sensitivity is assessed by means of the power – the probability of reaching a preset level of significance under the assumption that various alternative hypotheses are true. In the approach described here, sensitivity is assessed by means of the distribution of the random variable *P*, considered under the assumption of various alternatives. In confidence intervals, corresponding sensitivity assessments are not directly in terms of length of intervals but rather in terms of the probability of including false values of the parameter.

The two avenues to sufficient statistic *s* often lead to the same destination but with some differences of interpretation and justification. These differences can lead to more flexible specifications and uses of the same statistical tools. For example, it suffices for our purposes that the error probabilities are only approximate. Whereas Neyman–Pearson confidence intervals fix a single confidence level for a parameter of interest, in the current approach one would want to report several confidence limits at different levels. These benchmarks serve to more fully convey what the data are saying with respect to which values are, and are not, consistent with the data at different levels.

This interpretation of confidence intervals also scotches criticisms of examples where, due to given restrictions, it can happen that a $(1 - \alpha)$ estimate contains all possible parameter values. Although such an inference is "trivially true," it is scarcely vacuous in our construal. That all parameter values are consistent with the data is an informative statement about the limitations of the data to detect discrepancies at the particular level.

9 Conditioning for Separation from Nuisance Parameters

In most realistic situations there is a nuisance parameter λ in addition to the parameter of interest. In this section we consider the formulation of tests to accommodate such nuisance parameters – first from the current perspective and then in their relation to tests developed from the traditional Neyman–Pearson perspective. In order to take a small value of *p* as evidence that it is due to a discrepancy between the null hypothesis and the actual data-generating procedure, we need a test statistic with a distribution that is split off from that of the unknown nuisance parameter λ. The parameter θ may be partitioned into components $\theta = (\psi, \lambda)$ such that the null hypothesis is that $\psi = \psi_0$, where λ is an unknown nuisance parameter. Interest may focus on alternatives $\psi > \psi_0$.

In this case one aim is to achieve a factorization $s = (t, v)$, where t is one-dimensional such that

- the random variable V has a distribution depending only on λ and
- the conditional distribution of T given $V = v$ depends only on ψ.

In constructing significance tests, these conditions may sometimes be achieved by conditioning on a sufficient statistic V for the nuisance parameter, thereby reducing the null hypothesis to a simple hypothesis, where ψ is the only unknown. The test statistic is $T | V$ (i.e., T given $V = v$). Although the distribution of V depends on an unknown, the fact that it is disconnected from the parameter under test, ψ, allows values of p for a hypothesis about ψ (e.g., $H_0\colon \psi = \psi_0$) to be calculated from this conditional distribution, $T | V$. This may be called *technical conditioning for separation from nuisance parameters*. It has the additional advantage that it largely determines the appropriate test statistic by the requirement of producing the most sensitive test possible with the data at hand.

Example 6. *Conditioning for separation from nuisance parameters.* Suppose that Y_1 and Y_2 have independent Poisson distributions of means μ_1 and μ_2, respectively, but that it is only the ratio of the means, μ_1/μ_2, that is of interest; that is, the null hypothesis concerns $\psi = \mu_1/\mu_2$ and, therefore, $H_0\colon \psi = \psi_0$. Thus, the nuisance parameter λ is $\mu_1 + \mu_2$. In fact, for any given value of ψ, it can be shown that there is a sufficiency reduction to $V = y_1 + y_2$. That is, for any given value of ψ, the observed value v contains all the information about nuisance parameter λ. (V is a *complete sufficient statistic* for λ.) There is a factorization into information about λ and the complementary term of the distribution of, say, Y_1 given $V = v$, which depends only on ψ and, thus, contains all the information about ψ so long as there is no other information about the nuisance parameter. The variable Y, given $V = v$ has a binomial distribution with probability of success $\psi / (\psi + 1)$; accordingly, the test rejects the null hypothesis for large values of y. This conditional distribution serves as our test statistic for the hypothesis of interest. In addition to achieving separation from nuisance parameters, the observed value of V also indicates the precision of the inference to be drawn. The same test would emerge based on the goal of achieving a uniformly most powerful size α *similar* test.

9.1 Conditioning to Achieve UMP Size α Rejection Regions

In the most familiar class of cases, the aforementioned strategy for constructing appropriately sensitive tests, separate from nuisance parameters, produces the same tests entailed by Neyman–Pearson theory, albeit with a difference in rationale. In particular, when certain requirements are satisfied rendering the statistic V a "complete" sufficient statistic for nuisance parameter λ, there is no other way of achieving the Neyman–Pearson goal of an exactly α-level rejection region that is fixed regardless of nuisance parameters – exactly *similar* tests.[3] These requirements are satisfied in many familiar classes of significance tests. In all such cases, exactly similar size α rejection regions are equivalent to regions where the conditional probability of Y being significant at level α is independent of v:

$$\Pr(T(Y) \text{ is significant at level } \alpha \mid v; H_0) = \alpha,$$

where v is the value of the statistic V that is used to eliminate dependence on the nuisance parameter. Rejection regions where this condition holds are called regions of *Neyman structure*. Having reduced the null hypothesis to a simple hypothesis, one may then ensure the test has maximum power against alternatives to the null within the class of α-level tests. In the most familiar cases, therefore, conditioning on a sufficient statistic for a nuisance parameter may be regarded as an outgrowth of the aim of calculating the relevant *p*-value independent of unknowns, or, alternatively, as a by-product of seeking to obtain the most powerful similar tests.

However, requiring exactly similar rejection regions precludes tests that merely satisfy the weaker requirement of being able to calculate *p* approximately, with only minimal dependence on nuisance parameters; and yet these tests may be superior from the perspective of ensuring adequate sensitivity to departures, given the particular data and inference of relevance. This fact is especially relevant when optimal tests are absent. Some examples are considered in Section 10.

9.2 Some Limitations

The constructions sketched in the preceding sections reveal the underpinnings of a substantial part of what may be called elementary statistical methods, including standard problems about binomial, Poisson, and normal distributions, and the method of least squares for so-called linear models. When we go to more complicated situations, the factorizations that underlie the arguments no longer hold. In some generality, however, we may show that they hold approximately and we may use that fact to obtain *p*-values whose interpretation is only very mildly dependent on the values of nuisance parameters. The distributional calculations needed to

---

[3] For a discussion on the technical notion of (bounded) completeness, see Lehmann (1986).

find p often involve appeal to so-called asymptotic theory or, perhaps more commonly nowadays, involve computer simulation. The goal of ensuring minimal dependence of the validity of primary inferences on unknown nuisance parameters is thereby achieved. A contrasting way to assess hypothesis H_0 in the face of several parameters is to assign (prior) probability distributions to each and integrate out the uninteresting parameters to arrive at the posterior for the null hypothesis given the data. Then, however, the resulting inference depends on introducing probability distributions for the unknown nuisance parameters, and the primary inference may be vitiated by faulty priors. The corresponding Bayesian treatment does not involve mathematical approximations, except where forced by the numerical complexity of some applications, but it does depend, often relatively critically, on a precise formulation of the prior distribution. Even in so-called impersonal Bayesian accounts, it can depend on the particular ordering of importance of the nuisance parameters. (We return to this in Section 12.)

10 Conditioning to Induce Relevance to the Particular Inference

Being able to calculate the p-value under the null, split off from nuisance parameters, although a necessary accomplishment, does not by itself entail that the calculation will be appropriately relevant for purposes of inference from the data. Although conditioning on sufficient statistics for nuisance parameters is also to tailor the inference to the best estimates of the background or nuisance parameters, more may be required in certain cases to ensure relevance to the given question of interest. We now turn to this issue.

Suppose then that one can calculate the p-value associated with an observed difference t_{obs}, namely $P(T \geq t_{obs}; \psi = \psi_0)$. If $P(T \geq t_{obs}; \psi = \psi_0)$ is very low (e.g., .001), then t_{obs} is grounds to reject H_0 or to infer a discordance with H_0 in the direction of the specified alternative at the corresponding level .001. There are two main rationales for this interpretation:

1. It is to follow a rule with low error rates (i.e., erroneous rejections) in the long run when H_0 is true. In particular, we may give any particular value p the following hypothetical interpretation. Suppose that we were to treat the data as just decisive evidence against H_0; then, in hypothetical repetitions, H_0 would be rejected in a long-run proportion p of the cases in which it is actually true.

However, this theoretical calibration of a significance test may be used as a measuring instrument to make inferences about how consistent or inconsistent these data show this hypothesis to be in the *particular case* at hand. In such contexts the justification is that

2. It is to follow a rule where the low *p*-value corresponds to the *specific data set* providing evidence of inconsistency with or discrepancy from H_0.

This evidential construal follows the frequentist principle FEV in 7(I). This aim is accomplished only to the extent that it can be assured that the small observed *p*-value is due to the actual data-generating process being discrepant from that described in H_0. Moreover, the *p*-values (or corresponding confidence levels) associated with the inference should validly reflect the stringency and sensitivity of the actual test and the specific data observed.

Once these requirements in rationale 2 are satisfied, the low-error-rate rationale 1 follows, but the converse is not true. Many criticisms of frequentist significance tests (and related methods) are based on arguments that overlook the avenues open in frequentist theory for ensuring the relevancy of the sampling distribution on which *p*-values are to be based. It is one of the concerns addressed by the conditionality principle.

10.1 Weak Conditionality Principle (WCP)

Example 7. *Two measuring instruments of different precisions.* Suppose a single observation Y is made on a normally distributed random variable with unknown mean μ. A randomizing device chooses which of two instruments to use in measuring y: E' or E'', with probabilities v' or v''. The first instrument has known small variance, say 10^{-4}, whereas the second has known large variance, say 10^4. The full data indicate whether E' or E'' was performed, and the value of Y, y' or y'', respectively. The randomizer may be seen as an indicator of which experiment is performed to produce the data; for this purpose we typically consider an indicator statistic A, which takes values 1 and 2 with probabilities v' and v'', respectively; $S = (Y, A)$ is sufficient. Statistic A, being a subset of S whose distribution is independent of the parameter of interest, is an example of an *ancillary statistic* (Cox, 1958).

Using this setup, one may define a *mixture test*. First let the device (e.g., a coin toss) choose the instrument to use, then report the result of using it and calculate the *p*-value. In testing a null hypothesis, say, $\mu = 0$, the same *y* measurement would correspond to a much smaller *p*-value were it to have come from E' (Y is normal $N(\mu, 10^{-4})$) than if it had come from E'' (Y is normal $N(\mu, 10^4)$): denote them as $p'(y)$ and $p''(y)$, respectively. However, if one were to consider the overall type I error of the mixture corresponding to the observed y, one would average: $[p'(y) + p''(y)]/2$. This is the convex combination of the p-values averaged over the probabilities from E' and E'', chosen by the randomizer. The p-value associated with an inference – if calculated using (the unconditional distribution of) the mixture of tests, abbreviated as $\mathrm{Infr}_{E\text{-mix}}(y)$ – would be based on this average.

The point essentially is that the marginal distribution of a p-value averaged over the two possible configurations is misleading for a particular set of data. It would mean that an individual fortunate in obtaining the use of a precise instrument in effect sacrifices some of that information in order to rescue an investigator who has been unfortunate enough to have the randomizer choose a far less precise tool. From the perspective of interpreting the specific data that are actually available, this makes no sense. Once it is known whether E' or E'' has been run, the p-value assessment should be made conditional on the experiment actually run. In some other cases, the basis for conditioning may not be so obvious; therefore, there is a need for a systematic formulation.

Weak Conditionality Principle (WCP): If a mixture experiment (of the aforementioned type) is performed, then, if it is known which experiment produced the data, inferences about θ *are appropriately drawn in terms of the sampling behavior* in the experiment known to have been performed.

To avoid equivocation, it is important to understand what is being asserted. The WCP does not state a mathematical identity but it asserts that the *appropriate* way to draw the inference is not by means of the unconditional but rather by means of the conditional, sampling distribution of the experiment known to have produced the data. Once we know the data have been generated by E_j, given that our inference is about some aspect of E_j, our inference should not be influenced by whether a coin was tossed to decide which of two experiments to perform, and the result was to perform E_j. WCP is a *normative* epistemological claim about the appropriate manner of reaching an inference in the given context. We are assuming, of course, that all the stipulations in WCP are satisfied. Another example very often referred to in this context is the following.

Example 8. Suppose two independent and identically distributed random variables Y_1, Y_2 can each take values $\varphi - 1$ or $\varphi + 1$ with probability .5, φ unknown. The data take one of two possible configurations: either both values are the same, say $y_1 = y_2 = y'$, or there are two different values, say $y_1 = y'' - 1$ and $y_2 = y'' + 1$. Let A be an indicator of which of the two configurations obtains in a given sample. The minimal sufficient statistic S is (Y, A). In the case of the second configuration, $A = 2$, the sample values differ by 2 and, thus, ψ is exactly known to be y''. In the first configuration $(A = 1)$, the observed y-values are the same; thus, the two possible values for ψ, namely, $y' \pm 1$, are equally concordant with the data. Although the distribution of A is fixed independently of the parameter of interest, ψ, learning whether $A = 1$ or $A = 2$ is very relevant to the precision achieved; hence, the relevant inference would be conditional on its value.

As with Example 7, the sufficient statistic S being of dimension 2, while there is only one parameter, indicates the *incompleteness* of S. This opens the door to different p-values or confidence levels when calculated conditionally. In particular, the marginal distribution of a p-value averaged over the two possible configurations $(.5(0) + .5(.5) = .25)$ would be misleading for any particular set of data. Here the problem is generally given as estimating ψ with a confidence set with $n = 2$. If the two observed values are the same, then infer $\psi = y' - 1$; if they are different, infer ψ is y''. Overall, the probability of an erroneous inference is .25. If two distinct values have been observed, all but one of the parameter values are ruled out with highest severity (p-value is zero), whereas when both observed values are the same, the test fails to discriminate between the two logically possible parameter values (the p-value for either value is .25).

The general argument here is analogous to the ones seen earlier. We seek a factorization $s = (t, a)$, where t is one-dimensional, and we can write

$$f_S(s; \psi) = f_A(a) f_{T|A}(t; a, \psi),$$

where the first factor A has a fixed distribution that does not depend on θ. Now we argue that it is equivalent to obtain the data in two steps.

First, we observe that $A = a$ (either $A = 2$, we are lucky enough to have observed two different values, or $A = 1$, we are unlucky enough to have observed both the same).

Second, we observe, conditionally on the first step, that $T = t$, an observation from the conditional distribution $f_{T|A}$ (e.g., given E'' is performed, observe y''). In the case of the mixture in Example 7, observing A corresponds to applying the randomizer, indicating which experiment to perform. In other kinds of examples, the second step might correspond, in an analogous manner, to conditioning on a statistic A that is indicative of the level of precision achieved. The second step defines a unique p. We may call this process *technical conditioning to induce relevance* of the frequentist probability to the inference at hand.

Because it is given that the distribution of A does not involve ψ, merely learning the value of A at the first step tells us nothing directly about the value of ψ. However, it may, and indeed in general will, say something about the amount of information actually obtained; and thus is relevant to determining what is learned from the observed data as regards the actual data-generating procedure. If, for example, y' results from E', then it is the properties of E' that are relevant for evaluating warranted inferences about E'.

The concern about average p-values in mixtures and related examples that underwrites the need to condition often arises in relation to the pre-data emphasis typically associated with the behavioristic "accept/reject" accounts of testing from which we have already distinguished the present approach. The justification for the WCP is fully within the frequentist sampling philosophy for contexts of scientific inference. There is no suggestion, for example, that only the particular data set should be considered. That would entail abandoning altogether the sampling distribution as the basis for inference. It is rather a matter of identifying an appropriate sampling distribution for the inference goal at hand.

It is not uncommon to see statistics texts argue that in frequentist theory one is faced with the following dilemma: either to deny the appropriateness of conditioning on the precision of the tool chosen by the toss of a coin, or else to embrace the strong likelihood principle, which entails that frequentist sampling distributions are irrelevant to inference once the data are obtained. This is a false dilemma. Conditioning is warranted to achieve objective frequentist goals, and the conditionality principle coupled with sufficiency does not entail the strong likelihood principle. The "dilemma" argument is therefore an illusion (see Mayo, 7(III)).

11 Bayesian Alternatives

There are two further possible approaches to these issues. One involves a notion of probability as a personalistic degree of belief. It allows the incorporation of evidence other than that which can be modeled via a frequency concept of probability but, by its very nature, is not focused on the extraction and presentation of evidence of a public and objective kind. Indeed the founders of this approach emphasized its connection with individual decision making. Its main appeal in statistical work is some mixture of internal formal coherency with the apparent ability to incorporate information that is of a broader kind than that represented by a probabilistic model based on frequencies. The essential focus is too far from our concern with objectivity for this to be a generally satisfactory basis for statistical analysis in science.

The other approach, based on a notion of rational degree of belief, has in some respects similar objectives to the frequentist view sketched earlier and often leads to formally very similar or even numerically identical answers. There are, however, substantial difficulties over the interpretation to be given to the probabilities used to specify an initial state of knowledge, and hence also to the final, or posterior, probabilities. We now turn to this issue.

11.1 What Do Reference Posteriors Measure?

Attempts to develop conventional "default," "uninformative," or "reference" priors are deliberately designed to prevent prior opinions and beliefs from influencing the posterior probabilities, thereby attaining an "objective" or impersonal Bayesian formulation. The goal is to retain the benefits of the Bayesian approach while avoiding the problems posed by introducing subjective opinions into scientific inference. A classic conundrum, however, is that no unique "noninformative" flat prior exists that would be appropriate for all inference problems within a given model. (To assume one exists leads to inconsistencies in calculating posterior marginal probabilities.) Any representation of ignorance or lack of information that succeeds for one parameterization will, under a different parameterization, appear to entail having knowledge; so that special properties of particular parameterizations have to be appealed to (Dawid, A.P., Stone, M.M., and Zidek, J.V., 1973).

Rather than seek uninformative priors, the majority of contemporary reference Bayesian research is directed to finding priors that are to be regarded as *conventions* for obtaining reference posteriors. The priors are not to be considered expressions of uncertainty, ignorance, or degree of belief. Conventional priors may not even be probabilities in that a constant or flat prior for a parameter may not sum to 1 (improper prior). They are conventions intended to allow the data to be "dominant" in some sense.

The most elaborately developed versions of this are the reference priors chosen to maximize the contribution of the data to the resulting inference (Bernardo, 2005). However, if priors are not probabilities, what then is the interpretation of a posterior? It may be stipulated, by definition, that the posteriors based on a reference prior *are* objective degrees of belief in the parameters of interest. More is required to show that the calculated posteriors succeed in measuring a warranted strength of evidence afforded by data in the approximate truth or correctness of the various parameter values. Even if the reference prior research program succeeds in identifying priors that satisfy its own desiderata (Bernardo), it is necessary to show they satisfy this epistemic goal. Otherwise, it is not clear how to evaluate critically the adequacy of reference Bayesian computations for their intended epistemological measurement, in contrast to possibly regarding them as convenient mathematical procedures for deriving methods with good frequentist properties (Cox, 2006).

11.2 Problems with Nuisance Parameters

To ensure that unknown nuisance parameters exert minimal threats to the validity of p-value and other frequentist calculations, we saw how techniques for conditioning on sufficient statistics for such parameters are employed. By contrast, the Bayesian requires a joint distribution for all these unknowns, and the posterior will depend on how this is assigned. Not only may the calculation of a reference prior be relatively complicated but the prior for a particular parameter may depend on whether it is a parameter "of interest" or if it is a nuisance parameter, and even on the "order of importance" in which nuisance parameters are arranged. For example, if a problem has two nuisance parameters, the appropriate reference prior may differ according to which is considered the more important. The dependency on such apparently arbitrary choices tends to diminish the central goal of maximizing the contribution of the data to the resulting inference. The problem is not so much that different researchers can arrive at different posterior degrees with the same data; it is that such choices would appear to be inextricably bound up with the reported posteriors. As such, it would not be apparent which parts of the final inference were due to the data and which to the particular choice of ordering parameters.

11.3 Priors Depend on the Sampling Rule

Reference priors differ according to the sampling distribution associated with the model formulation. The result is to forfeit what is often considered a benefit of the Bayesian approach and to violate the strong likelihood principle (SLP), despite it often being regarded as the cornerstone of Bayesian coherency. Now the sampling distribution and the consideration of relevant hypothetical repetitions are at the heart of the frequentist objective assessment of reliability and precision, but violation of the SLP introduces incoherency into the reference Bayesian account. Reference Bayesians increasingly look upon the violation of the SLP as the "price" that has to be paid for objectivity. We agree. Violating the SLP is necessary for controlling error probabilities, but this alone is not sufficient for objectivity in our sense.

Granted, as some (e.g., Berger, 2004) have noted in practice, arriving at subjective priors, especially in complex cases, also produces coherency violations.[4] But there would seem to be an important difference between falling short of a formal principle (e.g., due to human limitations) and having its violation be required in principle to obtain the recommended priors.

11.4 Reference Posteriors with Good Frequentist Properties

Reference priors yield inferences with some good frequentist properties, at least in one-dimensional problems – a feature usually called *matching*. Although welcome, it falls short of showing their success as objective methods. First, as is generally true in science, the fact that a theory can be made to match known successes does not redound as strongly to that theory as did the successes that emanated from first principles or basic foundations. This must be especially so where achieving the matches seems to impose swallowing violations of its initial basic theories or principles.

Even if there are some cases where good frequentist solutions are more neatly generated through Bayesian machinery, it would show only their technical value for goals that differ fundamentally from their own. But producing identical numbers could only be taken as performing the tasks of frequentist inference by reinterpreting them to mean confidence levels and significance levels, not posteriors.

What some Bayesians seem to have in mind when pointing, as evidence of the success of reference priors, is that in some cases it is possible to match reference posteriors, construed as degrees of rational belief, with frequentist error probabilities. That is, the ability to match numbers helps to justify construing reference posteriors as objective degrees of belief in hypotheses. It is hard to know how to assess this, even in the very special cases where it

---

[4] If the prior were intended to represent external knowledge, a Bayesian might justify using different priors in the cases described in Example 5 – binomial versus negative binomial – by considering that the latter is often used when the probability of success is small.

holds. Frequentist performance, we have shown, may, if correctly specified, be used to obtain measures of consistency with hypothesized parameter values and to assess sensitivity and precision of inferences about the system or phenomena at hand. It is not clear how the reference Bayesian's stated aim – objective degree of belief assignments – is attained through long-run error rates or coverage probabilities, even where these are achieved.

It is important not to confuse two kinds of "error probabilities": Frequentist error probabilities relate to the sampling distribution, where we consider hypothetically different outcomes that could have occurred in investigating this one system of interest. The Bayesian allusion to frequentist "matching" refers to the fixed data and considers frequencies over different systems (that could be investigated by a model like the one at hand). Something would need to be said as to why it is relevant to consider other hypotheses, perhaps even in different fields, in reasoning about this particular *H*. The situation with frequentist priors considered as generating empirical Bayesian methods is distinct.

The ability to arrive at numbers that agree approximately and asymptotically with frequency-based measures in certain special cases does not seem sufficient grounds for the assurances often given that frequentist goals are being well achieved with reference priors, considering the cases of disagreement and the differences in interpretation in general. There is also the problem that distinct approaches housed under the impersonal Bayesian banner do not always agree with each other as to what method is to be recommended, even in fairly ordinary cases. Many seem to regard reference Bayesian theory to be a resting point until satisfactory subjective or informative priors are available. It is hard to see how this gives strong support to the reference prior research program.

12 Testing Model Assumptions

An important part of frequentist theory is its ability to check model assumptions. The use of statistics whose distribution does not depend on the model assumption to be checked lets the frequentist split off this task from the primary inference of interest.

Testing model adequacy formally within Bayesian formulations is not straightforward unless the model being used in initial analysis is itself embedded in some bigger family, typically putting fairly high prior probability on the initial model being close to the truth. That presupposes a reasonably clear notion of the possible departures that might arise.

13 Concluding Remarks

A statistical account, to be fairly assessed and usefully explicated, requires a clear understanding of both its aims and its methods. Frequentist methods, as here conceived, aim to learn about aspects of actual phenomena by testing the statistical consistency of hypotheses framed within statistical models of the phenomena. This approach has different tools, just as in science in general, for different questions and various stages of inquiries within its domain. Yet the frequentist standpoint supplies a unified argument and interconnected strategies that relate to achieving the goal of objective learning about the world. It achieves this goal by means of calibrations afforded by checkable standards. While there are certainly roles for other approaches, notably the use of personalistic Bayesian ideas in the context of personal decision making, we consider that a frequentist formulation provides a secure basis for the aims on which we are focusing: the analysis of data and the communication of conclusions in scientific research.

In our discussion of a frequency-based approach, the information in the data about the model is split into two parts: one captures all the information, assuming the model to be correct, and the other allows for checking the adequacy of the model. At least in simple problems this splitting is unique and unambiguous. The only other step in the development is the need in some, but not all, contexts to condition the probability calculations to achieve separation from nuisance parameters and to ensure their relevance to the issue under study. The appropriate choice for the statistic on which conditioning takes place depends on the particular aim of conditioning in the case at hand, taking account of the type of null hypothesis of interest. In cases with nuisance parameters, conditioning on a sufficient statistic V enables assessments of p-values and confidence levels that are free from threats to validity from these unknowns. In the most familiar class of cases, the strategies for constructing appropriately sensitive tests, separate from nuisance parameters, produces the same tests entailed by Neyman–Pearson theory, albeit with a difference in rationale.

In conditioning to induce relevance, the aim is to ensure inferences are constrained to reflect the actual precision of our test as an instrument for probing the underlying data-generation mechanisms. Tests that emerge to ensure relevance for the particular inference may differ from those developed based solely on Neyman–Pearson long-run behavior goals. The conditional perspective grows out of the desire that p-values (and confidence levels) reflect the relevant precision of particular inferences. It allows us to avoid well-known counterintuitive examples while remaining within the frequentist, sampling theory framework. Understanding the sampling properties of statistical tools, as well as attention to the data-collection process, is the key to inductive inference from data to underlying values of θ. The sampling account deliberately leaves a clear trail regarding the basis for inferences drawn, providing a valuable framework for an objective scrutiny and improvement of results.

There is a historical and philosophical basis for a different notion of "objectivity" – one that is satisfied by automatic, conventional, "a priori" measures. In the contemporary "impersonal Bayesian" accounts, much as in earlier conventional approaches, the stumbling block remains one of showing appropriateness for achieving empirical scientific goals. Although it is among the most promising accounts currently on offer, we have also seen that the impersonal Bayesian paradigm is at odds with two fundamental goals of much Bayesian discussion: incorporating background information via priors, and adhering to the SLP (and the associated freedom from having to consider sampling distributions and stopping rules). Taking stock of the implications for foundations of statistics is therefore especially needful.

References

Berger, J. (2004), "The Case for Objective Bayesian Analysis," *Bayesian Analysis*, 1: 1–17.
Bernardo, J.M. (2005), "Reference Analysis," *Handbook of Statistics*, vol. 35, Elsevier, Amsterdam.
Birnbaum, A. (1962), "On the Foundations of Statistical Inference," *Journal of the American Statistical Association*, 57: 269–306.
Cox, D.R. (1958), "Some Problems Connected with Statistical Inference," *Annals of Mathematical Statistics*, 29: 357–72.
Cox, D.R. (1990), "Role of Models in Statistical Analysis," *Statistical Science*, 5: 169–74.
Cox, D.R. (2006), *Principles of Statistical Inference*, Cambridge University Press, Cambridge.
Cox, D.R., and Brandwood, L. (1959), "On a Discriminatory Problem Connected with the Works of Plato," *Journal of the Royal Statistical Society B*, 21: 195–200.
Cox, D.R. and Hinkley, D.V. (1974), *Theoretical Statistics*, Chapman and Hall, London.
Dawid, A.P., Stone, M. and Zidek, J.V. (1973), "Marginalization Paradoxes in Bayesian and Structural Inference," (with discussion), *Journal of the Royal Statistical Society B*, 35: 189–233.
Fisher, R.A. (1935), *Design of Experiments*, Oliver and Boyd, Edinburgh.
Hacking, I. (1965), *Logic of Statistical Inference*, Cambridge University Press, Cambridge.
Lehmann, E.L. (1986), *Testing Statistical Hypotheses*, 2nd ed., Wiley, New York.
Mayo, D.G. and Cox, D.R. (2006), "Frequentist Statistics as a Theory of Inductive Inference," in J. Rojo (ed.), *Optimality: The Second Erich L. Lehmann Symposiun*, Lecture Notes-Monograph Series, Institute of Mathematical Statistics (IMS), 49: 77–97.

III An Error in the Argument from Conditionality and Sufficiency to the Likelihood Principle

Deborah G. Mayo

Cox and Mayo (7(II), this volume) make the following bold assertion:

> It is not uncommon to see statistics texts argue that in frequentist theory one is faced with the following dilemma, either to deny the appropriateness of conditioning on the precision of the tool chosen by the toss of a coin, or else to embrace the strong likelihood principle which entails that frequentist sampling distributions are irrelevant to inference once the data are obtained. This is a false dilemma: Conditioning is warranted in achieving objective frequentist goals, and the conditionality principle coupled with sufficiency does not entail the strong likelihood principle. The "dilemma" argument is therefore an illusion.

Given how widespread is the presumption that the (weak) conditionality principle (CP) plus the sufficiency principle (SP) entails (and is entailed by) the (strong) likelihood principle (LP), and given the dire consequence for error statistics that follows from assuming it is true, some justification for our dismissal is warranted. The discussion of the three principles (in 7(II)) sets the stage for doing this. The argument purporting to show that CP + SP entails LP was first given by Birnbaum (1962), although the most familiar version is that found in Berger and Wolpert (1988). To have a statement of the SLP in front of us for this discussion I restate it here:

The Strong Likelihood Principle (SLP): Suppose that we have *two* experiments E' and E'', with different probability models $f'_{Y'}(y'; \theta)$ and $f''_{Y''}(y''; \theta)$, respectively, with the same unknown parameter θ. If y'^* and y''^* are observed data from E' and E'', respectively, where the likelihoods of y'^* and y''^* are proportional, *then* y'^* and y''^* have the identical evidential import for any inference about θ.

This notation, commonly used in discussing this result, suggests a useful shorthand: we may dub those outcomes y'^* and y''^* that satisfy the requirements of the SLP "star pairs" from the experiments E' and E''.

Principles of inference based on the sampling distribution do not in general satisfy the (strong) likelihood principle; indeed, were the SLP to be accepted it would render sampling distributions irrelevant for inference once the data were in hand. Understandably, therefore, statisticians find it surprising that Birnbaum purports to show the SLP is entailed by the two principles accepted in most formulations of frequentist theory. (The SLP is, at least formally, incorporated in most formulations of Bayesian and purely likelihood-based accounts of statistical inference.)

Example 5 (p. 286) showed how frequentist theory violates the SLP; the following case is even more dramatic.

Example: Fixed versus sequential sampling. Suppose Y' and Y'' are sets of independent observations from $N(\theta, \sigma^2)$, with σ known, and p-values are to be calculated for the null hypothesis $\theta = 0$. In test E' the sample size is fixed, whereas in E'' the sampling rule is to continue sampling until $1.96\sigma_y$ is attained or exceeded, where $\sigma_y = \sigma/n^{.5}$. Suppose E'' is first able to stop with n_0 trials. Then y'' has a proportional likelihood to a result that could have occurred from E', where n_0 was fixed in advance, and it happened that after n_0 trials y' was $1.96\sigma_y$ from zero. Although the corresponding p-values would be different, the two results would be inferentially equivalent according to the SLP. According to the SLP, the fact that our subject planned to persist until he got the desired success rate – the fact that he *tried and tried again* – can make *no* difference to the evidential import of the data: the data should be interpreted in just the same way as if the number of trials was fixed at the start and statistical significance resulted. (Bayesians call this the stopping rule principle.) For those of us wishing to avoid misleading inferences with high or maximal probability, this example, "taken in the context of examining consistency with $\theta = 0$, is enough to refute the strong likelihood principle" (Cox, 1977, p. 54) because, with probability 1, it will stop with a "nominally" significant result even though $\theta = 0$.

By contrast, Savage declares:

The persistent experimenter can arrive at data that nominally reject any null hypothesis at any significance level, when the null hypothesis is in fact true.... These truths are usually misinterpreted to suggest that the data of such a persistent experimenter are worthless or at least need special interpretation... The likelihood principle, however, affirms that the experimenter's intention to persist does not change the import of his experience. (Savage, 1962a, p. 18)

Before anyone came forward with independent grounds for accepting the SLP, one merely had a radically different perspective on the goals for sensible accounts of statistics. This situation was to change dramatically with Birn-baum's argument that purported to provide just such grounds, using only premises apparently supported by frequentists. Doubtless this is why Savage (1962b) greeted Birnbaum's argument for the SLP as a "breakthrough." As it is often said, "the proof is surprisingly simple" (p. 467). However, it is precisely the overly quick series of equivalences so characteristic of each presentation that disguises how the vital terms shift throughout. Although important caveats to the Birnbaum argument have often been raised, the soundness of his results has generally been accepted by frequentists, like-lihoodists, and Bayesians alike (see Casella and Berger, 2002; Lee, 2004; Lehmann, 1981; Robins and Wasserman, 2000; Royall, 1997). Yet a close look at the argument reveals it to be deeply (and interestingly) flawed.

At the risk of belaboring key elements of the Birnbaum experiment, I first informally outline the ingredients. Then the more usual notation can be grasped without getting dizzy (hopefully).

1 Introduction to the Birnbaum Experiment: E-BB

Let E' and E'' be expements some of whose outcomes are "star pairs," i.e., some y'^* from E' and y''^* from E'' satisfy the proportional likelihood requirement in the SLP. An example would be E', the normal test with n fixed, the significance level set to .05 and E'' the corresponding test with the rule: stop when a 0.05 statistically significant result is achieved. There are two steps to the Birnbaum experiment.

Step 1: Use E' and E'' to define a special type of mixture experiment: Flip a coin to determine whether E' or E'' is to be performed. One can indicate which is performed using statistic J: in particular, $j = 1$ and $j = 2$ when E' and E'' are performed, respectively. *So far this is an ordinary mixture.*

Step 2: Once you toss the coin and perform the indicated experiment, you are to report the result in the following way:

If the result has a star pair in the experiment not performed, then report it came from E', whether or not it came from E' or E''. Say $j = 2$ so the optional stopping experiment is run, and let the experiment achieve .05 sig-nificance when $n = 100$. Does this outcome have a "star pair" in experiment E'? Yes, its star pair is the case where n was fixed at 100, and statistical signifi-cance at the .05 level was achieved. The Birnbaum experiment instructs you to report: (E', y'^*) that the result came from the fixed sample size test, even though the result actually came from E''. We can construe his rule about reporting results in terms of a statistic, call it T_{BB}, that erases the fact that a result came from E'' and writes down E' (in the case of starred outcomes):

$$T_{BB}(E'', y''^*) = (E', y'^*).$$

Let us abbreviate the special mixture experiment of Birnbaum's as E-BB. When (E', y'^*) is reported from E-BB, you know that the result could have come from E' or E'', but you do not know which.

We have not said what E-BB stipulates if the result does not have a star pair. In that case, report the results as usual. For example, let $j = 1$, so E' is performed, and say the results after the fixed number of trials is not statistically significant. Since these latter (case 2) results do not enter into the main proof we can focus just on the star pairs (case 1).

I have said nothing so far about the argument purporting to show the SLP, I am just describing the set-up. In setting out the argument, it will be useful to employ the abbreviation introduced in 7(II), p. 287:

Infr$_E(y)$ is the inference from outcome y from experiment E *according to the methodology of inference being used.*

Although this abbreviation readily lends itself for use across schools of inference, here we use it to discuss an argument that purports to be relevant for frequentist sampling theory; therefore, the relativity to the sampling distribution of any associated statistic is fundamental. Birnbaum is free to stipulate his rule for reporting results, so long as inferences are computed using the sampling distributions corresponding to T_{BB} in experiment E-BB. If this is done consistently however, the desired result (SLP) no longer follows – or so I will argue.

2 The Argument from the Birnbaum Experiment

I first trace out the standard version of the argument that purports to derive the SLP from sufficiency and conditionality (found, for example, in Berger and Wolpert, 1988; Casella and Berger, 2002). It begins with the antecedent of the SLP.

A. We are to consider a pair of experiments E' and E'' with f' differing from f'' where E' and E'' have some outcomes y'^* and y''^* with proportional likelihoods.

We may think of E' as the normal test with n fixed, and E'' the normal test with optional stopping as mentioned earlier. The argument purports to show the following:

$$\text{Infr}_{E'}(y'^*) = \text{Infr}_{E''}(y''^*).$$

B. We are to consider a mixture experiment whereby a fair coin is flipped and "heads" leads to performing E' and reporting the outcome y', whereas "tails" leads to E'' and reporting the outcome y''. Each outcome would have two components (E^j, y^j) $(j = 0, 1)$, and the distribution for the mixture would be sampled over the distinct sample spaces of E' and E''.

In the special version of this experiment that Birnbaum puts forward, E_{BB}, whenever an outcome from E'' has a "star pair" in E' we report it as (E', Y^*), (case 1), else we are to report the experiment performed and the outcome (case 2).

That is, Birnbaum's experiment, E_{BB}, is based on the statistic T_{BB}.

$$T_{BB}(E^j, y^j) = \begin{cases} (E', y'^*) & \text{if } j = 1 \text{ and } y' = y'^* \text{ or if } j = 2 \text{ and } y'' = y''^* \\ (E^j, y^j) \end{cases}$$

For example, if the result is (E'', y''^*), we are to report (E', y'^*).

Because the argument for the SLP is dependent on case 1 outcomes, we may focus only on them for now. (Note, however, that only case 2 outcomes describe the ordinary mixture experiment.) Now for the premises:

(1) The next step is to argue that in drawing inferences from outcomes in experiment E_{BB} the inference from (E'', y''^*) is or should be the same as if the result were (E', y'^*), as they both yield the identical output (E', y'^*) according to (sufficient) statistic T_{BB}. This gives the first premise of the argument:

(1) $\text{Infr}_{E\text{-BB}}(E', y'^*) = \text{Infr}_{E\text{-BB}}(E'', y''^*).$

(2) The argument next points out that WCP tells us that, once it is known which of E' or E'' produced the outcome, we should compute the inference just as if it was known all along that E^j was going to be performed. Applying WCP to Birnbaum's mixture gives premise (2):

(2) $\text{Infr}_{E\text{-BB}}(E^j, y^{j*}) = \text{Infr}_{Ej}(y^{j*}).$

Premises (1) and (2) entail the inference from y'^* is (or ought to be) identical to the inference from y''^*, which is the SLP.

Here I have written the argument to be formally valid; the trouble is that the premises cannot both be true. If one interprets premise (1) so that it comes out true, then premise (2) comes out false, and vice versa. Premise

(2) asserts: The inference from the outcome (E^j, y^{j*}) computed using the sampling distribution of E_{BB} is appropriately identified with an inference from outcome y^{j*} based on the sampling distribution of E^j, which is clearly false. The sampling distribution to arrive at $\text{Infr}_{E\text{-BB}}$ would be the convex combination averaged over the two ways that y^{j*} could have occurred. This differs from the sampling distributions of both $\text{Infr}_{E'}(y'^*)$ and $\text{Infr}_{E''}(y''^*)$.

3 Second Variation of the Birnbaum Argument

To help bring out the flaw in the argument, let us consider it in a different, equivalent, manner. We construe the premises so that they all are true, after which we can see the argument is formally invalid.

We retain the first premise as in the standard argument. To flesh it out in words, it asserts the following:

(1) If the inference from the outcome of the mixture, (E^j, y^{j*}), is computed using the unconditional formulation of E_{BB}, then the inference from (E', y'^*) is the same as that from (E'', y''^*).

$$(1)\ \ \text{Infr}_{E\text{-BB}}(E', y''^*) = \text{Infr}_{E\text{-BB}}(E'', y''^*).$$

We need to replace the false premise (2) of the standard formulation with a true premise (2)′:

(2)′ If the inference is to be conditional on the experiment actually performed then the inference from (E^j, y^{j*}) should be the same as $\text{Infr}_{E^j}(y^{j*})$.

Note we can speak of an inference from an outcome of E_{BB} without saying the inference is based on the sampling distribution of E_{BB} − avoiding the assertion that renders premise (2) of the standard variation false.

We have from the WCP: Once (E^j, y^{j*}) is known to have occurred, the inference should condition on the experiment actually performed, E^j. The conclusion is the SLP:

Therefore, the inference from y'^* is or should be the same as that from y''^*.

Because our goal here is to give the most generous interpretation that makes the premises true, we may assume it is intended for the phrase "once (E^j, y^j) is known to have occurred" is understood to apply to premise (1) as well. The problem now is that, even by allowing all the premises to be true, the conclusion does not follow. The conclusion could "follow" only if it is assumed that you both should and should not use the conditional formulation. The antecedent of premise (1) is the denial of the antecedent of premise (2)′. The argument is invalid; if made valid, as in the first rendition, it requires adding a contradiction as a premise. Then it is unsound.

4 An Explicit Counterexample

Because the argument is bound to seem convoluted for those not entrenched in the treatments of this example, it is illuminating to engage in the logician's technique of proving invalidity. We consider a specific example of an SLP violation and show that no contradiction results in adhering to both WCP and SP.

Suppose optional stopping experiment E'' is performed and obtains the 1.96 standard deviation difference when $n = 100$. Let σ be 1, $1.96\sigma_y = .196$. This value is sometimes said to be a "nominally" .05 significant result.

The calculation of the "actual" p-value would need to account for the optional stopping, resulting in a higher p-value than if the .196 had been obtained from a fixed-sample-size experiment. This leads to a violation of the SLP.

So we have

$$\text{Infr}_{E'}(y'^* = .196) = .05,$$

$$\text{Infr}_{E''}(y''^* = .196) = .37 \text{ (approximately).}$$

Although these two outcomes have proportional likelihoods, they lead to different inferences. This result gives a clear violation of the SLP because $.05 \neq .37$.

$$\text{Infr}_{E'}(y'^* = .196) \neq \text{Infr}_{E''}(y''^* = .196).$$

Now the Birnbaum argument alleges that, if we allow this violation of the SLP, then we cannot also adhere to the WCP and the SP, on pain of contradiction (Berger and Wolpert 1988, p. 28). But we can see that no contradiction results from not-SLP and WCP and SP.

Sufficiency Principle (SP)

Where is sufficiency to enter? We know that SP applies to a single experiment, whereas E' and E'' are distinct experiments. But a mixture of the two permits inference in terms of outcomes whose first component indicates the experiment run and the second the outcome from that experiment: (E^j, y^j). The mixture can have a single sampling distribution composed of the convex combination of the two.

For the mixed experiment we are discussing, the statistic associated with the Birnbaum experiment E_{BB} is T_{BB}, where

$$T_{BB}(E'', 1.96\sigma_y) = T_{BB}(E', 1.96\sigma_y) = (E', 1.96\sigma_y).$$

The unconditional p-value associated with the outcome from the mixed experiment would be based on the convex combination of the two experiments; it would be $.5(.05 + .37) = .21$. If we are to base inferences on the sampling distribution of T_{BB}, then in both cases the average p-value associated with .196 would be .21.

$$(1) \quad \text{Infr}_{E\text{-}BB}(E'', .196) = \text{Infr}_{E\text{-}BB}(E', .196) = .21.$$

This may be seen as an application of the SP within E_{BB}. Sufficiency reports on a mathematical equivalence that results, provided that it is given that inference is to be based on a certain statistic T_{BB} (and its sampling distribution) defined on experiment E_{BB}. Having defined E_{BB} in this way, the mathematical equivalence in (1) is straightforward. But the WCP, if it is applied here, asserts *we ought not* to base inference on this statistic.

WCP is Accepted

The WCP says that if a mixture experiment outputs $(E'', .196)$, the inference should be calculated conditionally, that is, based on the sampling distribution of E''. Applying WCP we have the following:

The inference from $(E'', y''^* = .196)$ should be based on $\text{Infr}_{E''}(.196) = .37$. Likewise, if the mixture experiment resulted in $(E', .196)$, the inference should be based on the sampling distribution of E': $\text{Infr}_{E'}(.196) = .05$.

As was argued in Chapter 7(II), an argument for applying WCP is based on normative epistemological considerations of the way to make error probabilities relevant for specific inferences. This is why it may be open to dispute by those who hold, for a given context, that only average error rates matter in the long run. It is not about a mathematical identity, and in fact it is only because the unconditional analysis differs from the conditional one that it makes sense to argue that the former leads to counterintuitive inferences that are remedied by the distinct analysis offered by the latter.

So no contradiction results! Equivalently the preceding argument is invalid.

5 The Argument Formally Construed

Philosophers may wish to see the formal rendering of the argument.

Let A be as follows: outcome (E^j, y^{j*}) should be analyzed unconditionally as in Birnbaum experiment E_{BB}.

Then the WCP asserts not-A:

Not-A: outcome (E^j, y^{j*}) should be analyzed conditonally using the sampling distribution of E^j.

Premise 1: If A, then the inference from (E', y'^*) should be equal to the inference from (E'', y'). (The converse holds as well.)

Premise 2: Not-A.

Premise 2a: If not-A then the inference from (E'', y''^*) should be equal to the inference from y'''^*.

Premise 2b: If not-A then the inference from (E', y'^*) should be equal to the inference from y'^*.

Therefore, the inference from y'''^* should be equal to the inference from y'^*, the SLP.

From premises 2, 2a, and 2b, we have the following:

The inference from (E'', y''^*) should be equal to the inference from y'''^*.
The inference from (E', y'^*) should be equal to the inference from y'^*.

But nevertheless we can have that the inference from y'''^* should NOT be equal to the inference from y'^*.

This demonstrates the argument is invalid.

The peculiarity of Birnbaum's experiment and its differences from the ordinary mixture have been duly noted (e.g., Cox and Hinkley, 1974, p. 41; Durbin, 1970), and many take it alone as grounds that Birnbaum's argument is not compelling. We see that an even stronger criticism is warranted.

My argument holds identically for derivation of the SLP based on a weaker condition than SP (see Evans et al., 1986), which I do not discuss here.

6 Concluding Remark

Chapter 7(II) grew out of numerous exchanges with David Cox in the 2 years since ERROR06. My goal was to extract and articulate the philosophical grounds underlying Cox's work on the nature and justification of types of conditioning in relation to the goal of ensuring that error statistical assessments are relevant for the tasks of objective statistical inference. It became immediately clear to me that I would have to resolve the Birnbaum SLP puzzle if 7(II) was to consistently fill in the pieces of our earlier work in 7(I), hence, the birth of this essay. Fortunately, spotting the fallacy proved easy (in July 2007). To explain it as simply as I would have liked proved far more challenging.

References

Berger, J.O., and Wolpert, R.L. (1988), *The Likelihood Principle*, California Institute of Mathematical Statistics, Hayward, CA.

Birnbaum, A. (1962), "On the Foundations of Statistical Inference" (with discussion), *Journal of the American Statistical Association*, 57: 269–326.

Birnbaum, A. (1970), "More on Concepts of Statistical Evidence," *Journal of the American Statistical Association*, 67: 858–61.

Casella, G., and Berger, R.L. (2002), *Statistical Inference*, 2nd ed., Duxbury, Pacific Grove, CA.

Cox, D.R. (1977), "The Role of Significance Tests" (with discussion), *Scandinavian Journal of Statistics*, 4: 49–70.

Cox, D.R. and Hinkley, D.V. (1974), *Theoretical Statistics*, Chapman and Hall, London.

Durbin, J. (1970), "On Birnbaum's Theorem and the Relation between Sufficiency, Conditionality and Likelihood," *Journal of the American Statistical Association*, 65: 395–8.

Evans, M., Fraser, D.A.S., and Monette, G. (1986), "Likelihood," *Canadian Journal of Statistics*, 14: 180–90.

Lee, P.M. (2004), *Bayesian Statistics: an Introduction*, 3rd ed., Hodder Arnold, New York.

Lehmann, E.L. (1981), "An Interpretation of Completeness in Basu's Theorem," *Journal of the American Statistical Association*, 76: 335–40.

Mayo, D.G. and Cox, D.R. (2006), "Frequentist Statistics as a Theory of Inductive Inference," in J. Rojo (ed.), *Optimality: The Second Erich L. Lehmann Symposium*, Lecture Notes-Monograph Series, Institute of Mathematical Statistics (IMS), Vol. 49: 77–97.

Robins, J., and Wasserman, L. (2000), "Conditioning, Likelihood, and Coherence: A Review of Some Foundational Concepts," *Journal of the American Statistical Association*, 95: 1340–6.

Royall, R.M. (1997), *Statistical Evidence: A Likelihood Paradigm*, Chapman & Hall, London.

Savage, L., ed. (1962a), *The Foundations of Statistical Inference: A Discussion*. Methuen, London.

Savage, L. (1962b), "'Discussion on Birnbaum (1962)," *Journal of the American Statistical Association*, 57: 307–8.

IV On a New Philosophy of Frequentist Inference
Exchanges with David Cox and Deborah G. Mayo

Aris Spanos

1. *Experimental Reasoning and Reliability*: How can methods for controlling long-run error probabilities be relevant for inductive inference in
   science? How does one secure error reliability in statistical inference?
   How do statistical inferences relate to substantive claims and theories?
2. *Objectivity and Rationality*: What is objectivity in statistical inference?
   What is the relationship between controlling error probabilities and
   objectivity? Can frequentist statistics provide an account of inductive
   inference? Is a genuine philosophy for frequentist inference possible?
   Does model validation constitute illegitimate double use of data?

1 Introduction

The twin papers by Cox and Mayo in this chapter constitute a breath of fresh
air in an area that has long suffered from chronic inattention. A renowned
statistician and a well-known philosopher of science and statistics have
joined forces to grapple with some of the most inveterate foundational
problems that have bedeviled frequentist statistics since the 1950s. The end
result is more than a few convincing answers pertaining to some of these
chronic problems, which are rarely discussed explicitly in either statistics
or philosophy of science. Although here I give my own conception of what
this amounts to, it is my hope that others recognize that the twin papers
put forward some crucial steps toward providing a genuine philosophy for
frequentist inference.

I need to profess at the outset that I am *not* an uninterested outsider
to these philosophical issues and discussions. I learned my statistics as an
undergraduate at the London School of Economics (LSE) from Cox and
Hinkley (1974); and over the years, both as a graduate student at LSE
and later, as a practicing econometrician, I have been greatly influenced

by Cox's books and papers (Hand and Herzberg, 2005). What appealed to me most in his writings was a dauntless inclination to raise as well as grapple with fundamental issues in statistical modeling and inference as they arise at the level of a practitioner, offering what seemed to me right-headed suggestions guided by a discerning intuition. From his writings, I learned to appreciate several subtle issues and problems in statistical modeling, including the value of preliminary data analysis and graphical techniques in learning from data, the importance of model adequacy, the difficulties in fusing statistical and substantive information, the distinctness of Fisher's inductive reasoning, and the perplexing variety of statistical testing. The challenge to seek more systematic accounts for some of his right-headed suggestions in a unifying framework fascinated me, and that objective greatly influenced my research agenda over the years.

Over the past eight years or so, I have collaborated with Mayo on several projects pertaining to various foundational issues. Our discussions often overlapped with some of the issues Mayo and Cox were grappling with, and I was aware of the exchanges that eventually gave rise to these twin papers. Most intriguing for me was the sense of discovery at how the frequentist procedures and principles Cox had long developed primarily on intuitive grounds obtained their most authentic and meaningful justification within a unified philosophy of learning from data. In what follows, I try to articulate that sense of discovery.

What is particularly interesting (and unique) about these twin papers is that they show us the weaving together of threads from Cox's perspective on frequentist inference, which has a distinct Fisherian undertone, and Mayo's error-statistical perspective, which appears to enjoy more affinity with the Neyman-Pearson (N-P) framework. The end result is a harmonious blend of the Fisherian and N-P perspectives to weave a coherent frequentist inductive reasoning anchored firmly on error probabilities. The key to this reconciliation is provided by recognizing that Fisher's p-value reasoning is based on a *post-data* error probability, and Neyman and Pearson's type I and II error reasoning is based on *pre-data* error probabilities. In the coalescing, both predata and post-data error probabilities fulfill crucial complementary roles, contrary to a prevailing view of critics, e.g. Savage (1962a).

Some of the particular problems and issues discussed in 7(I), (II), (III) include: frequentist inductive reasoning as it relates to the p-value and the accept/reject rules, the relevant error probabilities – especially when selection effects are involved, the different roles of conditioning in frequentist inference, model adequacy and its relationship to sufficiency, the use and abuse of the likelihood principle, objectivity in frequentist and Bayesian statistics, and Bayesian criticisms of frequentist inference. In addition, these papers raise interesting issues pertaining to the Popperian philosophy of science as it relates to frequentist inductive reasoning.

2 Statistics and Philosophy of Science since the 1950s: A Bird's-Eye View

To do even partial justice to the joint papers by Mayo and Cox, I need to place them in the proper context, which includes both statistics and philosophy of science, and, by necessity, I have to use very broad brushstrokes to avoid a long digression; I apologize for that at the outset.

The modern approach to frequentist (classical) statistics was pioneered by Fisher (1922) as model-based statistical induction, anchored on the notion of a *statistical model.* Fisher (1925, 1934), almost single-handedly, erected the current theory of "optimal" point estimation and formalized p-value significance testing. Neyman and Pearson (1933) proposed an "optimal" theory for hypothesis testing by modifying and extending Fisher's significance testing. Neyman (1937) proposed an "optimal" theory for interval estimation analogous to N-P testing.

Broadly speaking, the probabilistic foundations of frequentist statistics and the technical apparatus associated with statistical inference methods were largely in place by the late 1930s, but the philosophical foundations associated with the proper form of the underlying *inductive reasoning* were rather befuddled. Fisher was arguing for "inductive inference," spearheaded by his significance testing in conjunction with p-values and his fiducial probability for interval estimation. Neyman was arguing for "inductive behavior" based on N-P testing and confidence interval estimation in conjunction with predata error probabilities (see Mayo, 2006).

The last exchange between these pioneers of frequentist statistics took place in the mid 1950s (see Fisher, 1955; Neyman, 1956; Pearson, 1955) and left the philosophical foundations of the field in a state of confusion with many more questions than answers: What are the differences between a Fisher significance test and an N-P test? Does a proper test require the specification of an alternative hypothesis? What about goodness-of-fit tests like Pearson's? Are the notions of type II error probability and power applicable to Fisher-type tests? What about the use of error probabilities postdata? Is the p-value a legitimate error probability? What is the relationship between p-values and posterior probabilities? Does Fisher's fiducial distribution give rise to legitimate error probabilities? Can one distinguish between different values of the unknown parameter within an observed confidence interval

(CI)? Can one infer substantive significance from an observed CI? In what sense does conditioning on an ancillary statistic enhance the precision and data specificity of inference?

In addition to these questions, it was not at all obvious under what circumstances an N-P tester could safeguard the coarse *accept/reject decisions* against:

1. the *fallacy of acceptance*: interpreting *accept* H_0 [*no evidence against H_0*] as *evidence for H_0*, or
2. the *fallacy of rejection*: interpreting *reject* H_0 [*evidence against H_0*] as *evidence for H_1*.

A well-known example of the latter is the conflation of *statistical* with *substantive significance*.

Fisher's use of the *p*-value to reflect the "strength of evidence" against the null was equally susceptible to the fallacy of rejection because the *p*-value often goes to zero as the sample size $n \to \infty$.

Moreover, interpreting a *p*-value that is *not* "small enough" as evidence for H_0 would render it susceptible to the fallacy of acceptance.

The subsequent literature on frequentist statistics sheds very little additional light on these philosophical/foundational issues. The literature in philosophy of science overlooked apparent connections between the Popperian and Fisherian versions of falsification and more or less ignored the important developments in frequentist statistics.[1] In direct contrast to the extensive use of statistics in almost all scientific fields, by the early 1950s, logical empiricism had adopted combinations of hypothetico-deductive and Bayesian perspectives on inductive inference, with Carnap's confirmatory logics (logical relations between statements and evidence) dominating the evidential accounts in philosophy of science (see Neyman's [1957] reply to Carnap).

Not surprisingly, because of the absence of genuine guidance from statistics or philosophy of science, the practitioners in several disciplines, such as epidemiology, psychology, sociology, economics, and political science, came up with their own "pragmatic" ways to deal with the philosophical puzzles bedeviling the frequentist approach. This resulted in a hybrid of the Fisher and N-P accounts, criticized as "inconsistent from both perspectives and burdened with conceptual confusion" (Gigerenzer, 1993, p. 323). That these methods were open to unthinking use and abuse made them a

---

[1] Exceptions include Giere (1969), Hacking (1965), Seidenfeld (1979), philosophical contributors to Harper and Hooker (1976), and Godambe and Sprott (1971).

convenient scapegoat for the limits and shortcomings of several research areas, a practice that continues unabated to this day (see references). In my own field of economics, Ziliak and McCloskey (2008) propose their own "economic" way to deal with the statistical versus substantive significance problem; see Spanos (2008) for a critical review.

By the early 1960s, disagreements about the philosophical underpinnings of frequentist inductive reasoning were increasingly taken as prima facie evidence that it failed to provide a genuine account for inference or evidence. This encouraged the supposition of the philosophical superiority of Bayesian inference, which does away with error probabilities altogether and upholds foundational *principles* like the *likelihood* and *coherency*. Despite the vast disagreements between different Bayesian schools, this impression continues to be reiterated, largely unchallenged in both statistics (see Ghosh et al., 2006) and philosophy of science (see Howson and Urbach, 1993).

3 Inductive Reasoning in Frequentist Statistics

A pivotal contribution of Mayo and Cox is a general *frequentist principle for inductive reasoning*, which they motivate as a modification and extension of the *p*-value reasoning:

FEV (i): data z_0 provides (strong) evidence against the null H_0 (*for* a discrepancy from H_0), if and only if (iff) the *p*-value, $P(d(Z) > d(z_0); H_0) = p(z_0)$ is very low or, equivalently, $P(d(Z) \leq d(z_0); H_0) = (1 - p(z_0))$ is very high.

Corollary. Data z_0 do *not* provide (strong) evidence against H_0, if $P(d(Z) > d(z_0); H_0) = p(z_0)$ is *not* very low.

This is a formal version of our "minimal scientific principle for evidence" (p. 3). The question that naturally arises is whether the aforementioned conditions relating to the *p*-value can be strengthened enough to avoid the fallacies of acceptance and rejection. The answer provided by Mayo and Cox is that, in cases where one can quantify departures from H_0 using a discrepancy parameter $\gamma \geq 0$, one can strengthen the corollary to guard against the fallacy of acceptance in the following form:

FEV(ii): A moderate $p(z_0)$-value is evidence of the absence of a discrepancy γ from H_0, only if $P(d(Z) > d(z_0); \mu_0 + \gamma)$ is very high.

What about the fallacy of rejection? It is well known that a very low *p*-value establishes the existence of some discrepancy $\gamma \geq 0$ from H_0 but provides no information concerning the magnitude of γ licensed by data z_0. This magnitude can be established using an obvious modification of FEV(ii) to strengthen FEV(i) in safeguarding it against the fallacy of rejection:

FEV(iii): A very low $p(z_0)$-value is evidence *for* a discrepancy $\gamma \geq 0$ from H_0, only if $P(d(Z) \leq d(z_0); \mu_0 + \gamma)$ is very high.

The preceding principles constitute crucial extensions of post-data frequentist inductive reasoning in cases where one can quantify departures from H_0 using a *discrepancy parameter* γ. Under such circumstances, the FEV(ii) and FEV (iii) rules can be seen as special cases of the *severity evaluations* associated with N-P accept/reject decisions (Mayo 1996, Mayo and Spanos, 2006):

$$\text{SEV}(T_\alpha; z_0; \mu \leq \mu_1) = P(d(Z) > d(z_0); \mu > \mu_1), \quad \text{for } \mu_1 = \mu_0 + \gamma, \gamma \geq 0,$$
$$\text{SEV}(T_\alpha; z_0; \mu > \mu_1) = P(d(Z) \leq d(z_0); \mu \leq \mu_1), \text{ respectively.}$$

At first sight these evaluations give the impression that they stem exclusively from the Neyman-Pearson (N-P) testing perspective because they remind one of the evaluation of power and the probability of type II error. This first impression is misleading, however, because on closer examination the severity evaluations draw from both the N-P and Fisherian perspectives on testing; they constitute a harmonious reconciliation of the two that can be used to address several of the questions mentioned in Section 2. Like the *p*-value, but unlike the type II error probability and power, the severity evaluations constitute *post-data error probabilities*. They involve events in the sample space denoting *lesser* (or *greater*) accordance with H_0 than z_0 is. Like the type II error probability and power, the severity evaluations involve scenarios with specific discrepancies from the null, $\mu_1 = \mu_0 + \gamma$ (for some $\gamma \geq 0$), but, unlike them, the emphasis here is on evaluating the *post-data capacity* of the test in question. Notwithstanding Fisher's rhetoric against type II errors and power (see Fisher, 1955), enough evidence exists to suggest that he also viewed the optimality of tests in terms of their capacity (sensitivity) to detect discrepancies from the null hypothesis: "By increasing the size of the experiment, we can render it more sensitive, meaning by this that it will allow of the detection of... a quantitatively smaller departure from the null hypothesis" (Fisher, 1935, pp. 21–22).

This emphasis on capacity to detect discrepancies is exactly what is needed to provide an evidential construal of frequentist tests when combined with the following principle:

Severity Principle (SP): Data z_0 do *not* provide good evidence for hypothesis H (H_0 or H_1) if z_0 is used in conjunction with a test procedure that

Table 7.1. *Simple Normal Model*

$Z_t = \mu + u_t, \quad t \in \mathbb{N},$
[1] Normality: $Z_t \sim N(.,.),$
[2] Constant mean: $E(Z_t) = \mu,$
[3] Constant variance: $\mathrm{Var}(Z_t) = \sigma^2,$
[4] Independence: $\{Z_t, \ t \in \mathbb{N}\}$ is an independent process.

has a very low capacity to uncover discrepancies from H when present (see Mayo, 1996).

4 Model Adequacy

Another largely neglected area of frequentist statistics that Cox and Mayo discuss is that of *model adequacy*. The issues of statistical model specification and adequacy can be traced back to Cox's (1958) paper and constitutes a recurring theme in many of his writings, including Cox and Hinkley (1974) and Cox (1990, 2006).

4.1 Model Adequacy and Error Statistics

Since Fisher (1922) it has been known, but not widely appreciated, that the reliability of inductive inference depends crucially on the validity of the prespecified statistical model, which is generically denoted by $\mathcal{M}_\theta(\mathbf{z}) = \{f(\mathbf{z}; \boldsymbol{\theta}), \boldsymbol{\theta} \in \Theta\}$, $\mathbf{z} \in \mathbb{R}_z^n$, where $f(\mathbf{z}; \boldsymbol{\theta})$ denotes the distribution of the sample $\mathbf{Z} := (Z_1, Z_2, \ldots, Z_n)$, and Θ and $\mathbb{R}_z^n$ denote the parameter and sample spaces, respectively. The statistical model is chosen to render the data $\mathbf{z}_0 := (z_1, z_2, \ldots, z_n)$ a "typical realization" of the stochastic process $\{Z_t, \ t \in \mathbb{N} := (1, 2, \ldots, n \ldots)\}$, whose probabilistic structure is parameterized by $\mathcal{M}_\theta(\mathbf{z})$. A crucial first step in assessing model adequacy is to be able to specify a statistical model in terms of a complete set of probabilistic assumptions that are testable vis-à-vis data $\mathbf{z}_0$ (see Spanos, 1999).

The quintessential statistical model is given in Table 7.1.

Statistical model adequacy – that the assumptions underlying a statistical model $\mathcal{M}_\theta(\mathbf{z})$ (e.g., conditions [1]–[4] in Table 7.1) are valid for data $\mathbf{z}_0$ – is crucially important for frequentist inference because it secures *error reliability* – the nominal and actual error probabilities coincide (approximately) – which is fundamental in learning from data and is an important component of objectivity. Departures from model assumptions give rise to *error unreliability* – divergences between nominal and actual error probabilities – leading inductive inferences astray. Spanos (2005) showed that even seemingly minor departures (e.g., the presence of correlation, say, $\rho =$ .1, instead of assumption [4]), can give rise to sizeable divergences between nominal (.05) and actual (.25) error probabilities.

4.2 Misspecification (M-S) testing

The idea underlying model validation is to construct M-S tests using "distance" functions (test statistics) whose distribution under:

H_0: the probabilistic assumptions constituting $\mathcal{M}_\theta(\mathbf{z})$ hold for data $\mathbf{z}_0$,

is known, and at the same time they have adequate power against potential departures from the model assumptions. The logic of M-S testing is the one underlying Fisher's significance testing (Mayo and Cox, p. 259), where one identifies a test statistic $d(\mathbf{Z})$ to measure the distance between what is expected under H_0 with $d(\mathbf{z}_0)$. When the relevant p-value, $P(d(\mathbf{Z}) > d(\mathbf{z}_0)$; H_0 true) $= p(\mathbf{z}_0)$, is very small, then there is evidence of violations of the model assumption(s) entailed by H_0. If the p-value is not small enough, one is entitled to rule out only departures the test had enough capacity to detect. Mayo and Cox bring out the importance of combining *omnibus* and *directional* M-S tests because they shed different light on possible departures from the model assumptions.

4.3 Model Adequacy and Sufficiency

An important impediment to model validation using M-S testing has been the argument that it involves *illegitimate double-use of data*, a topic discussed in Chapter 4. The same data are used to draw inferences concerning θ as well as test the validity of assumptions [1]–[4]. Although nobody can deny that M-S testing involves double use of data, the charge of illegitimacy can be challenged on several grounds (see Mayo and Spanos, 2004). Cox and Mayo offer a purely statistical argument based on sufficiency. This argument arises in cases where there exists a sufficient statistic $S(\mathbf{Z})$ for θ, which gives rise to the following reduction:

$$f(\mathbf{z}; \theta) \propto f(\mathbf{z}|\mathbf{s}) \cdot f(\mathbf{s}; \theta) \quad \text{for all } \mathbf{z} \in \mathbb{R}_z^n \tag{1}$$

In this reduction, the information in the data about the model is split into two parts, one $[f(\mathbf{s}; \theta)]$ captures all the information assuming the model to be correct, and the other $[f(\mathbf{z}|\mathbf{s})]$ allows checking the adequacy of the model. By a simple *modus tollens* argument, $f(\mathbf{z}; \theta)$ implies $f(\mathbf{z}|\mathbf{s})$; thus, any departure from $f(\mathbf{z}|\mathbf{s})$ implies that $f(\mathbf{z}; \theta)$ is false. In the case of a simple Bernoulli model, $f(\mathbf{z}|\mathbf{s})$ is a discrete uniform distribution, and in the case of the simple Poisson model, $f(\mathbf{z}|\mathbf{s})$ is a multinomial distribution with identical cell probabilities. In both of these cases the form of $f(\mathbf{z}|\mathbf{s})$ gives rise to testable restrictions, which can be used to assess the validity of the original model, $\mathcal{M}_\theta(\mathbf{z})$.

This result can be extended to other situations where the structure of $f(\mathbf{z}; \theta)$ gives rise to reductions that admit statistics whose sampling distributions are free of θ.

4.4 Model Validation and the Role of Sufficiency and Ancillarity

Motivated by informal arguments of Mayo (1981) and Hendry (1995), Spanos (2007a) showed that under certain assumptions $f(\mathbf{z}; \theta)$ can be reduced into a product of two components as in reduction (1), but now it involves a *minimal sufficient* $S(\mathbf{Z})$ and a *maximal ancillary* statistic $R(\mathbf{Z})$ for θ:

$$f(\mathbf{z}; \theta) \propto f(\mathbf{s}|\theta) \cdot f(\mathbf{r}) \quad \text{for all } (\mathbf{r}, \mathbf{s}) \in \mathbb{R}_z^n. \tag{2}$$

This reduction is analogous to reduction (1), but here $\mathbf{s}$ is also independent of $\mathbf{r}$. The crucial argument for relying on $f(\mathbf{r})$ is that the probing for departures from $\mathcal{M}_\theta(\mathbf{z})$ is based on error probabilities that do not depend on the true θ.

Example. In the case of the simple Normal model with $\theta = (\mu, \sigma^2)$ (Table 7.1), reduction (2) holds with

$$\mathbf{s} = (\bar{z} = (1/n) \sum_{t=1}^{n} z_t, \quad s^2 = [1/(n-1)] \sum_{t=1}^{n} (z_t - \bar{z})^2), \quad \mathbf{r} = (r_3, \ldots, r_n)$$

$$r_t = \frac{\sqrt{n}\,(z_t - \bar{z})}{s}, \quad t = 3, 4, \ldots, n,$$

where $\mathbf{r}$ represents the *studentized residuals*. These results extend to the linear regression and hold approximately in many cases where asymptotic Normality is invoked.

From a methodological perspective, the separation in reductions (1) and (2) reflects the drastically different questions posed for inference and model adequacy purposes. The question posed for model adequacy purposes is as follows: "Do data $\mathbf{z}_0$ represent a truly typical realization of the stochastic mechanism specified by $\mathcal{M}_\theta(\mathbf{z})$?" The generic form of the model validation hypotheses is:

$$H_0: f^*(\mathbf{z}) \in \mathcal{M}_0(\mathbf{z}) \quad \text{vs.} \quad H_1: f^*(\mathbf{z}) \in [\mathcal{P}(\mathbf{z}) - \mathcal{M}_0(\mathbf{z})],$$

where $f^*(\mathbf{z})$ denotes the true (but unknown) distribution of the sample, and $\mathcal{P}(\mathbf{z})$ the set of all possible models that could have given rise to data $\mathbf{z}_0$. This form clearly indicates that M-S probing takes place *outside the boundaries* of $\mathcal{M}_\theta(\mathbf{z})$. Note that the generic alternative $[\mathcal{P}(\mathbf{z})-\mathcal{M}_\theta(\mathbf{z})]$ is intractable as it stands; thus, in practice one must consider different forms of departures from H_0, which can be as vague as a direction of departure or as specific as an encompassing model $\mathcal{M}_\varphi(\mathbf{z})$; $\mathcal{M}_\theta(\mathbf{z}) \subset \mathcal{M}_\varphi(\mathbf{z})$ (see Spanos, 1999).

In contrast, the questions posed by inferences concerning θ take the model $\mathcal{M}_\theta(\mathbf{z})$ as given (adequate for data $\mathbf{z}_0$) and probe their validity *within the model's boundaries* (see Spanos, 1999).

On a personal note, I ascertained the crucial differences between testing *within* (N-P) and testing *outside the boundaries* (M-S) of a statistical model and ramifications thereof, after many years of puzzling over what renders Fisher's significance testing different from N-P testing. What eventually guided me to that realization was the "unique" discussion of testing in Cox's writings, in particular the exposition in chapters 3–6 of Cox and Hinkley (1974). I also came to appreciate the value of preliminary data analysis and graphical techniques in guiding and enhancing the assessment of model adequacy from Cox's writings. After years of grappling with these issues, the result is a unifying modeling framework wherein these techniques become indispensable facets of statistical modeling and inference (see Spanos, 1986).

5 Revisiting Bayesian Criticisms of Frequentist Statistics

Of special importance in Cox and Mayo (7(II)) is their use of the new philosophy of frequentist inference to shed light on earlier philosophical debates concerning frequentist versus Bayesian inference and more recent developments in objective (O) Bayesianism.

5.1 Revisiting the Likelihood Principle (LP)

A crucial philosophical debate concerning frequentist versus Bayesian inference began with Birnbaum's (1962) result claiming to show that the (strong) LP follows from Sufficiency Principle (SP) and the Weak Conditionality Principle (WCP); if frequentists wish to condition on $\mathbf{z}_0$, they are faced with either renouncing sufficiency or renouncing error probabilities altogether. Cox and Mayo counter this argument as follows:

It is not uncommon to see statistics texts argue that in frequentist theory one is faced with the following dilemma, either to deny the appropriateness of conditioning on the precision of the tool chosen by the toss of a coin, or else to embrace the strong likelihood principle which entails that frequentist sampling distributions are irrelevant to inference once the data are obtained. This is a false dilemma: Conditioning is warranted in achieving objective frequentist goals, and the conditionality principle coupled with sufficiency does not entail the strong likelihood principle. The "dilemma" argument is therefore an illusion. (Cox and Mayo, p. 298)

Indeed, in 7(III) Mayo goes much further than simply raising questions about the cogency of the LP for frequentist inference. She subjects Birnbaum's "proof" to a careful logical scrutiny. On logical grounds alone, she brings out the *fallacy* that shows that those who greeted Birnbaum's paper as a "landmark in statistics" (see Savage, 1962b) with skepticism had good cause to withhold assent. By and large, however, Birnbaum's result is taken at face value (see Berger and Wolpert, 1988). As such, arguments for conditioning are taken as arguments for LP, which is just a short step away from Bayesianism; Ghosh et al. (2006) argue:

> Suppose you are a classical statistician and faced with this example [Welch's uniform] you are ready to make conditional inference as recommended by Fisher. Unfortunately, there is a catch. Classical statistics also recommends that inference be based on minimal sufficient statistics. These two principles, namely the conditionality principle (CP) and sufficiency principle (SP) together have a far reaching implication. Birnbaum (1962) proved that they imply one must then follow the likelihood principle (LP), which requires inference be based on the likelihood alone, ignoring the sample.... Bayesian analysis satisfies the LP since the posterior depends on the data only through the likelihood. Most classical inference procedures violate the LP. (p. 38)

Even frequentist statisticians who treated the WCP + S = LP equation with skepticism are legitimately challenged to give a principled ground for conditioning rather than using the unconditional error probabilities. When we consider the kinds of canonical examples that argue for conditioning, especially in light of FEV, it is not difficult to find a principled argument, and that alone attests to its value as a frequentist principle. According to Cox and Mayo, "[M]any criticisms of frequentist significance tests (and related methods) are based on arguments that overlook the avenues open in frequentist theory for ensuring relevancy of the sampling distribution on which p-values are to be based" (Chapter 7, p. 295). Indeed, one may explain the inappropriateness of the unconditional inference by appealing to the notion of *relevant error probabilities*, as elaborated in the twin Cox and Mayo papers, where "relevance" includes *error reliability* (stemming from model adequacy) and *inference specificity* (relating to the inference at hand).

5.2 Revisiting the Welch Example for Conditional Inference

I argue that the various examples Bayesians employ to make their case involve some kind of "rigging" of the statistical model so that it appears as though embracing the conditionality principle (CP) is the only way out, when in fact other frequentist principles invariably allow extrication. To illustrate the general point, I consider the case of the Welch uniform, which seems the most realistic among these examples; Cox and Mayo discuss two other examples (pp. 295–7). In the Welch (1939) uniform case where $Z_k \sim U(\theta - .5, \theta + .5)$, the rigging stems from the fact that this distribution is *irregular* in that its support depends on the unknown parameter θ (see Cox and Hinkley, 1974, p. 112). This irregularity creates a constraint between θ and the data z_0 in the sense that, whatever the data, $\theta \in A(z_0) = [z_{[n]} - .5, z_{[1]} + .5]$, where $z_{[n]} = \max(z_1, \ldots, z_n)$ and $z_{[1]} = \min(z_1, \ldots, z_n)$. Hence, post-data, the unconditional sampling distribution $f(\hat{\theta}; \theta)$, $[\theta - .5 \leq \hat{\theta} \leq \theta + .5]$, where $\hat{\theta} = [(Z_{[n]} + Z_{[1]})/2]$ is an estimator of θ, ignores the support information $\theta \in A(z_0)$, and thus a confidence interval may include *infeasible values* of θ. The CP argument suggests that the conditional distribution $f(\hat{\theta} \mid R; \theta)$, where $R = (Z_{[n]} - Z_{[1]})$ is an ancillary statistic, gives rise to much better inference results (see Berger and Wolpert, 1988; Ghosh et al., 2006; Young and Smith, 2005). Notwithstanding such claims, Cox and Hinkley (1974, pp. 220–1) showed that this conditional inference is also *highly problematic* because $f(\hat{\theta} \mid R; \theta)$ has *no* discriminatory capacity because it is uniform over $[-.5(1 - R), .5(1 - R)]$.

Using the notion of relevant error probabilities, Spanos (2007b) showed that, *post-data*, the truncated sampling distribution $f(\hat{\theta} \mid A(z_0); \theta)$, for $\theta \in A(z_0)$, provides the relevant basis for inference because it accounts for the deterministic support information $\theta \in A(z_0)$ – created by the irregularity of the model – without sacrificing the discriminatory capacity of $f(\hat{\theta}; \theta)$.

Learning from data can only result from using *relevant* error probabilities in the sense that they reflect faithfully (in an error-reliable sense) the mechanism that actually generated the data z_0.

5.3 Objective Bayesian Perspective

It is refreshing to see Cox and Mayo give a hard-nosed statement of what scientific objectivity demands of an account of statistics, show how it relates to frequentist statistics, and contrast that with the notion of "objectivity" used by O-Bayesians (see Berger, 2004). They proceed to bring out several weaknesses of this perspective. The question that naturally arises from their discussion is "if one renounces the likelihood, the stopping rule, and the coherence principles, marginalizes the use of prior information as largely untrustworthy, and seeks procedures with 'good' error probabilistic properties (whatever that means), what is left to render the inference Bayesian, apart from a *belief* (misguided in my view) that the only way to provide an evidential account of inference is to attach probabilities to hypotheses?"

6 Error Statistics and Popperian Falsification

The philosophy of frequentist inference growing out of 7(I) bears much fruit in 7(II). Among the most noteworthy is the stage it sets for philosophical debates concerning Popperian falsification. Although recognizing the importance of a "new experimentalist" focus on the details of obtaining, modeling, and making inferences about data, a reluctance exists among Popperian philosophers of science to see that full-fledged ampliative or inductive inferences are involved. Part of the reason is the supposition that inductive inference is a matter of assigning post-data degrees of probability or belief to hypotheses (Musgrave, Achinstein, this volume), whereas a true Popperian ("progressive" Popperian, as Mayo calls him) would insist that what matters is not *highly probable*, but rather *highly probed* hypotheses resulting from severe tests – genuine attempts at refutation using tests with enough capacity to detect departures. In a sense, the error-statistical interpretation of tests in these joint papers offers solutions to Popper's problems about deductive "falsification" and "corroboration," wherein, by essentially the same logic as that of statistical falsification, one may warrant claims that pass severe tests.

This suggestion is importantly different from using *p*-values and other error probabilities as post-data degrees of confirmation, support, or the like – in contrast with the most familiar "evidential" interpretations of frequentist methods, or the well-known O-Bayesian "reconciliations" (see Berger, 2003). Where in philosophy of science we might say this enables us to move away from what Musgrave (in this volume) calls "justificationist" approaches, in statistics, it allows us to answer those skeptical of regarding "frequentist statistics as a theory of inductive inference."

7 Conclusion

My hope is that I was able to convey to the reader the sense of discovery that I felt in coming to see how several of the frequentist procedures and principles developed by Cox over many years obtained their philosophical justification within a coherent statistical framework, the central goal of which is "to extract what can be *learned from data* that can also be vouched for" (p. 278).

The series of exchanges between Cox and Mayo exemplifies the central goals of the "two-way street," wherein statistical ideas inform philosophical debates and problems while philosophical and foundational analysis offer unification and justification for statistical methodology. In the back-and-forth exchanges that gave rise to 7(I) and 7(II), it is apparent that both scholars have moved away from their usual comfort zones to some degree; yet by taking the two papers together, the shared principles of the error-statistical philosophy begin to crystallize and the essential pieces of the puzzle fall into place.

Aside from their specific contributions, the Cox and Mayo papers are noteworthy for taking some crucial steps toward legitimating the philosophy of frequentist statistics as an important research program in its own right. I hope that others, statisticians and philosophers of science, will join them in what may properly be described as an expansion of contemporary work in the foundations of statistics.

References

Berger, J. (2003), "Could Fisher, Jeffreys and Neyman Have Agreed on Testing?" *Statistical Science*, 18: 1–12.

Berger, J. (2004), "The Case for Objective Bayesian Analysis," *Bayesian Analysis*, 1: 1–17.

Berger, J., and Wolpert, R. (1988), *The Likelihood Principle*, 2nd ed., Institute of Mathematical Statistics, Hayward, CA.

Birnbaum, A. (1962), "On the Foundations of Statistical Inference" (with discussion), *Journal of the American Statistical Association*, 57: 269–306.

Cox, D.R. (1958), "Some Problems Connected with Statistical Inference," *Annals of Mathematical Statistics*, 29: 357–72.

Cox, D.R. (1990), "Role of Models in Statistical Analysis," *Statistical Science*, 5: 169–74.

Cox, D.R. (2006), *Principles of Statistical Inference*, Cambridge University Press, Cambridge.

Cox, D.R., and Hinkley, D.V. (1974), *Theoretical Statistics*, Chapman & Hall, London.

Fisher, R.A. (1922), "The Mathematical Foundations of Theoretical Statistics," *Philosophical Transactions of the Royal Society A*, 222: 309–68.

Fisher, R.A. (1925), "Theory of Statistical Estimation," *Proceedings of the Cambridge Philosophical Society*, 22: 700–25.

Fisher, R.A. (1934), "Two New Properties of Maximum Likelihood," *Proceedings of the Royal Statistical Society A*, 144: 285–307.

Fisher, R.A. (1935), *The Design of Experiments*, Oliver and Boyd, Edinburgh.

Fisher, R.A. (1955) "Statistical Methods and Scientific Induction," *Journal of the Royal Statistical Society B*, 17: 69–78.

Ghosh, J.K., Delampady, M., and Samanta, T. (2006), *An Introduction to Bayesian Analysis: Theory and Methods*, Springer, New York.

Giere, R.N. (1969), "Bayesian Statistics and Biased Procedures," *Synthese*, 20: 371–87.

Gigerenzer, G. (1993), "The Superego, the Ego, and the Id in Statistical Reasoning," pp. 311–39 in G. Keren and C. Lewis (eds.), *A Handbook of Data Analysis in the Behavioral Sciences: Methodological Issues*, Lawrence Erlbaum Associates, Hillsdale, NJ.

Godambe, V.P., and D.A. Sprott (eds.) (1971), *Foundations of Statistical Inference: a Symposium*, Holt, Rinehart and Winston, Toronto.

Hacking, I. (1965), *Logic of Statistical Inference*, Cambridge University Press, Cambridge.

Hand, D.J., and Herzberg, A.M. (2005), *Selected Statistical Papers of Sir David Cox*, vols. 1–2, Cambridge University Press, Cambridge.

Harper, W.L., and C.A. Hooker (eds.) (1976), *Foundations of Probability Theory, Statistical Inference, and Statistical Theories of Science, Vol. II: Foundations and Philosophy of Statistical Inference*, Reidel, Dordrecht, The Netherlands.

Hendry, D.F. (1995), *Dynamic Econometrics*, Oxford University Press, Oxford.

Howson, C., and Urbach, P. (1993), *Scientific Reasoning: The Bayesian Approach*, 2nd ed., Open Court, Chicago.

Mayo, D.G. (1981), "Testing Statistical Testing," pp. 175–230 in J. Pitt (ed.), *Philosophy in Economics*, D. Reidel, Dordrecht.

Mayo, D.G. (1996), *Error and the Growth of Experimental Knowledge*, University of Chicago Press, Chicago.

Mayo, D.G. (2005), "Philosophy of Statistics," pp. 802–15 in S. Sarkar and J. Pfeifer (eds.), *Philosophy of Science: An Encyclopedia*, Routledge, London.

Mayo, D.G., and Spanos, A. (2004), "Methodology in Practice: Statistical Misspecification Testing," *Philosophy of Science*, 71: 1007–25.

Mayo, D.G. and Spanos, A. (2006), "Severe Testing as a Basic Concept in a Neyman-Pearson Philosophy of Induction," *British Journal for the Philosophy of Science*, 57: 323–57.

Neyman, J. (1937), "Outline of a Theory of Statistical Estimation Based on the Classical Theory of Probability," *Philosophical Transactions of the Royal Statistical Society of London A*, 236: 333–80.

Neyman, J. (1956), "Note on an Article by Sir Ronald Fisher," *Journal of the Royal Statistical Society B*, 18: 288–94.

Neyman, J. (1957), "Inductive Behavior as a Basic Concept of Philosophy of Science," *Revue de L'Institut International de Statistique*, 25: 7–22.

Neyman, J., and Pearson, E.S. (1933), "On the Problem of the Most Efficient Tests of Statistical Hypotheses," *Philosophical Transactions of the Royal Society A*, 231: 289–337.

Pearson, E.S. (1955), "Statistical Concepts in Their Relation to Reality," *Journal of the Royal Statistical Society B*, 17: 204–7.

Savage, L., ed. (1962a), *The Foundations of Statistical Inference: A Discussion*. Methuen, London.

Savage, L. (1962b), "Discussion" on Birnbaum (1962), *Journal of the American Statistical Association*, 57: 307–8.

Seidenfeld, T. (1979), *Philosophical Problems of Statistical Inference: Learning from R.A. Fisher*, Reidel, Dordrecht, The Netherlands.

Spanos, A. (1986), *Statistical Foundations of Econometric Modelling*, Cambridge University Press, Cambridge.

Spanos, A. (1999), *Probability Theory and Statistical Inference: Econometric Modeling with Observational Data*, Cambridge University Press, Cambridge.

Spanos, A. (2005), "Misspecification, Robustness and the Reliability of Inference: The Simple t-Test in the Presence of Markov Dependence," Working Paper, Virginia Tech.

Spanos, A. (2007a), "Sufficiency and Ancillarity Revisited: Testing the Validity of a Statistical Model," Working Paper, Virginia Tech.

Spanos, A. (2007b), "Revisiting the Welch Uniform Model: A Case for Conditional Inference?" Working Paper, Virginia Tech.

Spanos, A. (2008), "Stephen Ziliak and Deirdre McCloskey's The Cult of Statistical Significance: How the Standard Error Costs Us Jobs, Justice, and Lives," *Erasmus Journal for Philosophy and Economics*, 1: 154–64.

Welch, B.L. (1939), "On Confidence Limits and Sufficiency, and Particular Reference to Parameters of Location," *Annals of Mathematical Statistics*, 10: 58–69.

Young, G.A., and Smith, R.L. (2005), *Essentials of Statistical Inference*, Cambridge University Press, Cambridge.

Ziliak, S.T., and McCloskey, D.N. (2008), *The Cult of Significance: How the Standard Error Costs Us Jobs, Justice and Lives*, University of Michigan Press, Ann Arbor, MI.

Related Exchanges

Bartz-Beielstein, T. (2008), "How Experimental Algorithmics Can Benefit From Mayo's Extensions To Neyman-Pearson Theory Of Testing," *Synthese (Error and Methodology in Practice: Selected Papers from ERROR 2006)*, vol. 163(3): 385–96.

Casella, G. (2004), "Commentary on Mayo," pp. 99–101 in M.L. Taper and S.R. Lele (eds.), *The Nature of Scientific Evidence: Statistical, Philosophical, and Empirical Considerations*, University of Chicago Press, Chicago.

Mayo, D.G. (2003), "Severe Testing as a Guide for Inductive Learning," pp. 89–118 in H. Kyburg, Jr. and M. Thalos (eds.), *Probability Is the Very Guide of Life: The Philosophical Uses of Chance*, Open Court, Chicago and La Salle, Illinois.

Mayo, D.G. (2004), "An Error-Statistical Philosophy of Evidence," pp. 79–97 and "Rejoinder" pp. 101–15, in M.L. Taper and S.R. Lele (eds.), *The Nature of Scientific Evidence: Statistical, Philosophical, and Empirical Considerations*, University of Chicago Press, Chicago.

Taper, M.L. and Lele, S.R. (eds.) (2004), *The Nature of Scientific Evidence: Statistical, Philosophical, and Empirical Considerations*, University of Chicago Press, Chicago.

Additional References on the Significance Test Debates (see webpage for updates)

Altman, D.G., Machin D., Bryant T.N., and Gardner M.J. (2000), *Statistics with Confidence*, (eds.), British Medical Journal Books, Bristol.

Barnett, V. (1982), *Comparative Statistical Inference*, 2nd ed., John Wiley & Sons, New York.

Cohen, J. (1994), "The Earth is Round (p < .05)," *American Psychologist*, 49:997–1003.

Harlow L.L., Mulaik, S.A., and Steiger, J.H. (1997) *What If There Were No Significance Tests?* Erlbaum, Mahwah, NJ.

Lieberman, B. (1971), *Contemporary Problems in Statistics: a Book of Readings for the Behavioral Sciences*, Oxford University Press, Oxford.

Morrison, D.E., and Henkel, R.E. (1970), *The Significance Test Controversy: A Reader*, Aldine, Chicago.

EIGHT

Causal Modeling, Explanation and Severe Testing

I Explanation and Truth

Clark Glymour

1 The Issues

There are at least two aspects to understanding: comprehensibility and veracity. Good explanations are supposed to aid the first and to give grounds for believing the second – to provide insight into the "hidden springs" that produce phenomena and to warrant that our estimates of such structures are correct enough. For Copernicus, Kepler, Newton, Dalton, and Einstein, the virtues of explanations were guides to discovery, to sorting among hypotheses and forming beliefs. The virtues were variously named: simplicity, harmony, unity, elegance, and determinacy. These explanatory virtues have never been well articulated, but what makes them contributions to perspicacity is sometimes apparent, while what makes them a guide to truth is obscure.

We know that under various assumptions there is a connection between *testing* – doing something that could reveal the falsity of various claims – and the second aspect of coming to understand. Various forms of testing can be stages in strategies that reliably converge on the truth, and some cases even provide probabilistic guarantees that the truth is not too far away. But how can *explaining* be any kind of reliable guide to any truth worth knowing? *What structure or content of thoughts or acts distinguishes explanations, and in virtue of what, if anything, are explanations indications of truth that should prompt belief in their content, or provide a guide in inquiry? How, if at all, do explanatory virtues aid in selecting what to believe so that the result is a true understanding?*

My answer is that explanations and the explanatory virtues that facilitate comprehension also facilitate testing the claims made in explanation. I do not claim that every aspect and device of explanation marks a test or testability – but I will give particular historical examples in which the tests corresponding to good scientific explanations made a case for the truth of theories. I claim the explanations specify tests whose results, if in accord with the explanations, converge on the truth *conditionally*. In one of my examples, as far as is known, the conditions are met, whereas in another they proved false. But more important than the history mongering, the much-neglected methodological project is threefold: first, to articulate general conditions – assumptions – under which the tests corresponding to explanations in various kinds of theories can converge on the truth; second, to articulate and prove methods for finding theories, as the data change, whose various explanations converge on the truth under general assumptions; and third, to articulate, as much as possible, means to test whether the general conditions hold in particular circumstances. To illustrate these projects, I describe work that shows that the connection between testability and explanation is intrinsic to causal explanations of regularities, for which a variety of conditions sufficient for convergence to the truth have been established, and a variety of feasible, convergent methods of discovery have been proved. First, I give two brief dismissals.

2 Hempel

According to Carl Hempel, the link between explanation and testing is provided by an equivalence known to philosophers as the symmetry thesis: premises explain a phenomenon if and only if the phenomenon could be predicted from the premises. Because he held that to explain a particular fact or regularity was to deduce a statement of it from general, "lawlike" premises and singular statements of fact, the symmetry thesis was a natural, even necessary, consequence.

Explanation in Hempel's formulation cannot be much of a guide to truth. If hypothesis H explains statement E, whether singular or general, then so does the hypothesis $H\&A$, where A is any "lawlike" claim consistent with H. It is natural to try to obviate the "tacking on" of intuitively irrelevant hypotheses by requiring that the explanation be in some sense "simple," but no satisfactory account of logical simplicity that could help the problem has ever been produced. The simplest – in the sense of logically weakest – Hempelian explanation of any lawlike regularity is that regularity itself.

Hempel's failure to link explanation and truth has sometimes been taken as an indication that no intimate connection exists among logic, explanation, and truth. The inference is too facile: not every logical relation, and not every explanation that exploits logical relationships, is a matter of a deduction of Hempelian form. I consider some other relationships in due course, but I first turn to two other approaches to the issues I have posed.

3 Musgrave

Posing these questions as I do, and suggesting the answers I will, puts me at loggerheads with the views of a distinguished philosopher, Alan Musgrave, who thinks that whether explanations demonstrably *are* guides to truth is orthogonal to whether we should believe them to be. Musgrave can here serve as representative for Peter Strawson and for many others who have advanced similar views, however differently put.

In a discussion that is admirably direct, Musgrave writes:

It is reasonable to believe that the best available explanation of any fact is true.

> *F* is a fact.
> Hypothesis *H* explains *F*.
> No available competing hypothesis explains *F* as well as *H* does.
> Therefore, *it is reasonable to believe that H* is true.

This scheme is valid and instances of it might well be sound. (Chapter 3, this volume, p. 92)

But the scheme can be sound in any instance only if the first premise is true in every instance because the first claim is a general proposition about rationality. Judged by that premise, almost all of the aforementioned Greats were unreasonable men: they did not believe the best explanations available in their time and they sought better. More than better explanations, they sought true explanations. I want to know how any explanatory virtues *could* guide the search for truth. For two reasons, Musgrave seems to think this sort of quest is quixotic. First, he argues that the success of science shows that reasonableness is a pretty good guide to the truth. No need exists to investigate the conditions under which various explanatory virtues are reliable guides to the truth: whatever those conditions are, the conduct of science embodies them through a process of conjecture and criticism. Second, he argues, explanatory virtues distinguish between hypotheses and their fictionalist parodies when no possible evidence exists that could do so. Because the fictionalist or Berkeleyan parodies could – in a purely logical sense – be true, explanatory virtues cannot in every logically possible case be a reliable guide to truth.

Were Musgrave correct, obedience to the counsel to be reasonable would stop theoretical innovation in science. But good reasons exist to think he is not right. Science is a social enterprise and social enterprises wander, some-times into cul-de-sacs. Lots of physicists think that is what has happened with string theory; I think that is what has happened with psychometrics, with a good deal of cognitive psychology, and with a great deal of quan-titative social science. What methodologists ought to provide, if they can, are effective methods justified by demonstrations of their reliability – if not in every possible circumstance, then in circumstances as general as can be found. On the second count, granted that explanatory virtues cannot be reliable marks of truth in deciding between each substantive theory and a fictionalist parody, it does not follow that they cannot be reliable marks of truth in other choices among other alternatives, and if so – if we do not take fictionalist parodies seriously (and I do not) – it is important to know what those conditions could be and how the explanatory virtues work within those conditions.

4 Probability and Explanation

Bayesian statistics is one thing (often a useful thing, but not germane to my topic); Bayesian epistemology is something else. The idea of putting proba-bilities over hypotheses delivered to philosophers a godsend, an entire pack-age of superficiality. I consider for illustration just one of several attempts to explain explanatory unification with subjective probabilities.

Wayne Myrvold (2003) says that theories unify by making the subjective probabilities of phenomena *dependent*. The virtue of having phenomena unified by a theory is that the result is greater "support" for the theory than for a theory that does not unify the phenomena. Here then is a possible connection between explanation and testing for the truth of the explanans – the explanation leads to the prediction of one phenomenon from another. I think a charitable way of putting Myrvold's claim is this: a grasp of explana-tory unification prompts the rational understanding to increase the credence given to a unifying theory and provides that theory a support or rationale measured by the ratio of the degrees of belief given the theory after, to before, the unified phenomena were recognized.

In motivation, Myrvold remarks that Kepler observed that the motions of the planets are closely related to the motion of the sun: "A system, such as Tycho's or Copernicus', which centers the orbits of the planets on the sun, makes one set of phenomena – the mean apparent motions of the planets, and the Sun, along the ecliptic – carry information about what, on the Ptolemaic system, are independent phenomena, the deviations of the planets from their mean apparent motions." What Myrvold has in mind here, I think, is that as the sun and a superior planet, such as Mars, approach opposite positions in the celestial sphere, the daily motion of Mars against that sphere slows down, at opposition stops, and then reverses direction for a time. The position of the sun on the celestial sphere when opposition to a superior planet occurs changes from year to year, but the value of one variable – the apparent motion of a superior planet – gives information about the value of another variable – the position of the sun.

The example does not quite fit his scheme, for two reasons. First, with appropriate parameter values as estimated by Ptolemy, Ptolemaic theory gives the same relationships. Second, the relations are purely mathematical in both theories. Probability has nothing to do with it. The unification in Myrvold's example is entirely derivative from logical and semantic implications of what the theories *say* about the phenomena, not from anyone's personal degrees of belief about what the theories say and about the phenomena. Subjective probabilistic theories of explanatory virtues always mistake epiphenomena (degrees of belief) for phenomena (content).

What is true for Myrvold's account holds as well, I think, for other exercises in Bayesian accounts of scientific coherence and scientific explanation, although the details vary.[1] The Bayesian epistemology literature is filled with shadows and illusions: for example, Bayesian philosophers solve the difficulty that logical relations constituting an explanation can be conjoined with irrelevancies – the tacking on problems – by just *saying*, ad hoc for each case, that tacking on reduces degree of belief. For understanding, we need to look behind the phony probability numbers to the mathematical relationships they shadow.

5 Explanation, Unification, and Equivalences

Some of the mathematical connections that lie behind Myrvold's shadow unification are illustrated by Kepler's arguments for the Copernican framework, by Dalton's explanation of the law of definite proportions, by the equivalence of Joule's law for the expansion of gases and the ideal gas law, by the general relativistic explanation of conservation laws, and no doubt by many other cases. The idea is that unifying explanations transform independently established regularities into one another or into necessary truths.

---

[1] It is not only subjective Bayesians who connect probability and explanation in very odd ways. Consider propensities. Population biologists are said to use "fitness" of organisms or of types of organisms to explain their emergence or spread. Susan Mills and John Beatty (1979) have this to say: fitness cannot be used to explain evolutionary phenomena if fitness cannot be measured. But if fitness = number of children, then number of children cannot be explained by fitness without circularity. So let fitness = *expected* number of children, where the expectation is from a "propensity." *Now*, they say, fitness can explain the number of children.

Kepler argued that the Copernican theory transforms an apparently contingent regularity of the superior planets already noted by Ptolemy (if in a number of solar years, a superior planet passes through a number of oppositions to the sun and a number of revolutions with respect to the fixed stars, then the number of solar years is the sum of the other two numbers) into a mathematical necessity (if point 1 moves in a closed continuous curve around a center, and point 2 does as well, and the orbit of point 2 is everywhere inside the orbit of point 1, and the angular velocity of point 2 is greater than that of point 1, then the number of revolutions of 1 is the number of revolutions of 2 plus the number of overtakings (collineations on the same side of the center) of 2 by 1).

The transformation of the empirical regularity into a necessary feature of a heliocentric astronomy – necessary, at least, given the empirical fact that the time required for a revolution of longitude is always greater than a solar year – is central to the hypothesis of a circumsolar system. If the empirical regularity is false, then the entire Copernican framework is wrong; nothing is salvageable except by ad hoc moves (as in, *oh well, Mars isn't really a* "*planet*").

The same regularity is accounted for in Ptolemaic theory by adjusting parameters so that the line from each superior planet to the center of its epicycle on the deferent is always parallel to the line from the sun to the Earth. If one or another superior planet should not satisfy the regularity, no harm: the Ptolemaic parameters can be adjusted to save the phenomenon. Ptolemy wrote that he had no explanation for the empirical regularity, and so far as explanations point to evidence, he was right.[2]

Clerk Maxwell's theory reduces to mathematical identities both the fact that magnetic fields are closed and that time-changing magnetic fields constrain electrical fields. The falsity of these equations would destroy the theory.

$$\operatorname{div} B = 0 \tag{1}$$

$$\operatorname{curl} E = -\frac{1}{c}\frac{\partial H}{\partial t} \tag{2}$$

$$E = -\operatorname{grad}'\phi - \frac{1}{c}\frac{\partial A}{\partial t} \tag{3}$$

$$B = \operatorname{curl} A \tag{4}$$

---

[2] Musgrave is interestingly wrong when he writes: "The detailed content of the Copernican theory, and the fact that some of the detail of the Ptolemaic theory is similar to it, is essential to the explanation of the success of Ptolemaic theory" (this volume, p. 97). The success of Ptolemaic theory can also be explained by Harald Bohr's theorem: any semiperiodic function can be approximated arbitrarily closely by iterations of the epicycle on deferent construction.

It follows that

$$\mathrm{div\,curl\,} A = 0 \tag{1*}$$

$$-\mathrm{curl\,grad\,}\phi - \frac{1}{c}\mathrm{curl}\frac{\partial t}{\partial t} = -\frac{1}{c}\mathrm{curl}\frac{\partial A}{\partial t} \tag{2*}$$

Both (1*) and (2*) are identities.

In other cases – thermodynamics and Joule's law and Charles's law, for example, or the atomic theory and the law of definite proportions and the additivity of mass – the identifications of quantities provided by a theory merely transform one regularity into another. But that is enough to provide a test of those parts of the theory that identify measured quantities with thermodynamic quantities.

6 Tests, Assumptions, and Causal Explanations

Deborah Mayo advocates "severe testing" of hypotheses, meaning "[t]he probability is high that test T would not yield so high a score for H as e, given that H is false."

A common objection to Mayo's emphasis on severe tests is that only "low-level" statistical hypotheses can be so tested. Hypotheses of full-blooded theories that talk about properties not recorded in the data are beyond severe testing. I show to the contrary in what follows that something like severe testing is available for a broad class of causal explanations, including causal explanations that postulate unobserved quantities. Some philosophers – the late David Lewis for example – have claimed that all scientific explanation is causal explanation. Whether or not that is so, the class of cases I consider should put to rest the claim that serious explanations cannot be severely tested.

A further objection to Mayo's emphasis on severe testing is that tests, especially tests of hypotheses that are "high level," require assumptions, and whatever those assumptions may be, they themselves cannot be tested, at least not without further assumptions that are themselves untested, or that if tested require still other assumptions, and so forth. Because total evidence is always finite, the entire testing procedure must, so the informal argument goes, be unfounded and depend on assumptions for which there is no test. And whether or not it is a good objection, the conclusion is correct.[3] If assumptions are needed to reliably sort among explanations,

---

[3] Many years ago I attempted an account of theory testing in which a theory could be tested with respect to itself – using some of its claims as assumptions in tests of other claims, and other claims of the theory in tests of the assumptions, and so on, back and forth,

which assumptions should be used? No convincing, single answer exists because there are trade-offs: the weaker the assumptions, the less informative the conclusions that can be drawn from data. The appropriate methodological enterprise is not to shout "So there!" citing Duhem and Quine and such, but to investigate what can be tested and reliably discovered under varying assumptions.

For causal explanations it seems appropriate to begin by generalizing and formalizing the assumptions that are implicit in causal inference from experiments. In experiments to determine whether X causes Y, three possible results exist: yes, no, and no decision, allowing Bayesian inferences as cases of no decision. Other things equal, if an experiment finds no association, that is evidence for the negative; if an association is found, for the positive, hence the principles: prima facie, if X does not cause Y, Y does not cause X, no common cause of X and Y exists, and observation of cases is independent of values of X, Y, then X and Y are independent. Conversely, if X causes Y, then X and Y are dependent. I generalize the first and the contrapositive of the second to systems of variables, whether observed experimentally or not, whose causal relations are represented by a directed graph without cycles (hereafter, a DAG) whose directed edges connect variables, as follows:

Causal Markov Condition: Consider a causal system with variables V represented by a DAG G, and joint probability distribution P on the member of V. If the variables V in a system of variables are causally sufficient – for each external cause acting on the system, that cause directly affects only one variable in the system – each variable, X, in V is independent of variables in V that are not effects of X, conditional on the direct causes of X in V (the parents of X in G).

Faithfulness Condition: All conditional independence relations in P are consequences of the Markov Condition applied to G.

The Markov condition can be generalized, and the faithfulness condition can be strengthened and weakened. The faithfulness condition is really a kind

---

so that each of a set of axioms sufficient to entail the entire theory would be tested, as it were, with respect to their joint consequences. The formal theory was something of a disaster and became a makework target for philosophical critics – although Ken Gemes has recently remedied some of the formal problems. Writing the conclusion of the book, I realized it had entirely failed to show that any formalization of the basic idea – essentially severe testing (I did not use the phrase) of a theory without assumptions beyond the theory itself – would reliably lead to truth, whatever the truth might be. Fifteen years earlier, Hilary Putnam had proved a result that implies, among other things, that no such procedure can be reliable. Perhaps the idea of severe testing without assumptions is so naïve that no proof of its shallowness is necessary.

of simplicity assumption that excludes some logically possible hypotheses and that may not be true in some particular cases, but if it is assumed to be true severe tests of causal hypotheses using experimental data are possible. Zhang and Spirtes (2007) have broken faithfulness into components and shown how to test every component except when three variables may form an unfaithful graphical triangle.

A causal hypothesis, positive or negative, can explain data from a randomized experiment, and the randomized experiment in reciprocation tests the hypothesis. The Markov and faithfulness assumptions, one a matter of simplicity, are what tie the explanations and the tests together in causal inference from randomized experiments. Suppose X is randomized and Y and Z are two outcome variables, where X is associated with Y but not with Z. If faithfulness were not to hold, the data may be explained by the hypothesis that X causes Y and Y causes Z, producing an association between X and Z, but also X causes Z directly, not through Y, and the two mechanisms of influence of X on Z perfectly cancel one another so that in the data X and Z are not associated. Could we test this unfaithful explanation? Sure, on that explanation X and Z should be associated in the data *conditional* on values of Y, and that can be tested. But the appropriateness of the test depends on the Markov condition.

For a substantive example, consider a recent genetics experiment. Gene alleles L are randomized (never mind how) and traits T_1 and T_2 (messenger RNA concentrations) are measured. Traits T_1 and T_2 are both found to be associated with L. To determine the structure of the dependency, the scientists tested whether T_2 is independent of L conditional on T_1; finding that it is, it was concluded that $L \rightarrow T_1 \rightarrow T_2$. The inference requires both Markov and faithfulness assumptions. The principles are the ties that bind explanation and testing of causal hypotheses together.

But is this "severe testing"? Without trying to reconstruct precisely what Mayo had in mind, I propose to follow a mathematically precise notion that seems to me related:

Let G be a class of objects $<G, P>$ where G is a DAG and P is a distribution both Markov and faithful to G. Let $\mathbf{O}$ be a subset of the vertices of G, and let $\mathbf{O}_n$ range over sequences of samples drawn independently from the $\mathbf{O}$ marginal $P_\mathbf{o}$ of P; hence, with joint distribution, $P_\mathbf{o} \times \cdots \times P_\mathbf{o} = P_o^n$. Let θ be any "parameter" whose values range over some feature determined by $<G, P>$ for each $<G, P>$ in G. Strictly, θ is a function from $<G, P>$ to the real line. For example, if G consists of linear models whose vertex set includes variables X, Y, the parameter θ might represent the linear coefficient corresponding to an $X \rightarrow Y$ edge (with the convention that is zero if the edge does not occur in the graph). Or θ might represent the proposition that an edge $X \to Y$ does not exist (coded as 0, 1), or that a partial correlation is not equal to zero, and so on.

Following Zhang (2002), we consider tests of a null hypothesis that θ has value θ_o versus the alternative $\theta \neq \theta_o$ and define the following:

$\Omega_0 = \{P$: there exists $<G, P> \in G$ such that $\theta(<G.P>) = \theta_0)$.

$\Omega_1 = \{P$: there exists $<G, P> \in G$ such that $\theta(<G.P>) \neq \theta_0)$.

$\Omega_{1\delta} = \{P$: there exists $<G, P> \in G$ such that $|\theta(<G.P>) - \theta_0| \geq \delta)$.

A *test* τ is a partial function from values of O_n for each $n > 0$ to $\{0, 1\}$, corresponding to P in Ω_0 or in Ω_1. A test so defined is essentially a restricted estimator of subsets of distributions.

Now we can define various sorts of tests.

A test is pointwise consistent if and only if (iff)

1. for all P in Ω_0, $\lim_n P_n (\tau(O_n) = 1) = 0$
2. for all P in Ω_1, $\lim_n P_n(\tau(O_n) = 0) = 0$

A test is nontrivial iff for some P in $\Omega_0 \cup \Omega_1$, $\lim_n P_n(\tau(O_n) = 0) = 1$ or $\lim_n P_n(\tau(O_n) = 1) = 1$.

Nontrivial tests cannot take a pass everywhere, but consistent tests must take a pass eventually for P in $\Omega_0 \cap \Omega_1$. A test that is nontrivial for all P, θ, and pointwise consistent, has the property that if $\theta = \theta_0$, then for each alternative, θ_1, and for any positive probability p, a sample size exists for which (and for larger) the probability that the test says θ_1 is less than p. That sample size may vary with the alternative; therefore, there may never be a finite sample size that excludes (up to probability p) *every* alternative hypothesis at once. Likewise, if the null hypothesis is false, the same kind of convergence holds.

Pointwise consistent tests on finite samples are not what I think Mayo intends by a "severe test" because there need be no finite sample on which the probability of data for which a null (or alternative) hypothesis is "passed" by a test is improbable on every alternative (or every distribution for the null). What she has in mind seems closer to a uniformly consistent test, which Zhang defines as

3. $\lim_n \sup_{P \in \Omega_0} P_n (\tau(O_n) = 1) = 0$ and
4. for every $\delta > 0$, $\lim_n \sup_{P \in \Omega_{1\delta}} P_n(\tau(O_n) = 0) = 0$.

With uniform consistency, however close you want an alternative to be to the truth, if the null is true then for every $0 < p < 1$ there is a sufficient sample size after which the null passes the test with probability at least $1 - p$, and on *every* alternative that close to the null, the probability is less than or equal to p that the null passes the test.

Define the *causal effect* of variable X on variable Y in $<G, P>$ if $<G, P>$ is a linear model, to be the sum over all directed paths in G of the product, for each path, of the linear coefficients associated with edges on that path. The causal effect so defined predicts the effect on Y of exogenous interventions that force values on X. If $<G, P>$ is a multinomial model, define the causal effect on Y of an intervention that sets $X = x_1$, in contrast to an intervention that sets $X = x_2$, as the difference in expected values of Y when the probability $P(Y|X = x)$ is calculated conditional on X but leaving the probability of all ancestors of X as defined unconditionally by P.[4]

Zhang notes the following. If θ ranges over values of the causal effect of X on Y, and θ_0 is any value of θ, then given a complete time order and causal sufficiency, uniformly consistent tests are defined everywhere for the hypothesis $\theta = \theta_0$ against the alternative $\theta \neq \theta_0$.

The conditions – time order and causal sufficiency (no unobserved common causes) – are exactly the conditions that experimental randomization is intended to provide.

With data that are obtained without experimental controls, we may not know time order, and causal sufficiency may not hold – unrecorded common causes may be present that are responsible in whole or part for the measured associations among variables. Why assume faithfulness, or even the Markov condition? Zhang has provided (in his master's thesis!) a collection of proofs of conditions for the existence – or nonexistence – of uniform tests of causal hypotheses under alternatives to faithfulness. The idea of the alternative is that strong dependencies produce strong correlations. It is an open question if, for linear systems in which the measured variables are not Gaussian, faithfulness is even needed at all, with or without latent variables, for uniform tests of causal effects of measured variables (Shimizu et al., 2006b). But the Markov condition is indispensable, and no test exists for it in a particular nonexperimental data set. We must rely on the fact that, in macroscopic systems, whenever we are able to neatly randomize each variable in a system, the condition is found to hold.

There is little point in knowing the true hypothesis is testable if one cannot find it. It has been known for fifteen years that certain graphical model estimation procedures – search procedures – exist that are pointwise

---

[4] For details about this and other issues of graphical causal models, see Spirtes et al. (2001).

convergent estimators of the Markov equivalence class of causal graphs. Zhang also shows that uniformly convergent procedures are possible under his modification of faithfulness using an algorithm called Conservative PC (CPC). (For an appropriate finite sample size, a uniformly convergent search procedure finds (with probability 1) a hypothesis for which a uniform test exists that the hypothesis passes.) For non-Gaussian linear systems it appears likely (as yet no proof exists) that uniformly convergent procedures are possible assuming only indeterminism (Shimizu et al., 2006a).

7 An Illustration: Neural Cascades from Functional Magnetic Resonance Imaging

This is all rather dry and abstract, which is to say real mathematical philosophy, but it has practical, scientific implications. I give an example from recent unpublished work by my colleague Joe Ramsey and me.

Depolarization of cortical nerve cells produces a surge of oxygenated blood flow in the area of the cells, followed rapidly by deoxygenation, the blood oxygen level-dependent (BOLD) response, which is captured by functional magnetic resonance imaging. BOLD signals rise after a dip delayed as much as 2 seconds from signal onset at a (usually) unknown delay from firing of neurons in a recorded voxel, and continue for 8 to 12 seconds or more after onset. Onset of the BOLD signal can vary among recorded cell volumes, called voxels. Neural signals between cortical regions are passed in ~100 ms or less. In not atypical experimental designs, a voxel is sampled every 2 seconds, with longer delays between stimulus events. The difference between the speed of neural signals and the slower sampling rate of BOLD signals in functional magnetic resonance imaging (fMRI) measurements presents a problem for determining the structure of neural cascades in response to stimuli, because signals recorded at a particular time are approximately a sum over BOLD effects at previous times. Proposals have been made to deconvolute the BOLD signal using staggered measurement intervals and general linear models (e.g., Miezin et al., 2000). We pursue a different approach. The inference problem as we conceive it is as follows: (1) to estimate clusters of voxels (otherwise known as regions of interest) that have a distinctively, and collectively, raised BOLD signal in response to the stimuli; (2) to form an aggregate measure from the BOLD intensities of the several voxels within each cluster; (3) to estimate a time-series description of the BOLD signals aggregated by clusters; (4) from (3), to estimate the causal relations among clusters in the activation cascade the stimuli produce; (5) to determine the stability of those relations within subjects in experimental repetitions; and (6) to determine the invariance of the cascade structure between subjects in experimental structures.

1. Several clustering procedures are in the literature and built into software. Two sources of validation exist: known functional roles of cortical regions, and whether the clustering yields informative, robust, and invariant inferred cascade structures. The latter can only be determined post hoc from solutions to the other five issues.

2. Aggregate measures can be formed from averages over voxels in a cluster, from maximum signal intensity within a cluster, from maximum likelihood estimation of a single latent variable source of voxel signals within a cluster, or perhaps other methods.

3. Because different cortical tissue may have different BOLD response delays to neural activity, the time order of onset of BOLD signals from different areas cannot be reliably used to constrain the causal structure of a neural cascade. Nonetheless, using any of a variety of new, consistent, graphical model search methods that have appeared recently (Chickering, 2003; Chu and Glymour, 2008; Ramsey et al., 2006; Shimizu et al., 2005; Spirtes et al., 2000), a structural Variable Auto-Regressive time series with independent noises can be formed from the fMRI records, with lags depending on the sampling rate. With samples at 2-second intervals, for example, lags of 2 to 6 might be relevant. Statistical indicators are available for when further lags are irrelevant. The time series cannot be expected to be stationary, but that is not of principal concern because we do not propose to use the series directly to model features of the cascade – the time series is a step in the analysis, not a conclusion. If, however, the data series involves repeated presentations of similar stimuli, a stable time series would indicate within-subject invariance, but a better test would be similar structure across repetitions of the entire experimental series, well separated in time, with the same subject.

4. The neural influences, and the resultant BOLD variations, between clusters occur more rapidly than the sampling rate, and the aim is to identify the structure of these processes occurring *between* fMRI registrations of BOLD signals. Recent methods (Demiralp and Hoover, 2003; Moneta and Spirtes, 2006; Swanson and Granger, 1997) developed for graphical models allow the possibility of such inferences. The idea is to remove the influences of previous values' recorded variables on current values. The procedure is roughly as follows: (a) find an appropriate structural VAR model for the time series of measurements;

(b) regress each variable X_t on its self-lags and on its lagged ancestors in the VAR model; (c) apply any of the graphical model specification procedures referenced earlier to the residuals to obtain an estimated contemporaneous causal structure; (5), (6) within subjects, for repeated trials of the same brief stimulus type, locations of hemodynamic response are thought be nearly invariant; across subjects in the same experimental design they are more variable. It is to be expected that variation in clusters should also be found in such cases. Cross-subject cluster identification may be aided by roughly mapping nearby areas in two brains to one another to maximize overlap and to maximize isomorphism of inferred cascade structure as much as possible. We propose no formal measure for this kind of comparison.

Under quite general assumptions, for most of the algorithms we deploy, we have proofs of convergence to correct information about structure. Nonetheless, no guarantee exists that these procedures would find all (or only) the dependencies between regions of interest. The procedures may mislead or fail because of significantly nonlinear dependencies or because cellular influences not clustered act as common causes of cluster activity. Chu and Glymour (2008) develop graphical model methods that address both of these possible problems. Even without such methods, an important indication that the time series has captured a more rapid process is that the contemporaneous graphical structure found a repeated structure in the time series.

Analysis of unpublished data illustrates the behavior of the procedure. In the experiment, a participant is presented with a stimulus every few seconds, over 734 seconds. A brain region is scanned at 2-second intervals over 734 seconds. A standard procedure clusters voxels that produce large signals. The distribution of voxel signals at a time within each cluster are left skewed; the distribution of maximum signals with a cluster over time is roughly symmetrical and roughly single-peaked, but not clearly Gaussian.

We use the maximum voxel signal in a cluster at a time as our variables. To identify the cascade structure we use the CPC algorithm and parts of the LiNGAM algorithm (which is described later).

The CPC algorithm uses sample data to construct a Markov equivalence class of DAGs. Two DAGs are Markov equivalent if (1) they have the same vertices and (2) every pair of vertices that are adjacent (connected by a directed edge) in one DAG are also adjacent (connected by a directed edge, not necessarily in the same direction) in the other DAG, and vice versa; if a graphical structure of the form $X \rightarrow Y \leftarrow Z$ with X, Z not adjacent occurs in one DAG, the same structure occurs in the other. In systems with independent noises, statistical models built on Markov equivalent graphs imply the same conditional independence relations (Pearl, 1988) and in linear models imply the same vanishing partial correlations (Spirtes et al., 2000). As we deploy it here, the CPC algorithm uses statistical decisions (based on normal distribution theory) about vanishing partial correlations to construct a Markov equivalence class. Simulation studies (Druzdzel and Voortman, 2008) indicate that the search procedure is accurate when variables are skewed or show kurtosis. Our own simulations find it reasonably accurate with bimodal distributions as well. Any number of other procedures are possible, as noted earlier, including exhaustive tests of every linearized DAG on the variables. We choose CPC for illustration because it is reliable in simulation studies, scales up (for sparse graphs) to a hundred variables or more, and has been proven to be uniformly consistent. The algorithm has one parameter, a p-value level used in tests for rejecting hypotheses of vanishing correlation. We use .03. Variations between .05 and .02 make little difference, except that at .05 an additional influence is found.

The output of CPC applied to the fMRI records – treating each time measurement as a unit – is shown in Figure 8.1.

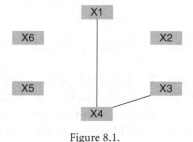

Figure 8.1.

The undirected graph represents all three DAGs in which no two edges are directed into X_4. We would like to have a reliable way to choose among these alternative DAGs, and for that we use a modification of the LiNGAM procedure. (Increasing the p-value to .05 would add another undirected edge between X_4 and X_5.)

LiNGAM (Shimizu et al., 2005) is a procedure for finding a unique DAG from data sampled from linear systems in which the noise terms are all, or all but one, not Gaussian. Each measured variable can be equivalently expressed as a function of a set of unrecorded noise terms – in what econometricians call the reduced form. If at most one of the noise terms is Gaussian, independent component procedures can estimate the noises. Each measured variable is returned as a function of all of the noises, but dependencies are pruned via a scoring procedure. Permutations of the resulting matrix of dependencies of variables on noises then return a unique lower triangular matrix with nonzero diagonal, corresponding to a unique DAG. The entries below the diagonal are the values of the linear coefficients. We use the scoring procedure as a postprocessor to CPC (Hoyer et al., in preparation), scoring each DAG in the CPC output on each of twenty bootstrapped samples and returning the DAG that most often has the highest score. Other scoring procedures could be used, including Bayesian posteriors and maximum p-values on a goodness-of-fit test. We use the LiNGAM score because it should be more robust to non-Gaussian distributions. The complexity of the procedure is determined by the number of DAGs in the CPC output. The result in our example is shown in Figure 8.2.

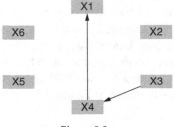

Figure 8.2.

Increasing the p-value cutoff to .05 would add a further edge directed from X_4 into X_5.

Our confidence in the CPC-LiNGAM output is limited by the fact that our sample units are patently not independent and identically distributed, because the BOLD signal at a sampled time would be an approximately linear combination of earlier BOLD effects within 8 to 12 seconds or more.

Next we lag the variables and apply CPC to the lags. The results for lags 2 and 5 are shown in Figure 8.3.

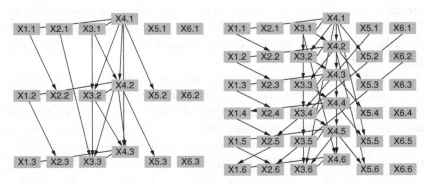

Figure 8.3.

The lag 5 graph is busier, presumably due to BOLD signal summation, but all of the lag 2 dependencies are retained. We use the LiNGAM scoring procedure for the lag 2 CPC output in Figure 8.4.

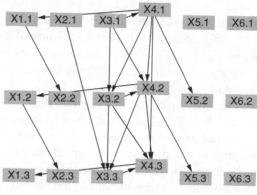

Figure 8.4.

Similar results are found with lag 5. The structure found among X_1, X_3, and X_4 from the first analysis with pooled data is retained among variables at the same time step, but two additional lagged dependencies show up repeatedly: $X_{1,t} \rightarrow X_{2,t+1}$ and $X_{4t} \rightarrow X_{5,t+1}$. At larger lags, as illustrated earlier, more dependencies appear. One question is whether these dependencies are the result of BOLD signal summation, chance, or real effects. To address that, we regress each variable, X_t on all variables V_{t-1}, V_{t-2} (including X_{t-1} and X_{t-2}) and apply the CPC–LiNGAM procedure to the residuals. The result is the structure we found with the pooled data earlier and is shown in Figure 8.5.

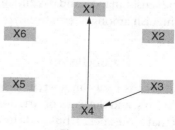

Figure 8.5.

Regressing on five lags and applying CPC–LiNGAM produces the same result, as does regression of each variable only on its parents and their lags. If we repeat the procedure with $p = .05$, an $X_{4,t} \rightarrow X_{5,t+1}$ connection appears in the time series and in the concurrent graph obtained after regression. Our tentative conclusion is that the neural cascade involves the aforementioned structure and, possibly, also an $X_4 \rightarrow X_5$ pathway. It is consistent with the data that the association of X_3 and X_4 in the contemporaneous model is due in whole or in part to unrecorded common causes, although we think that unlikely, but it is not consistent that the association of X_4 and X_1 or of X_4 and X_5 would be significantly confounded by an unobserved common cause. Such confounding is inconsistent with the independence of X_3 and X_1 or X_5 conditional in X_4.

We have not given the linear coefficients, but they can be produced automatically by the CPC-LiNGAM procedure or by other estimators. We note that the procedure can be modified to include other preprocessors to LiNGAM, and following Chu and Glymour (2008) can be adapted to nonlinear dependencies and to detect unrecorded common causes of variables in the contemporaneous graph obtained after regression. Our choice of maximum signal value as the measure for a cluster at a time was quite arbitrary, and it would be natural to investigate the results with different clustering methods and different measures for cluster and times.

The statistical procedures do not of themselves guarantee that the contemporaneous causal structure would be the same as the structure found from treating the fMRI records as independent events, but in this case it is. Should that be a robust finding of fMRI studies, the search procedure could of course be abbreviated. In our analysis, the variables in the contemporaneous causal structure obtained from the BOLD signals are proxies for the activity of the most active voxel in the respective clusters. The relationship between depolarization events and the BOLD signal could be modeled, but we do not attempt to do so – first because with present knowledge such modeling would be hypothetical, and second because almost any such model could be appended to the causal analysis and it is difficult to see how the fMRI data could help to distinguish among alternative hypotheses of this kind.

8 Conclusion

Mayo's focus on severe tests is really an aperçu. A good aperçu points to deep projects, as does Mayo's, and the work of Aris Spanos, an economist, is a deep development of that same aperçu, but aside from Spanos and Mayo's joint papers, the only deep response I know of in the entire philosophical literature is Zhang's extension of work on methods of inferring causal relationships.

Causal inference illustrates one common circumstance in which explanations and explanatory virtues are essential to producing testable hypotheses and to testing them. It illustrates how the explanatory virtues function as assumptions and how assumptions can be tested, and tested severely. And it also illustrates that not every assumption essential to testing a hypothesis by a set of data can be tested by the same data set.

References

Chickering, D. (2003), "Optimal Structure Identification with Greedy Search," *Journal of Machine Learning Research*, 3: 507–54.

Chu, T., and Glymour, C. (2008), "Search for Additive Nonlinear Time Series Causal Models," *Journal of Machine Learning Research*, 9: 967–91.

Demiralp, S., and Hoover, K. (2003), "Searching for the Causal Structure of a Vector Autoregression," *Oxford Bulletin of Economics*, 65: 745–67.

Druzdzel, M., and Voortman, M. (2008), "Insensitivity of Constraint-Based Causal Discovery Algorithms to Violations of the Assumption of Multivariate Normality," *Proceedings of the Florida Artificial Intelligence Research Conference (FLAIRS)* (to come).

Miezin, F., Maccotta, L., Ollinger, J.M., Petersen, S.E., and Buckner, R.L. (2000), "Characterizing the Hemodynamic Response: Effects of Presentation Rate, Sampling Procedure, and the Possibility of Ordering Brain Activity Based on Relative Timing," *NeuroImage*, 11: 735–59.

Mills, S., and Beatty, J. (1979), "The Propensity Interpretation of Fitness," *Philosophy of Science*, 46: 263–86.

Moneta, A., and Spirtes, P. (2006), "Graphical Models for Identification of Causal Structures in Multivariate Time Series," *Joint Conference on Information Sciences Proceedings*, Atlantis Press, http://www.atlantis-press.com/publications/aisr/jcis-06/.

Myrvold, W. (2003), "A Bayesian Account of the Virtue of Unification," *Philosophy of Science*, 70: 399–423.

Pearl, J. (1988), *Probabilistic Reasoning in Intelligent Systems Networks of Plausible Inference*, Morgan Kaufmann, San Mateo.

Ramsey, J., Zhang, J., and Spirtes, P. (2006), "Adjacency-Faithfulness and Conservative Causal Inference," pp. 401–8 in *Proceedings of the 22nd Conference on Uncertainty in Artificial Intelligence 2006*, Boston.

Shimizu, S., Hoyer, P.O., Hyvärinen, A., and Kerminen, A. (2006a), "A Linear Non-Gaussian Acyclic Model for Causal Discovery," *Journal of Machine Learning Research*, 7: 2003–30.

Shimizu, S., Hyvärinen, A., Kano, Y., and Hoyer, P.O. (2005), "Discovery of Non-Gaussian Linear Causal Models Using ICA," pp. 526–33 in *Proceedings of the 21st Conference on Uncertainty in Artificial Intelligence (UAI-2005)*.

Shimizu, S., Hyvarinen, A., Kano, Y., Hoyer, P.O., and Kerminen, A. (2006b), "Testing Significance of Mixing and Demixing Coefficients in ICA," pp. 901–8, *Independent Components Analysis and Blind Signal Separation: 6th International Conference, ICA 2006*, Springer, Berlin.

Spirtes, P., Glymour, C., and Scheines, R. (2000), *Causation, Prediction, and Search*, 2nd ed., Lecture Notes in Statistics, Springer-Verlag, New York.

Swanson, N.R., and Granger, C.W.J. (1997), "Impulse Response Functions Based on a Causal Approach to Residual Orthogonalization in Vector Autoregressions," *Journal of the American Statistical Association*, 92: 357–67.

Zhang, J. (December 2002), "Consistency in Causal Inference under a Variety of Assumptions," M.S. Thesis, Department of Philosophy, Carnegie Mellon University.

Zhang, J., and Spirtes, P. (2007), "Detection of Unfaithfulness and Robust Causal Inference," philsci-archive.pitt.edu.

Zhang, J. (2008), "Error Probabilities for Inference of Causal Directions," *Synthese (Error and Methodology in Practice: Selected Papers from ERROR 2006)*, 163(3): 409–18.

II Explanation and Testing
Exchanges with Clark Glymour

Deborah G. Mayo

Clark Glymour's contribution explores connections between explanation and testing and some applications of these themes to the discovery of causal explanations through graphical modeling. My remarks touch on a number of subthemes that have emerged from the back-and-forth exchanges Glymour and I have had since ERROR06,[1] especially insofar as they connect with problems and arguments of earlier chapters. As we get close to the end of the volume, I want also to provide the reader with some directions for interconnecting and building on earlier themes.

1. *Experimental Reasoning and Reliability*: How do logical accounts of explanation link with logics of confirmation and testing? When does *H*'s successfully explaining **x** warrant inferring the truth or correctness of *H*?

2. *Objectivity and Rationality*: Do explanatory virtues promote truth or do they conflict with well-testedness? How should probabilistic/statistical accounts enter into scrutinizing methodological desiderata (e.g., promote explanatory virtues) and rules (e.g., avoid irrelevant conjunction, varying evidence)?

3. *Metaphilosophical Themes*: How should probabilistic/statistical accounts enter into solving philosophical problems? What roles can or should philosophers play in methodological problems in practice? (Should we be in the business of improving practice as well as clarifying, reconstructing, or justifying practice?)

---

[1] Especially beneficial was the opportunity afforded by the joint course taught by Glymour and Spanos at Virginia Tech, Fall 2007.

1 Explanation and Testing

The philosophical literature on explanation is at least as large as the literature on testing, but the two have usually been tackled separately. I have always felt that the latter is more fundamental than the former and that, once an adequate account of inference is in hand, one may satisfactorily consider key explanatory questions. I find Glymour's thoughts on linking explanation and testing sufficiently insightful to enable me to take some first steps to connect the severity account of testing to some of the classic issues of explanation.

Glymour's insight is "that explanations and the explanatory virtues that facilitate comprehension also facilitate testing the claims made in explanation" (this volume, p. 331). Although he does not claim explanation and testability always go hand in hand, he considers examples where "the explanations specify tests whose results, if in accord with the explanations, converge on the truth *conditionally*" (p. 332). It will be interesting to see what happens if we replace his "test" with "severe test" as I define it. Recall our basic principle about evidence:

Severity Principle (weak): Data **x** (produced by G) do *not* provide good evidence for hypothesis H if **x** results from a test procedure with a very low probability or capacity of having uncovered the falsity of H (even if H is incorrect).

A test "uncovers" or signals the falsity of H by producing outcomes that are discordant with or that fail to "fit" what is expected were H correct. This weak notion of severity suffices for the current discussion.[2]

To Glymour's question, When does H's successfully explaining **x** warrant inferring the truth or correctness of H?, our answer would be "only when H is severely tested by **x**." (H could still be viewed as an explanation of **x** without anyone claiming that H is warranted *because* H explains **x**. Here I only consider the position that H is warranted *by dint* of H's explaining **x**.) I want to consider how much mileage may accrue from this answer when it comes to a key question raised by Glymour's discussion. For a model or theory H:

- How do/can explanatory virtues (possessed by H) promote H's well-testedness (or H's capacity for being subject to stringent tests)?

---

[2] Note that satisfying this weak principle does not mean the test is severe, merely that it is not blatantly insevere.

1.1 A Few Terminological Notes

Some informal uses of language in this exchange warrant a few notes. "*H* is severely tested" is a shorthand for "*H* passes a severe test," and to say that "*H* is (statistically) falsified" or "rejected" is the same as saying "not-*H* passes the test." Although this wording may seem nonstandard, the advantage is that it lets us work with the single criterion of severity for assessing any inference (even with test procedures built on controlling both erroneous acceptances and erroneous rejections). "*H* is false" may be seen to refer to the existence of a particular error that *H* is denying. The correctness of *H* may generally be seen to assert the absence of a specific error or discrepancy. To allude to examples in earlier exchanges, *H* may be viewed as the alternative hypothesis to a null hypothesis that asserts not-*H*. (For example, "*H*: $\mu > 2$ passes severely" may be identified with rejecting a null hypothesis "H_0: $\mu \leq 2$".)

2 Is There a Tension between Explanation and Testing?

2.1 A Goal for Science versus a Criterion for Inference

Pondering Glymour's question of the relationship between explanation and testing lets me elucidate and strengthen the responses I have given to my "high-level theory" critics in previous chapters. According to Chalmers, "[a] tension exists between the demand for severity and the desire for generality. The many possible applications of a general theory constitute many possible ways in which that theory could conceivably break down." (Chapter 2, p. 61). This leads him to recommend a weakened notion of severity. But there is a confusion here. The aim of science in the account I favor is not severity but *finding things out*, increasing understanding and learning. Severity is a criterion for determining what has and has not been learned, what inferences are warranted, and where our understanding may be in error. From the start I emphasized the *twin goals* of "severity and informativeness" (Mayo, 1996, p. 41). Trivial but severely passed claims do not advance learning. So we have, first, that the goal is finding things out, and second, that this demands learning from error and controlling error (which is not the same as error freedom). Does this mean that we shortchange "explanatory virtues"? My high-level theory critics say yes.

Musgrave (this volume) alludes to an earlier exchange in which Laudan (1997) urges me to factor "explanatory power back into theory evaluation."

There Laudan claims that "in the appraisal of theories and hypotheses, what does (and what should) principally matter to scientists is not so much whether those hypotheses are true or probable. What matters, rather, is the ability of theories to solve empirical problems – a feature that others might call a theory's explanatory or predictive power" (p. 306). Like Glymour, I deny that what matters is explanatory power *as opposed to* truth – truth does matter. Like Laudan I deny that one is after "highly probable" hypotheses in the sense of epistemic probabilists (e.g., Achinstein's Bayesian, or Musgrave's "justificationist"). What we want are hypotheses that have successfully been highly probed. What I wish to explore – taking advantage of Glymour's aperçu – is whether increasing predictive power is tied to increasing the probativeness of tests.

Laudan's "problem-solving" goal is covered by the twin informativeness/severity demands: we want informative solutions to interesting problems, but we also want to avoid procedures that would, with high probability, erroneously declare a problem solved by a hypothesized solution *H*. Far from showing "bland indifference to the issue of a theory's ability to account for many of the phenomena falling in its domain" (Laudan, 1997, pp. 306–7), when a theory fails to account for phenomena in its domain, that is a strong indication of errors (and potential rivals) not yet ruled out – concerns that are close to the heart of the severe tester. Moreover, as we saw in the case of GTR, such gaps are springboards for creating new hypothesized solutions of increased scope and depth, as well as developing tests that can reliably distinguish them.

2.2 Powerful Explanations and Powerful Tests

Glymour asks: "We know that under various assumptions there is a connection between *testing* – doing something that could reveal the falsity of various claims – and . . . coming to understand. Various forms of testing can be stages in strategies that reliably converge on the truth. . . . But how can *explaining* be any kind of reliable guide to any truth worth knowing?" (p. 331)

One answer seems to drop out directly from the severe testers' goals: to reliably infer "truth worth knowing," we seek improved tests – tests that enable us to corroborate severely claims beyond those already warranted. This leads to identifying interconnected probes for constraining claims and controlling error probabilities, which simultaneously results in frameworks and theories that unify. Starting from severe testing goals, we are led to stress an initial hypothesis by subjecting it to probes in ever-wider domains because that is how we increase the probability of revealing flaws. We thereby are led to a hypothesis or theory with characteristics that render it "explanatory."

3 When the "Symmetry" Thesis Holds

According to Carl Hempel, the link between explanation and testing is provided by an equivalence known to philosophers as the symmetry thesis: premises explain a phenomenon if and only if the phenomenon could be predicted from the premises. (Glymour, this volume p. 332)

Although the symmetry thesis is strictly false, Glymour, in our exchanges, tries to trace out when a symmetry does seem to hold. This led to my own way of linking the two (which may differ from what Glymour intended or holds). To begin with, the entanglement between successful explanations and probative tests is likely to remain hidden if we insist on starting philosophical analysis with a given theory and ignore the processes by which explanatory theories grow (or are discovered) from knowledge gaps.

3.1 The Growth of Explanatory Theories: The Case of Kuru

To move away from GTR, I consider some aspects of learning about the disorder known as Kuru found mainly among the Fore people of New Guinea in the 1960s. Kuru, and (what we now know to be) related diseases (e.g., Mad Cow, Crutzfield-Jacobs disease [CJD]) are known as "spongiform" diseases because the brains of their victims develop small holes, giving them a spongy appearance. (See Prusiner, 2003.)

For H to successfully explain $\mathbf{x}$ – as I am using that term – there must be good evidence for H.[3] For example, we would deny that witchcraft – one of the proposed explanations for Kuru – successfully explains Kuru among the Fore. Still, $\mathbf{x}$ may be excellent evidence for H (H may pass a severe test with $\mathbf{x}$) even though one would not normally say that H explains $\mathbf{x}$: hence the asymmetry. As with the familiar asymmetric examples (e.g., barometer-weather), the evidence that a patient is afflicted with Kuru warrants inferring H:

H: the patient has a hole-riddled cortex,

while the cortex holes do not explain her having Kuru.

---

[3] The data that is explained, $\mathbf{x}$, need not constitute this evidence for H.

Perhaps this example may be scotched by our (usual) stipulation that the hypothesis H refers to an aspect of the procedure generating **x**. But other examples may be found where we would concur that H is well tested by **x**, and yet H does not seem to explain **x**. For example, from considerable data in the 1950s, we had evidence **x**: instances of Kuru cluster within Fore families, in particular among women and their children, or elderly parents. Evidence **x** warrants, with severity,

H: there is an association of Kuru within families

but to say that H explains **x** sounds too much like explaining by "dormative properties."

These are the kinds of cases behind the well-known asymmetry between explanation and tests. Focus now on the interesting activity that is triggered by the recognition that H does not explain **x**: Prusiner, one of the early Kuru researchers, was well aware of this explanatory gap even without a clue as to what a satisfactory theory of Kuru might be. He asked: What causes Kuru? Is it transmitted through genetics? Infection? Or something else? Are its causes similar to those of other amyloid diseases (e.g., Alzheimer's)? How can it be controlled or irradicated? This leads to conjectures to fill these knowledge gaps and how such conjectures can be in error – which, in turn, provides an incentive to create and test more comprehensive theories.

That Kuru was a genetic disorder seemed to fit the pattern of observed cases, but here we see why the philosopher's vague notions of "fit" will not do. By 1979 it was recognized that a genetic explanation actually did not fit the pattern of data at all – Kuru was too common and too fatal among the Fore to be explained as a genetic disorder (it would have died out of the gene pool). Rather it was determined with severity that the correct explanation of the transmission in the Fore peoples was through mortuary cannibalism by the maternal kin (this was a main source of meat permitted women):

H: Kuru is transmitted through eating infected brains in funeral rites.

Ending these cannibalistic practices all but eradicated the disease, which had been of epidemic proportions. The fifty or so years it has taken to arrive at current theories – of which I am giving only the most sketchy glimpse – is a typical illustration of the back-and-forth movement between

A. the goal of attaining a more comprehensive understanding of phenomena (e.g., the dynamics of Kuru and related diseases, Mad Cow, CJD) and

B. the exploitation of multiple linkages to cross-check, and subtract out, errors.

Despite the inaccuracies of each link on its own, they may be put together to avoid errors. A unique aspect of this episode was the identification of a new entity – a prion – the first infectious agent containing no nucleic acid. (Many even considered this a Kuhnian revolution, involving, as it does, a changed metaphysics.) This revolutionary shift was driven by local experimental probes, most notably the fact that prions (e.g., from scrapie-infected brains) remain infectious even when subjected to radiation and other treatments known (through prior severe tests) to eradicate nucleic acids (whereas they are inactivated by treatments that destroy proteins). Only in the past ten years do we have theories of infectious prion proteins, and know something of how they replicate despite having no nucleic acid, by converting normal proteins into pathological prions.

Prion theories earned their badges for (being warranted with) high severity by affording ever stronger arguments from coincidence through interrelated checks (e.g., transmitting Kuru to chimpanzees). Moreover, these deeper prion theories were more severely corroborated than were the early, and more local, hypotheses such as *H*: Kuru is transmitted among the Fore through eating infected brains. (It does not even seem correct to consider these local hypotheses as "parts of" or entailed by the more comprehensive theories, but nothing turns on this.) However, and this is my main point, the same features that rendered these theories better tested, simultaneously earned them merit badges for deepening our understanding – for explaining the similarities and differences between Kuru, CJD, and Mad Cow – and for setting the stage to learn more about those aspects of prions that continue to puzzle microbiologists and epidemiologists. (Prusiner received the Nobel prize in 1997; see Prusiner, 2003.)

The position that emerges, then, is not that explanatory virtues are responsible for well-testedness, nor even that they are reliable signs of well-testedness *in and of themselves*; it is rather that these characteristics will (or tend to) be possessed by theories that result from fruitful scientific inquiries. By fruitful scientific inquiries I mean those inquiries that are driven by the goal of *finding things out* by reliable probes of errors.

4 Explanation, Unification, and Testing

Hempel's logical account of explanation inherits the problems with the corresponding hypothetico-deductive (HD) account of testing: *H* may entail, predict, or otherwise accord with **x**, but so good a fit may be highly probable (or even guaranteed) when *H* is false. The problem with such an HD account of confirmation is not only that it is based on the invalid affirming the consequent – that, after all, is only problematic when the hypothesis fails to be falsified. Even in the case of a deductively valid falsification, warranting the truth of the first premise "if H then $\mathbf{x}$" is generally problematic. The usual assumption that there is some conjunction of "background conditions" B so that H together with B entails $\mathbf{x}$ has scant relationship to the ways hypotheses are linked to actual data in practice. Furthermore, the truth of the conditional "if H (and B) then $\mathbf{x}$" does not vouchsafe what even HD theorists generally regard as a minimal requirement for a good test – that something has been done that could have found H false. This requires not mere logic but evidence that anomalies or "falsifying hypotheses" (as Popper called them) are identifiable and not too easily evadable. This, at any rate, would be required for a failure to falsify to count as any kind of evidence for H. This leads to my suggested reading of Glymour's point in his discussion of Copernicus. Ensuring that anomalies for H are recognizable goes hand in hand with H possessing explanatory attributes, such as the ability to transform a contingent empirical regularity into a necessary consequence. Says Glymour, "If the empirical regularity is false, then the entire Copernican framework is wrong; nothing is salvageable except by *ad hoc* moves (as in, *oh well, Mars isn't really a 'planet'*)." By contrast, "The same regularity is accounted for in Ptolemaic theory by adjusting parameters" (this volume, p. 336). I take the point to be that the former and not the latter warrants taking the agreement between H and the observed empirical regularity as H having passed a genuine test (even if weak). In the former case, the hypothesis or theory sticks its neck out, as it were, and says such-and-such would be observed, *by necessity*, so we get a stronger test.

The history of attempted improvements on HD accounts has been to add some additional requirement to the condition that H "accord with" evidence $\mathbf{x}$ in order to avoid too-easy confirmations. We have seen this with requiring or preferring theories that make novel predictions (e.g., Chapter 4). The same role, I suggest, is behind advocating certain explanatory virtues. However, because such attributes are neither necessary nor sufficient for meeting the "weak severity" requirement for a genuine test, it is more effective to make the severity requirement explicit. To be clear, I am not saying explanatory virtues are desirable only to vouchsafe genuine tests – they are desirable to achieve understanding and other goals, both epistemic and pragmatic. Here, my aim has been limited to exploring *how explanatory virtues may simultaneously promote grounds for inferring or believing the explanation*. It would be of interest to go further in the direction in which Glymour is valuably pointing us: toward analyzing the connection between explanatory power on the one hand and powerful tests on the other.

5 Irrelevant Conjunction

The issue of irrelevant conjunction is one on which Glymour and I have had numerous exchanges. On Hempel's logical account of explanation, Glymour notes, anything "lawlike," true or false, can be tacked on to a Hempelian explanation (or statistical relevance explanation) and generate another explanation (this volume, p. 332). That is,

If *H* explains **x**, then (*H* and *J*) explain **x**, where *J* is any "irrelevant" conjunct tacked on (the Pope's infallibility).

How would such a method fare on the severity account? From our necessary condition, we have that (*H* and *J*)'s explaining **x** cannot warrant taking **x** as evidence for the truth of (*H* and *J*) if **x** counts as a highly insevere test of (*H* and *J*). (See also Chapter 3, pp. 110, 123.)

A scrutiny of well-testedness may proceed by denying either condition for severity: (1) the fit condition, or the claim that (2) it is highly improbable to obtain so good a fit even if *H* is false. Here we are only requiring *weak* severity – as long as there is some reasonable chance (e.g., .5 or more) that the test would yield a worse fit, when *H* is false, then weak severity holds. Presumably, we are to grant that (*H* and *J*) fit **x** because *H* alone entails **x** (never mind how unrealistic such entailments usually are). Nevertheless condition 2 is violated. Say we start with data **x**, and that *H* explains **x**, and then irrelevant hypothesis *J* is tacked on. The fact that (*H* and *J*) fits **x** does not constitute having done anything to detect the falsity of *J*. Whether **x** or not-**x** occurs, the falsity of *J* would not be detected, and the conjunction would pass the test. Because this permits inferring hypotheses that have not been well tested in the least, the HD account is highly unreliable. (In a statistical setting, if the distribution of random variable *X* does not depend on hypothesis *J*, then observing *X* is uninformative about *J*, and in this informal context we have something similar.)

To go further, we should ask: when would we come across an assertion that a conjunction of hypotheses (*H* and *J*) explains **x**. Most commonly, saying *H* and *J* explain(s) **x** would be understood to mean either:

1. together they explain **x**, although neither does by itself, or
2. each explains **x** by itself.

A common example of case 1 would be tacking onto *H* an explanation of an *H*-anomaly (e.g., the Einstein deflection of light together with a mirror distortion explains the eclipse results at Sobral, (see Chapter 4)); an example of case 2 arises when *H* and *J* are rival ways of explaining **x**.

But the problem case describes a situation where *H* explains **x** and hypothesis *J* is "irrelevant." Although this is not defined, pretty clearly we can dream up the kind of example that the critics worry about. Consider

H: GTR and *J:* Kuru is transmitted through funerary cannibalism.

Let data **x** be a value of the observed deflection in accordance with GTR. The two hypotheses do not make reference to the same data models or experimental outcomes, so it is not clear that one can even satisfy the "fit" condition for a severe test.

That hypothesis $K = (H \text{ and } J)$ fits **x** requires, minimally, that $P(\mathbf{x}; K) > P(\mathbf{x}; \text{not-}K)$, and this would not seem to be satisfied (at least for an error statistician). Perhaps sufficient philosophical rigging can define something like a "Kuru or gravity experiment." However, the main force for rejecting the well-testedness of the conjunction of *H* and *J* is clearly that *J* has not been probed in the least by the deflection experiment.

We should emphasize a point regarding the weak severity requirement: it is not merely that the test needs to have some reasonable probability of detecting the falsity of *H*, if it is false. The indication of falsity (or discrepancy) has to be *because* of *H*'s falsity. For example, we would not consider that a GTR hypothesis *H* had been well probed by a "test" that rejected *H* whenever a coin landed heads (of course the fit condition would also fail). In a good test, moreover, the *more false H* is, the higher should be the probability that the test detects it (by producing a worse fit, or a failing result). With the irrelevant conjunct, however, the falsity of *J* does not increase the detection ability; the test is *not registering J*'s falsity at all.

Someone may ask, but what if one is given a bundle like conjunction *K* at the start, rather than creating *K* by tacking *J* onto *H*? Which part is well tested? With this question we are back to where we began in our discussions of theory testing (Chapter 1) and in responding to Chalmers and Musgrave. What is well tested is what has passed severely, and a good part of scientific inquiry involves figuring this out. For example, we saw that it was determined that warranting the equivalence principle did not count as severely testing all of GTR, but only metric versus nonmetric theories. The evidence for the equivalence principle did not have the ability to discriminate between the class of metric theories.

6 Metaphilosophical Notes: The Philosophical Role of Probabilistic Inference

"The idea of putting probabilities over hypotheses delivered to philosophers a godsend, an entire package of superficiality" (Glymour, this volume,

p. 334). I share Glymour's indictment of what often goes under the heading of "Bayesian epistemology" (to be distinguished from Bayesian statistics) – at least if the aim is solving rather than merely reconstructing. In Mayo (1996) I described the shortcomings of claims to "solve" problems about evidence, such as Duhem's problem, by means of a probabilistic reconstruction: "Solving Duhem comes down to a homework assignment of how various assumptions and priors allow the scientific inference reached to be in accord with that reached via Bayes's theorem" (p. 457). They do not tell us either how the assignments are arrived at or, more important, how to determine where the error really lies. The same problem arises, Worrall notes, in treating the issue of "use-novelty" among Bayesian philosophers. "The fact that every conceivable position in the prediction vs. accommodation debate has been defended on the basis of some Bayesian position is a perfect illustration of the fact that 'the' Bayesian position can explain everything and so really explains nothing" (Worrall, 2006, pp. 205–6).

Most ironic about this practice is that rather than use statistical ideas to answer questions about methodology, the Bayesian epistemologist starts out assuming the intuition or principle to be justified, the task then being the "homework problem" of finding assignments and/or selecting from one of the various ratio or difference probability measures to capture the assumed intuition. Take the "tacking problem" discussed in Section 5:

The Bayesian epistemology literature is filled with shadows and illusions: for example, Bayesian philosophers solve the difficulty that logical relations constituting an explanation can be conjoined with irrelevancies – the tacking on problems – by just *saying*, ad hoc for each case, that tacking on reduces degree of belief. (Glymour, this volume, p. 335)

At best, they are able to say that the conjunction gets less support than the conjunct, when what we want to say, it seems to me, is that there is no evidence for the irrelevant conjunct, and the supposed "test by which the irrelevant conjunct is inferred" is a terrible (zero-severity) test. Any account that cannot express this forfeits its ability to be relevant to criticizing even egregious violations of evidence requirements.

The same problem, Glymour observes, occurs in subjective Bayesian attempts to show why explanatory unification supplies greater degrees of belief. Here, too, we are to *start out* assuming some methodological principle (about unification and belief). The task is to carry out the Bayesian computation of hammering out probabilities to accord with the assumed principle. In fact, however, *H*'s unifying power and the warrant for believing or inferring *H* need not go hand in hand. Surely a theory that incorrectly "unifies" phenomena ought not to earn higher belief: the similarities between Kuru and other amyloid diseases such as Alzheimer's, for instance, should not give extra credence to a hypothesis that unified them rather than one that posited distinct mechanisms for Kuru and for Alzheimer's. The less-unified theory, in this case, is better tested (i.e., passed more severely).

I call on the philosopher of science with a penchant for probabilistic analysis to join in moving away from analytic reconstructions to link up to statistical inference (both for the problems and promise it affords).

7 Do Tests of Assumptions Involve Us in a Regress of Testing?

Because total evidence is always finite, the entire testing procedure must, so the informal argument goes, be unfounded and depend on assumptions for which there is no test. (Glymour, this volume, p. 337)

The issue of assumptions is fundamental and has already poked its head into several of the contributions in the form of experimental and model assumptions, and assumptions needed to link experimental and statistical hypotheses to substantive hypotheses and theories. For the error statistician, what matters – indeed, what makes an inferential situation "experimental" – is the ability to sustain reliability or severity assessments. Philosophers often point up assumptions that would be *sufficient* to warrant an inference without adequate consideration of whether they are *necessary* to warrant the inference. If a large-scale theory or paradigm is assumed, then "use-constructed" hypotheses are warranted; but it is a mistake, or so I have argued, to suppose the hypothesis cannot be warranted by other means. Experimental relativists could have arrived at inferences about the deflection effect by assuming GTR and estimating parameters within it, or they could have instead warranted deflection inferences without assuming any one metric theory – as they did! Or, to go back to my homely example in Chapter 1, I could have arrived at my inference about George's weight gain by assuming the first scale used was reliable, or I could have done what I did do: use several different weighing machines, calibrate them by reference to known standard weights, and reach a reliable inference about George's weight that did not require assuming the reliability of any particular scale used. In these remarks, in fact, I have been arguing that the impetus to evaluate severity without depending on unknown assumptions simultaneously leads to hypotheses and theories with good explanatory characteristics.

Glymour's work on causal modeling exemplifies the "can-do" attitude of the error statistician: "The appropriate methodological enterprise is not to shout 'So there!' citing Duhem and Quine and such, but to investigate what can be tested and reliably discovered under varying assumptions"

(p. 338). If any single theme should be attached to what I have referred to as the "new experimentalism," it is the idea that the secret to avoiding the skeptical upshots of classic problems of underdetermination is to make shrewd use of experimental strategies. Experimental strategies for me need not involve literal control or manipulation but are strategies for controlling and evaluating severity, at least qualitatively. How far this may be achieved in the observational contexts in which Glymour works is an open question, but I see no reason why a combination of literal experiments (including randomized trials), simulations, and nonobservational inquiries could not be used to address the kind of assumptions he worries about in causal modeling. But looking deliberately at "nonexperimental" or observational data may actually be the best way to understand why certain kinds of experimental controls enable reliable probes of causal connections. Glymour suggests that the conditions that vouchsafe no unobserved common causes "are exactly [those] that experimental randomization is intended to provide" (p. 341). This may offer an intriguing path to explaining how (and when) randomization works, and it also suggests ways to mimic the results when literal randomization is not possible. The advances achieved by Zhang and many others – their utilization in a vast array of social sciences, computer science, and technology – speak to the valuable aperçu of Glymour, as does the work on testing statistical assumptions by Aris Spanos. This takes us to the next exchange by Spanos.

References

Laudan, L. (1997), "How about Bust? Factoring Explanatory Power Back into Theory Evaluation," *Philosophy of Science*, 64: 306–16.

Mayo, D.G. (1996), *Error and the Growth of Experimental Knowledge*, University of Chicago Press, Chicago.

Mayo, D.G. (1997a), "Duhem's Problem, the Bayesian Way, and Error Statistics, or 'What's Belief Got to Do with It?'" *Philosophy of Science*, 64: 222–44.

Prusiner, S.B. (2003), *Prion Biology and Diseases*, 2nd ed., Cold Spring Harbor, Laboratory Press, Woodbury, NY.

Related Exchanges

Mayo, D.G. and Miller J., (2008), "The Error Statistical Philosopher As Normative Naturalist," *Synthese (Error and Methodology in Practice: Selected Papers from ERROR 2006)*, 163(3): 305–14.

Parker, W.S. (2008), "Computer Simulation Through An Error-Statistical Lens," *Synthese (Error and Methodology in Practice: Selected Papers from ERROR 2006)*, 163(3): 371–84.

III Graphical Causal Modeling and Error Statistics
Exchanges with Clark Glymour

Aris Spanos

1. *Experimental Reasoning and Reliability*: Can "experimental" virtues (e.g., reliability) be attained in "nonexperimental" modeling? How can the gap between statistical and structural (including causal) models be bridged?
2. *Objectivity and Rationality:* Can one objectively test assumptions linking actual data to statistical models, and statistical inferences to substantive questions?

My discussion focuses on aspects of Clark Glymour's contribution pertaining to empirical modeling in general and graphical causal (GC) modeling in particular. In addition to several constructive exchanges with Glymour in forums prior to and during the ERROR 06 conference, the remarks that follow reflect the back-and-forth dialogue during our teaching of a joint course at Virginia Tech in the fall of 2007 (as we were completing this volume).

One of the initial goals of undertaking the joint course was to explore whether developments in econometrics regarding *statistical model validation* using *Mis-Specification (M-S) testing* could bear fruit for Glymour's GC modeling program. As Glymour's contribution to this volume suggests, the *error-statistical* modeling framework can be used to clarify and address a number of methodological issues raised in this literature. At the very least, this framework affords the clarity needed to bring out the assumptions linking *statistical* and *structural models* (see Figure 6.1, p. 212) and, correspondingly, can delineate several issues pertaining to GC modeling.

In my view, GC modeling (see Glymour and Cooper, 1999; Edwards, 2000) constitutes one of the most important developments in empirical modeling of the past fifteen years or so. The considerable literature to which

Table 8.1. *The Simple Multivariate Normal Model*

$$X_t = \boldsymbol{\mu} + \mathbf{u}_t, \, t \in \mathbb{N}$$

[1] Normality: $X_t \sim N(.,.)$,

[2] Constant mean: $E(X_t) = \boldsymbol{\mu}$,

[3] Constant covariance: $\text{Cov}(X_t) = V > 0$,

[4] Independence: $\{X_t, \, t \in \mathbb{N}\}$ is an independent process.

---

work on GC modeling has given rise exemplifies the type of interdisciplinary work that holds significant promise for a more dynamic and effective philosophy of science. While engaging philosophers interested in modeling and causality (Hausman, 1998; Williamson, 2005; Woodward, 2003), it is relevant to practitioners who are grappling with numerous philosophical/methodological problems when modeling with observational data.[1]

The very fact that the error-statistical perspective broadens "experimental" inquiries to include those where the relevant error probabilities can be potentially ascertained challenges us to consider how analogous inferential goals and objectives can be met in contexts that lack literal manipulation and control, or even the implementation of experimental design techniques.

1 Statistical Model

The basic *statistical model* that underlies the overwhelming majority of structural GC models (see Spirtes et al., 2000) is the *simple multivariate Normal model* based on a vector stochastic process $\{X_t, \, t \in \mathbb{N}\}$, where $\mathbb{N} := \{1, \ldots, n, \ldots\}$, $X_t := (X_{1t}, X_{2t}, \ldots, X_{mt})$, assumed to be *Normal* (N), *Independent* (I) and *Identically Distributed* (ID), denoted by $X_t \sim$ NIID$(\boldsymbol{\mu}, V)$, $t \in \mathbb{N}$, where $E(X_t) = \boldsymbol{\mu} = [\mu_i]_{i=1,\ldots,m}$, $\text{Cov}(X_t) = V = [v_{ij}]_{ij=1,\ldots,m}$. A complete specification of the statistical model, denoted by $\mathcal{M}_\theta(\mathbf{x}) = \{f(\mathbf{x}; \boldsymbol{\theta}), \, \boldsymbol{\theta} \in \Theta\}$, $\mathbf{x} \in \mathbb{R}^{mn}$ (see Chapter 6), in terms of *testable* probabilistic assumptions [1]–[4], is given in Table 8.1.

The importance of such a specification stems from the fact that the reliability of any inference based on $\mathcal{M}_\theta(\mathbf{x})$ depends crucially on its *statistical adequacy*: assumptions [1]–[4] are valid for data $\mathbf{X}_0 := (X_1, \ldots, X_m)$, where each X_i, $i = 1, \ldots, m$, represents an $(n \times 1)$ column of data for X_{it}.

In particular, statistical adequacy secures the *error reliability* of statistical inferences based on $\mathcal{M}_\theta(\mathbf{x})$ in the sense that the *actual* error probabilities approximate closely the *nominal* ones.

---

[1] GC modeling has been applied with some success in several disciplines; see Glymour (2001), Hoover (2001, 2005), and Shipley (2000).

In practice, assumptions [1]–[4] can (and should) be assessed both *informally*, using graphical techniques, as well as *formally*, using thorough Mis-Specification (M-S) testing (see Spanos, 1999, chs. 5, 15).

Statistical adequacy is of crucial importance for GC modeling because its algorithms (see Pearl, 2000; Spirtes et al., 2000) rely heavily on testing *independence* and *conditional independence* between the observables X_t in selecting the appropriate GC model. Any departure from assumptions [1]–[4] is likely to lead these algorithms astray; applying a .05 significance-level test, when the actual type I error is closer to .90, can easily give rise to erroneous inferences.

Let me illustrate the *misspecification error* argument using a classic example in causal inference.

2 Certain Counterexamples to the Principle of Common Cause

The Principle of Common Cause (PCC), attributed to Reichenbach (1956), states that if P [two variables X_{1t} and X_{2t} are probabilistically dependent], then Q [either X_{1t} causes X_{2t}, X_{2t} causes X_{3t}, or X_{1t} and X_{2t} are joint effects of a common cause X_{3t}].

The PCC is crucial for GC modeling because the latter's two core presuppositions, the *Causal Markov* (CM) and the *Causal Faithfulness* (CF) conditions (see Spirtes et al., 2000) are closely related to it; the CM is closely linked to PCC and the CF provides a kind of *necessary* condition for Q.

Irrespective of the merits or demerits of the PCC, statistical adequacy can be used to rule out some of the counterexamples proposed in the philosophy of science literature – cases where there is apparent probabilistic dependence between X_{1t} and X_{2t} without any of the causal connections obtaining. In particular, because the use of the sample correlation $\hat{\rho}$, as a reliable measure of ρ (the probabilistic dependence between X_{1t} and X_{2t}), depends *crucially* on the validity of assumptions [1]–[4], certain violations of these assumptions would render $\hat{\rho}$ an artifact, thus voiding such counterexamples. This is the case with a familiar numerical counterexample (see Sober, 2001), where X_{1t} represents sea levels in Venice and X_{2t} represents prices of bread in Britain (Figure 8.6). Looking at Figure 8.6, it is clear that, at least one assumption, assumption [2] – possibly more – is *false* because the mean of both series is changing with t. The falsity of assumption [2] implies that $\hat{\rho}$ is *not* a reliable estimate of the correlation $\rho = [v_{12}/\sqrt{(v_{11} \cdot v_{22})}]$ parameter,[2] or

---

[2] The sample correlation $\hat{\rho} = \frac{\sum_{t=1}^{n}(X_{1t}-\overline{X}_1)(X_{2t}-\overline{X}_2)}{\sqrt{\sum_{t=1}^{n}(X_{1t}-\overline{X}_1)^2 \sum_{k=1}^{n}(X_{2t}-\overline{X}_2)^2}}$, when used as an estimator of $\rho = [v_{12}/\sqrt{(v_{11} \cdot v_{22})}]$, implicitly assumes that the mean of both variables, X_{1t} and X_{2t}, is constant; that is, $E(X_{1t}) = \mu_1$, $E(X_{2t}) = \mu_2$, and can be reliably estimated by

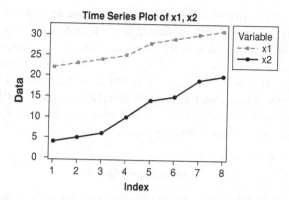

Figure 8.6. The t-plot of X_{1t} and X_{2t}.

any other notion of probabilistic dependence for that matter (see Section 3.1 of Chapter 6, this volume). Hence, the *claim* that X_{1t} and X_{2t} are probabilistically dependent,[3] that is, $\rho \neq 0$ (P is instantiated by the data in Figure 8.6) because $\hat{\rho} = .990$, with a *p*-value $p = .00000$, is *false*.

A simple way to demonstrate the falsity of this claim is to use the *deterministic trend* $t = 1, \ldots, 8$ to detrend the two series in an attempt to render assumption [2] approximately valid[4] and re-estimate ρ. The detrended data yield $\tilde{\rho} = .629$, with $p = .095$, indicating that ρ is statistically *insignificant*[5] at $\alpha = .05$.

$\overline{X}_1 = 1/n \sum_{t=1}^{n} X_{1t}$, $\overline{X}_2 = 1/n \sum_{t=1}^{n} X_{2t}$, respectively. But when the two means are changing with t, that is, $E(X_{1t}) = \mu_1(t)$, $E(X_{2t}) = \mu_2(t)$, $\overline{X}_1$ and $\overline{X}_2$ are "bad" (inconsistent) estimators, and they give rise to unreliable (spurious) statistical results (see Section 3.1 of Chapter 6, this volume).

[3] Note that *all* (estimation and testing) inferences and claims concerning *probabilistic dependence* always pertain to certain unknown *parameters* in $\mathcal{M}_\theta(\mathbf{x})$.

[4] It is important to emphasize that detrending does not constitute a "solution" to the spurious correlation (due to misspecification) problem, but a way to expose it, establishing what *cannot* be reliably inferred. Hence, suggestions to "clean up the data" using detrending, differencing, and so on, prior to analysis (Reiss, 2007; Steel, 2003) are misplaced, because they constitute ad hoc "fixes" that often distort the systematic information in the data and give rise to erroneous inferences. A proper solution (establishing what *can* be reliably inferred) requires *respecifying* $\mathcal{M}_\theta(\mathbf{x})$ with a view to attain a *statistically adequate* model, say $\mathcal{M}_\varphi(\mathbf{x})$, which *accounts for all departures* (systematic information) from the original assumptions [1]–[4], including *t*-heterogeneity or *t*-dependence. $\mathcal{M}_\varphi(\mathbf{x})$ can then be used to draw inferences about probabilistic dependence in its context. Such a respecification provides a natural home for several suggestions in Hoover (2003).

[5] After "subtracting out" both the *t*-heterogeneity and *t*-dependence, $\breve{\rho} = .698$, with $p = .292$, rendering ρ even less significant. However, in view of the small sample size ($n = 8$), caution should be exercised in interpreting all the results in this example; that does not affect the validity of the general argument.

This way of dealing with the alleged counterexample avoids the sticky problems that emerge in trying instead to explain it as a case where either (1) the apparent correlation is the result of "mixed populations" or (2) t can be treated as the *common cause* for X_{1t} and X_{2t} (see Cartwright, 2007, pp. 77–8). In relation to the latter, it is not at all clear what probabilistic sense can be given to the very notions of correlation with t: $\mathrm{Corr}(X_{it}, t)$, and conditioning on t: $\mathrm{D}(X_{it} \mid t; \boldsymbol{\psi}_1)$, $i = 1,2$, when the joint $\mathrm{D}(X_{it}, t; \boldsymbol{\varphi})$ and marginal $\mathrm{D}(t; \boldsymbol{\psi}_2)$ distributions are degenerate.[6]

3 Structural Model

It is important to emphasize that the notion of a *structural model*, as used in the error-statistical framework, is much broader than the notion of a graphical model as used in the GC literature, because it includes any *estimable* (in view of $\mathbf{X}_0$) model that is built primarily on *substantive* information. The primary (but not the only) family of *structural models* used in the GC literature is that of directed acyclic graphs (DAGs), denoted by $G(\mathrm{A}) := \{G(\boldsymbol{\alpha}), \boldsymbol{\alpha} \in \mathrm{A}\}$.

4 Selecting a DAG Model

The basic idea in GC modeling is to infer causal (substantive) relationships among the set of variables $\boldsymbol{X}_t$ using the statistical information contained in $\mathbf{X}_0$. GC modeling achieves that end by imposing two *core presuppositions*, the CM and CF conditions, which interrelate the causal and statistical information (see Pearl, 2000; Spirtes et al., 2000).

Example. For simplicity let us assume $\boldsymbol{\mu} = \mathbf{0}$ and consider the simple DAG model $G(\boldsymbol{\alpha}_0)$ among six observable variables (Figure 8.7), which for $\boldsymbol{\alpha}_0 = (a_{41}, a_{42}, a_{54}, a_{62}, a_{63}, a_{64}, v_1, \ldots, v_6)$ can be expressed in the form of the following structural model:

$$X_{1t} = \varepsilon_{1t}, \; X_{2t} = \varepsilon_{2t}, \; X_{3t} = \varepsilon_{3t},$$
$$X_{4t} = a_{41} X_{1t} + a_{42} X_{2t} + \varepsilon_{4t},$$
$$X_{5t} = a_{54} X_{4t} + \varepsilon_{5t},$$
$$X_{6t} = a_{62} X_{2t} + a_{63} X_{3t} + a_{64} X_{4t} + \varepsilon_{6t},$$

where $\boldsymbol{\varepsilon}_t := (\varepsilon_{1t}, \varepsilon_{2t}, \ldots, \varepsilon_{6t}) \sim N(\mathbf{0}, \boldsymbol{D})$, $\boldsymbol{D} = \mathrm{diag}(v_1, \ldots, v_6)$.

---

[6] Note that $\mathrm{D}(X_{it} \mid t; \boldsymbol{\psi}_1) = \mathrm{D}(X_{it}, t; \boldsymbol{\varphi}) / \mathrm{D}(t; \boldsymbol{\psi}_2)$, rendering the restriction $\mathrm{D}(t; \boldsymbol{\psi}_2) > 0$ necessary.

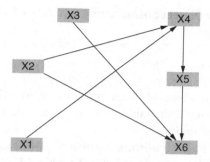

Figure 8.7. The graph of the DAG model, $G(\alpha_0)$.

4.1 Where Do Such DAG Models Come from?

The selection process giving rise to $G(\alpha_0)$ can be summarized as follows. The "acyclic" nature of $G(A)$ imposes a strict *causal ordering* on the variables $X_t := (X_{1t}, \ldots, X_{6t})$, which is either determined on a priori grounds or suggested by the data. Without any loss of generality, let the ordering for the preceding example be $X_{1t} \to X_{2t} \to X_{3t} \to X_{4t} \to X_{5t} \to X_{6t}$. Now their joint distribution, $D(X_t; \theta)$, simplifies into a product of univariate distributions (see Spanos, 2005):

$$D(X_t; \theta) = D(X_{1t}; \psi_1) \prod_{i=2}^{m} D(X_{it} | X_{(i-1)t}, \ldots, X_{1t}; \psi_i) \qquad (1)$$

The CM and CF conditions ensure the two-way equivalence: $D(X_t; \theta) \rightleftarrows$ DAG; the CM demarcates the type of causal relations admitted in $G(A)$, and the CF restricts probabilistic independencies to the ones entailed by the CM condition.

A particular DAG model, such as $G(\alpha_0)$, is selected from within $G(A)$ on the basis of data X_0 using one of several GC algorithms, which are guided by sequential testing relating to three primary types of hypotheses concerning zero *statistical* parameters in the context of $\mathcal{M}_\theta(x)$.

a. Testing independence using the parameters in $V = [v_{ij}]_{ij=1,\ldots,m}$, $v_{ij} := \mathrm{Cov}(X_{it}, X_{jt})$:

$$H_0: v_{ij} = 0 \text{ vs. } H_1: v_{ij} \neq 0, \text{ for some i, j} = 1, \ldots, m. \qquad (2)$$

b. Testing *conditional independence* using the inverse of the submatrices $V_k := [v_{ij}]_{ij=1,\ldots,k}$ for $k = 3, \ldots, m$, denoted by $V_k^{-1} = [\omega_{ij.k}]_{ij=1,\ldots,k}$, where $\omega_{ij.k}$ is the *partial covariance* between X_{it} and X_{jt} conditional on the rest of the variables in V_k:

$$H_0: \omega_{ij.k} = 0 \text{ vs. } H_1: \omega_{ij.k} \neq 0,$$

$$\text{for some i, j} = 1, \ldots, m \text{ and } k = 3, \ldots, m. \qquad (3)$$

c. Testing for the presence of *vanishing tetrads* (see Spirtes et al., 2000):

$$H_0: [v_{gh}v_{ij} - v_{gi}v_{hj}] = 0 \text{ vs. } H_1: [v_{gh}v_{ij} - v_{gi}v_{hj}] \neq 0,$$
$$\text{for some g, h, i, j} = 1, \ldots, m. \tag{4}$$

The nature of hypotheses (2)–(4) renders the various GC model selection algorithms potentially vulnerable to three inveterate problems that can undermine their reliability.

The first relates to the *statistical adequacy* issue raised earlier. If any of the statistical model assumptions [1]–[4] (Table 8.1) is false for data X_0, then the reliability of the selection algorithms would be seriously undermined because, unknown to the modeler, the nominal error probabilities are likely to be very different from the actual ones. The second potential source of unreliability stems from the sequential and multiple testing nature of hypotheses (2)–(4) and concerns the evaluation of the *relevant error probabilities*; this is beyond the scope of this paper (but see Chapter 7, this volume). The third source pertains to the potential susceptibility of the accept/reject decisions to the *fallacies of acceptance/rejection*. Identifying statistical significance/insignificance with the substantive significance/insignificance can easily lead these algorithms astray. It may be the case that the null hypotheses in hypotheses (2)–(4) are *not* rejected because the particular test does not have high enough power to detect the existing discrepancies. This is an instance of the *fallacy of acceptance*: no evidence against H_0 is (mis)interpreted as evidence for H_0. On the other hand, in cases of a large sample size *n*, using small *p*-values as evidence for a substantive discrepancy from the null might be unwarranted, which is an instance of the *fallacy of rejection*. The sequential nature of these testing procedures suggests that any erroneous decisions at one stage are likely to have serious cumulative effects on the reliability of subsequent inferences.

5 Severe Testing

The error-statistical perspective guards against the aforementioned fallacies using a post-data evaluation of inference based on severe testing reasoning. Assuming a threshold of, say, .90 to be the *severity* with which an inference (H_0 or H_1) "passes" with test T gives rise to a discrepancy from zero, say $\delta \neq 0$, warranted by data X_0. The severity evaluation can be used to establish the *smallest* discrepancy from H_0 warranted by data X_0, when H_1 'passes,' and the *largest* discrepancy when H_1 "passes" (see Mayo and Spanos, 2006). The warranted discrepancy δ for each of the testing results (accept or reject H_0) in hypotheses (2)–(4) can then be used to relate the statistical to the substantive significance or insignificance at each stage of sequential testing to make more informed decisions concerning the presence or absence of a substantive discrepancy.

In practice, this approach can be integrated into the selection algorithm to enhance its reliability.

6 Appraising the Empirical Validity of Explanatory (GC) Models

An additional strategy to safeguard against unreliable inferences is to assess the empirical validity (vis-à-vis data $\mathbf{X}_0$) of a DAG model $G(\boldsymbol{\alpha}_0)$ *after* it has been selected. Given the aforementioned vulnerabilities threatening the reliability of any selection procedure, it is inadvisable to assume that $G(\boldsymbol{\alpha}_0)$ is empirically valid just because it was chosen by an algorithm with "good" *asymptotic* properties. In practice, judicious modeling calls for a thorough probing of possible errors in selecting $G(\boldsymbol{\alpha}_0)$ on the basis of data $\mathbf{X}_0$. In particular, for declaring $G(\boldsymbol{\alpha}_0)$ the "best" causal model, one needs to safeguard against the possibility that there may exist much better DAG models, say $\{G(\boldsymbol{\alpha}_i), i = 1, \ldots, p\}$, within or even outside the $G(A)$ family, that are "equally consistent" with data $\mathbf{X}_0$; see Spanos (2009). This problem is known as the *underdetermination* of theory by data in philosophy of science.

The *error-statistical* modeling framework views this problem in terms of bridging the gap between data $\mathbf{X}_0$ and a structural model $G(\boldsymbol{\alpha}_0)$, using the statistical model $\mathcal{M}_\theta(\mathbf{x})$ as a link to eliminate structural models that are *statistically inadequate*; schematically the bridge is $\mathbf{X}_0 \curvearrowright \mathcal{M}_\theta(\mathbf{x}) \curvearrowright G(\boldsymbol{\alpha}_0)$ (see Figure 6.1, p. 212). The primary objective at the initial link, $\mathbf{X}_0 \curvearrowright \mathcal{M}_\theta(\mathbf{x})$, is to specify a statistical model that "accounts for the *statistical regularities*" (NIID) in data $\mathbf{X}_0$, as summarized by $D(\mathbf{X}_t; \boldsymbol{\theta})$. Its appropriateness is evaluated solely on statistical adequacy grounds: eliminate *statistical specification errors*. In this sense, the estimated $\mathcal{M}_\theta(\mathbf{x})$ might *not* necessarily have any explanatory content.

The introduction of $\mathcal{M}_\theta(\mathbf{x})$ changes the nature of the original underdetermination problem in the sense that it provides a more organized framing of the systematic information in $\mathbf{X}_0$ (without any loss of information) in the form of a *parametric model* (Table 8.1) that nests several potential structural models: $\{G(\boldsymbol{\alpha}_i), i = 0, 1, \ldots, p\}$. Indeed, the statistical *generating mechanism* (GM), $\mathbf{X}_t = \boldsymbol{\mu} + \mathbf{u}_t$ $(t = 1, \ldots, n, \ldots)$ (Table 8.1) treats all the variables in $\mathbf{X}_t$ *symmetrically*, leaving room for structural models to provide an adequate account of any *causal* information – which, by definition, treats the variables in $\mathbf{X}_t$ asymmetrically – underlying the *actual* GM.

Therefore, the underdetermination problem, viewed in the context of the link $\mathcal{M}_\theta(\mathbf{x}) \curvearrowright G(\boldsymbol{\alpha}_0)$, is addressed by imposing *structural (causal) restrictions,* rendering $G(\boldsymbol{\alpha}_0)$ more *informative* than $\mathcal{M}_\theta(\mathbf{x})$. This finding can be affirmed in terms of the unknown parameters in the sense that fewer parameters in $\boldsymbol{\alpha}_0$ than $\boldsymbol{\theta}$; $\boldsymbol{\theta}$ includes the unknown parameters in $V = [v_{ij}]_{ij=1,...,m}$, which amount to $(\frac{1}{2})m(m+1)$, that is, twenty-one *statistical parameters,* but only twelve *structural parameters* in $\boldsymbol{\alpha}_0 = (a_{41}, a_{42}, a_{54}, a_{62}, a_{63}, a_{64}, v_1, \ldots, v_6)$.

In econometrics, the underdetermination problem is viewed as a problem of *identifying* a particular *structural model* on the basis of data $\mathbf{X}_0$. In the context of the link $\mathcal{M}_\theta(\mathbf{x}) \curvearrowright G(\boldsymbol{\alpha}_0)$, identification of $G(\boldsymbol{\alpha}_0)$ is tantamount to the existence of an implicit function $H(\boldsymbol{\alpha}_0, \boldsymbol{\theta}) = 0$ determining $\boldsymbol{\alpha}_0$ in terms of $\boldsymbol{\theta}$ *uniquely*; $H(.,.)$ can be derived explicitly in practice (see Spanos, 2005). The difference between the number of parameters in $\boldsymbol{\alpha}_0$ and $\boldsymbol{\theta}$ suggests that there are nine *overidentifying restrictions.* But how adequate (vis-à-vis data $\mathbf{X}_0$) are these restrictions in accounting for the underlying causal structure? Formally their empirical validity can be tested using the following generic Neyman-Pearson (N-P) hypotheses:

$$H_0: H(\boldsymbol{\alpha}_0, \boldsymbol{\theta}) = 0, \text{ vs. } H_1: H(\boldsymbol{\alpha}_0, \boldsymbol{\theta}) \neq 0. \tag{5}$$

What do these overidentifying restrictions represent, and how do they relate to $\boldsymbol{\theta}$? To answer that question, we need to unpack $H(\boldsymbol{\alpha}_0, \boldsymbol{\theta}) = 0$ and bring out these restrictions explicitly by relating $G(\boldsymbol{\alpha}_0)$ to $\mathcal{M}_\theta(\mathbf{x})$ using a *mediating model* $G(\boldsymbol{\varphi})$; schematically, $\mathcal{M}_\theta(\mathbf{x}) \curvearrowright G(\boldsymbol{\varphi}) \curvearrowright G(\boldsymbol{\alpha}_0)$.

A by-product of this unpacking is to shed light on the notion of reparameterization/restriction.

The link $\mathcal{M}_\theta(\mathbf{x}) \curvearrowright G(\boldsymbol{\varphi})$ involves (1) the imposition of the strict causal ordering on the variables in X_t and (2) the "acyclicity" giving rise to the *unrestricted* DAG, say $G(\boldsymbol{\varphi})$, where $\boldsymbol{\varphi} := (b_{21}, b_{31}, b_{32}, b_{41}, b_{42}, b_{43}, b_{51}, b_{52}, b_{53}, b_{54}, b_{61}, b_{62}, b_{63}, b_{64}, b_{65}, \omega_1, \ldots, \omega_6)$:

$$X_{1t} = v_{1t},$$
$$X_{2t} = b_{21}X_{1t} + v_{2t},$$
$$X_{3t} = b_{31}X_{1t} + b_{32}X_{2t} + v_{3t},$$
$$X_{4t} = b_{41}X_{1t} + b_{42}X_{2t} + b_{43}X_{3t} + v_{4t},$$
$$X_{5t} = b_{51}X_{1t} + b_{52}X_{2t} + b_{53}X_{3t} + b_{54}X_{4t} + v_{5t},$$
$$X_{6t} = b_{61}X_{1t} + b_{62}X_{2t} + b_{63}X_{3t} + b_{64}X_{4t} + b_{65}X_{5t} + v_{6t},$$

where $v_t := (v_{1t}, v_{2t}, \ldots, v_{6t})$ denote uncorrelated Normal, white-noise error terms.

This is essentially a set of recursive *regression equations* whose probabilistic assumptions are entailed by assumptions [1]–[4] (Table 8.1; see Spanos,

2005). It is important to note that the number of parameters in φ is equal to that in θ (twenty-one in both), hence the link $\mathcal{M}_\theta(\mathbf{x}) \curvearrowright G(\varphi)$ constitutes a simple *reparameterization* (no restrictions), and $G(\varphi)$ is said to be *just identified*.

In contrast, going from $G(\varphi)$ to $G(\boldsymbol{\alpha}_0)$, involves the imposition of the following nine overidentifying restrictions:

$$H_0: b_{21} = 0,\ b_{31}= 0,\ b_{32}= 0,\ b_{43}= 0,\ b_{51}= 0,$$
$$b_{52} = 0,\ b_{53} = 0,\ b_{61}= 0,\ b_{65}= 0. \tag{6}$$

These restrictions represent an instantiation of $H(\boldsymbol{\alpha}_0, \theta) = 0$ in hypothesis (5), which, in some sense, encapsulates the explanatory capacity of $G(\boldsymbol{\alpha}_0)$, because these restrictions represent the *substantive* information over and above the statistical information contained in $\mathcal{M}_\theta(\mathbf{x})$ or $G(\varphi)$. Hence, $\mathcal{M}_\theta(\mathbf{x})$ provides the relevant statistical benchmark (with a "life of its own") against which one can assess the validity of any substantive information, including the restrictions in hypothesis (5) (see Chapter 6, this volume).

In view of the preceding discussion, rejection of H_0 in hypothesis (6) with high severity *does* provide evidence against the explanatory capacity of $G(\boldsymbol{\alpha}_0)$ vis-à-vis data $\mathbf{X}_0$. Indeed, the severity evaluation associated with generic hypotheses (5) can (and should) perform a dual role:

1. Assess the explanatory capacity of any $G(\boldsymbol{\alpha}_i)$ vis-à-vis data $\mathbf{X}_0$ by testing the hypotheses

$$H_0: H(\boldsymbol{\alpha}_i, \theta) = 0, \text{ vs. } H_1: H(\boldsymbol{\alpha}_i, \theta) \neq 0. \tag{7}$$

2. Compare alternative structural models from the same or different families including (i) DAG models with latent variables, (ii) undirected acyclic graphs, (iii) chain graphs, (iv) covariance graphs, and (v) reciprocal graphs (simultaneous equations models; see Edwards, 2000).

Model comparison 2 is easily implementable in practice because each structural model in $\{G(\boldsymbol{\alpha}_i),\ i = 0, 1, 2, \ldots, p\}$ from any of the families (i)–(v) can be embedded into the same statistical model $\mathcal{M}_\theta(\mathbf{x})$ (Table 8.1), its own restrictions $H(\boldsymbol{\alpha}_i, \theta) = 0, i = 0, 1, \ldots, p$, can be tested separately, and the results can be compared to assess which model $G(\boldsymbol{\alpha}_i)$ accords better with data $\mathbf{X}_0$. Hence, once a structural model $G(\boldsymbol{\alpha}_i)$ is *embedded* into $\mathcal{M}_\theta(\mathbf{x})$, the severity evaluation depends on assumptions [1]–[4] and *not* the nature of the structural model under consideration; this result stems from the N-P setup and might not obtain in other contexts.

Does *passing* the overidentifying restrictions $H(\boldsymbol{\alpha}_0, \theta) = 0$ *severely* with data $\mathbf{X}_0$ establish the *substantive adequacy* of the selected model $G(\boldsymbol{\alpha}_0)$? In general, the answer is *no* because there remains the problem of probing for possible errors in assessing how well the estimated DAG model $G(\boldsymbol{\alpha}_0)$ explains other aspects of the phenomenon of interest (see Figure 6.1, p. 212) this is sometimes referred to as assessing *external validity* (see Guala, 2005). Probing for potential errors in bridging the gap between an empirically validated model $G(\boldsymbol{\alpha}_0)$ and the phenomenon of interest can be severely tested and its substantive adequacy secured, under certain circumstances, including when one is facing an "experimental" situation in the broad sense: "Any planned inquiry in which there is a deliberate and reliable argument from error" (Mayo, 1996, p. 7).

7 Testing Faithfulness

Another important implication of the distinction between the structural DAG model, such as $G(\boldsymbol{\alpha}_0)$, and the underlying statistical model $\mathcal{M}_\theta(\mathbf{x})$ (Table 8.1) is that, sometimes, potential violations of the faithfulness condition can be tested. Faithfulness is primarily an issue that concerns the relationship DAG $\rightarrow D(\boldsymbol{X}_t; \boldsymbol{\theta})$ (reflecting the true Markov structure). A violation of faithfulness occurs when a particular set of values of the parameters $\boldsymbol{\alpha}_0$ is such that it undermines the true Markov structure (see Hoover, 2001). One can test for potential violations of faithfulness in the context of $\mathcal{M}_\theta(\mathbf{x})$ as follows:

1. Parametrize the joint distribution $D(\boldsymbol{X}_t; \boldsymbol{\theta})$ in different ways (specify possible regression models) to isolate the parameters involved in a potential faithfulness violation,
2. Test and establish that each of those parameters is nonzero; and then
3. Embed the faithfulness condition into a different parameterization (regression model) and test whether the particular combination of parameters is zero or not (see Spanos, 2006).

In contrast, the CF condition cannot be tested within the DAG family of models $G(\mathrm{A})$.

References

Cartwright, N. (2007), *Hunting Causes and Using Them*, Cambridge University Press, Cambridge.

Edwards, D. (2000), *Introduction to Graphical Modelling*, Springer, New York.

Glymour, C. (2001), *The Mind's Arrow: Bayes Nets and Graphical Causal Models in Psychology*, MIT Press, Cambridge, MA.

Glymour, C., and Cooper, G.F., eds. (1999), *Computation, Causation, & Discovery*, MIT Press, Cambridge, MA.

Guala, F. (2005), *The Methodology of Experimental Economics*, Cambridge University Press, Cambridge.

Hausman, D.M. (1998), *Causal Asymmetries*, Cambridge University Press, Cambridge.

Hoover, K.D. (2001), *Causality in Macroeconomics*, Cambridge University Press, Cambridge.

Hoover, K.D. (2003), "Non-stationary Time Series, Cointegration, and the Principle of the Common Cause," *British Journal for the Philosophy of Science*, 54: 527–51.

Hoover, K.D. (2005), "Automatic Inference of the Contemporaneous Causal Order of a System of Equations," *Econometric Theory*, 21: 69–77.

Mayo, D.G. (1996), *Error and the Growth of Experimental Knowledge*, University of Chicago Press, Chicago.

Mayo, D.G., and Spanos, A. (2006), "Severe Testing as a Basic Concept in a Neyman–Pearson Philosophy of Induction," *British Journal for the Philosophy of Science*, 57: 323–57.

Pearl, J. (2000), *Causality: Models, Reasoning and Inference*, Cambridge University Press, Cambridge.

Reichenbach, H. (1956), *The Direction of Time*, University of California Press, Berkeley, CA.

Reiss, J. (2007), "Time Series, Nonsense Correlations and the Principle of the Common Cause," pp. 179–96 in F. Russo and J. Williamson (eds.), *Causality and Probability in Science*, College Publications, London.

Shipley, B. (2000), *Cause and Correlation in Biology*, Cambridge University Press, Cambridge.

Sober, E. (2001), "Venetian Sea Levels, British Bread Prices, and the Principle of the Common Cause," *British Journal for the Philosophy of Science*, 52: 331–46.

Spanos, A. (1999), *Probability Theory and Statistical Inference: Econometric Modeling with Observational Data*, Cambridge University Press, Cambridge.

Spanos, A. (2005), "Structural Equation Modeling, Causal Inference and Statistical Adequacy," pp. 639–61 in P. Hajek, L. Valdes-Villanueva and D. Westerstahl (eds.), *Logic, Methodology and Philosophy of Science: Proceedings of the Twelfth International Congress*, King's College, London.

Spanos, A. (2006), "Revisiting the Omitted Variables Argument: Substantive vs. Statistical Adequacy," *Journal of Economic Methodology*, 13: 179–218.

Spanos, A. (2009), "Akaike-type Criteria and the Reliability of Inference: Model Selection vs. Statistical Model Specification," forthcoming, *Journal of Econometrics*.

Spirtes, P., Glymour, C., and Scheines, R. (2000), *Causation, Prediction and Search*, 2nd edition, MIT Press, Cambridge, MA.

Steel, D. (2003), "Making Time Stand Still: A Response to Sober's Counter-Example to the Principle of the Common Cause," *British Journal for the Philosophy of Science*, 54: 309–17.

Williamson, J. (2005), *Bayesian Nets and Causality: Philosophical and Computational Foundations*, Oxford University Press, Oxford.

Woodward, J. (2003), *Making Things Happen: A Theory of Causal Explanation*, Oxford University Press, Oxford.

NINE

Error and Legal Epistemology

Anomaly of Affirmative Defenses

Larry Laudan

1 Introduction

In *any* sophisticated system of inquiry, we expect to find several features intimately tied to questions of error and error avoidance. Above all, we want a system (1) that produces relatively few erroneous beliefs (without resort to the skeptical gimmick of avoiding error by refusing to believe anything) and (2) that, when it does make mistakes, commits errors that tend to be of the less egregious kind than of the more egregious kind (supposing that we can identify some error types as more serious than others). Finally, (3) we want to have mechanisms in place with the capacity to eventually identify the errors that we have made and to tell us how to correct for them. In short, we want to be able to *reduce* errors, to *distribute* those errors that do occur according to our preferences, and to have a self-correction device for identifying and *revising our erroneous beliefs*. This is, of course, an unabashedly Peircean view of the nature of inquiry, although one need not be (as I confess to being) a card-carrying pragmatist to find it congenial.

Most of the papers in this volume deal with the problem of error as it arises in the context of scientific research. That approach is fair enough because most of us are philosophers of science. In my remarks, however, I want to leave science to one side and look instead at some of the ways in which the *epistemology of error* intrudes into legal fact-finding. Like science, the law is an institutionalized, collective effort to find out the truth about matters of high interest to human beings. Unlike science, however, the law has been at the outer periphery of the field of vision of epistemologists in the twentieth century. I'd like to bring it closer to the center of focus because I believe it raises genuine epistemological issues not posed, or at least not explicitly addressed, by conventional scientific research. I have been struck in particular by the key role played by the burden of proof in the law. Scientific inquiry, as least in its sanitized textbook versions, subscribes to a disinterested, nonadversarial picture of inquiry. Because the "parties" or advocates are not clearly identified, scientific research generally leaves it wholly unspecified who is charged with establishing the bona fides of a hypothesis up for discussion. The law, by contrast, goes to great pains to make it very clear on whom falls the responsibility for proving some assertion or other. One party's failure to discharge its burden of proof means that the judgment or verdict favors the other party. In this sense, the burden of proof rules provide an articulated closure mechanism for terminating a controversy (something that science generally lacks).

It thus becomes a matter of high concern on whom falls the burden of proof. In civil trials, the burden moves repeatedly back and forth between the parties, if they are successful in satisfying the trier of fact that a proof burden temporarily placed on one party has been successfully met. In criminal trials, however, which are my primary focus, the burden of proof, at least in theory, falls squarely and invariably on the prosecutor. The presumption of innocence puts it there, instructing jurors that they must acquit the defendant until and unless the prosecutor has proved the defendant's guilt beyond a reasonable doubt. Model federal jury instructions are very clear about this point:

As a result of the defendant's plea of not guilty the burden is on the prosecution to prove guilt beyond a reasonable doubt. This burden never shifts to a defendant for the simple reason that the law never imposes upon a defendant in a criminal case the burden or duty of calling any witness or producing any evidence.[1]

It is the prosecution's burden to prove the defendant guilty beyond a reasonable doubt. That burden remains with the prosecution throughout the entire trial and never shifts to the defendant. The defendant is never required to prove that he is innocent.[2]

[T]he defendant in a criminal case never has any duty or obligation to testify or come forward with any evidence. This is because the burden of proof beyond a reasonable doubt remains on the government at all times, and the defendant is presumed innocent.[3]

These statements would appear to be unambiguous: "the burden never shifts to a defendant," "the defendant is never required to prove that he is innocent," and the defendant "never has any duty . . . [to] come forward

---

[1] 1–1 Modern Federal Jury Instructions-Criminal P 1.02.
[2] 1–4 Modern Federal Jury Instructions-Criminal P 4.01.
[3] 1–5 Modern Federal Jury Instructions-Criminal P 5.07.

with any evidence." As I have learned painfully over the past few years, "never" almost never means never when it comes to the law.

In this chapter, I want to look at a class of situations in which our untutored hunches about how proof functions in the law are stretched to, even beyond, the breaking point. What we come to see is that defendants are frequently asked to come forward with evidence, that they are often obliged to prove their innocence, and that the burden of proof is as apt to fall on the defendant as it is on the prosecutor. More than that, we see that various epistemic mechanisms exist in the law – especially those associated with the notion of the burden of proof – that find no obvious counterparts in other, more familiar, activities where the truth is sought, such as the natural sciences. This is one reason why I think that legal epistemology, as I call it, ought to be of much broader general interest to epistemologists than it is usually thought to be.

The class of cases where these surprises routinely occur forms what are usually known as *affirmative defenses* (hereafter ADs). Put succinctly, an AD arises whenever the defendant, instead of contesting the claim that he harmed someone else, usually concedes that point but insists that his actions were nonetheless legally blameless. ADs, I hasten to add, are not some *recherché* or esoteric area of the law. Fully one-third of all criminal trials alleging violent acts turn on an affirmative defense.[4] By focusing on this set of cases, I hope to persuade you that the law poses some epistemological puzzles that are both intriguing in themselves and are quite unlike the epistemic conundrums that those of us who grew up as philosophers of science are accustomed to dealing with.

It is doubtless helpful if, before moving to the technicalities, I mention some typical examples of affirmative defenses. Because the Congress and every state legislature can invent its own list of defenses that the courts must recognize, no comprehensive list can be made of them. Indeed, they vary dramatically from one jurisdiction to another. Here, however, the more common and familiar defenses are shown in Table 9.1. (I leave unexplained for now the rationale for my division of defenses into these two classes. Its significance emerges toward the end of the chapter.)

2 The Two Paradigmatic Forms of Affirmative Defenses

In the list given in Table 9.1, the ADs were divided between those defenses that provide a justification for the actions of the accused and those defenses

---

[4] Kalven and Zeisel, p. 221.

Table 9.1. *Affirmative defenses*

Justificatory defenses	
Defense of others, property, or self	Provocation
Insanity	Mistake of fact
Consent	Involuntary intoxication
Impossibility	Third-party guilt
Policy defenses	
Diplomatic, executive, or legislative immunity	Statute of limitations
Double jeopardy	Military orders
Entrapment	*De minimis* infraction
Plea-bargained immunity	Incompetency

that, although they also get the defendant off the prosecutorial hook, are grounded purely in reasons of public policy and expedience rather than being in any way morally exculpatory. For our purposes, however, a more epistemically salient way exists for dividing up ADs. This bifurcation reflects the fact that different jurisdictions impose quite different sorts of probatory burdens on those who invoke an affirmative defense. Essentially these burdens fall into two classes.

There is, for starters, what I call the *full-blown affirmative defense*. It has the following characteristics:

- The state must prove all elements of the crime beyond a reasonable doubt (hereafter BARD).
- If defendant asserts an AD, he must prove to a demanding standard (usually the preponderance of the evidence, sometimes more) that his alleged defense is true.
- A failure by the defendant to prove his defense more likely than not (as judged by the jury) obliges the jury to convict him.

Full-blown affirmative defenses are the rule in at least twelve states,[5] all of which have statutes requiring the defendant to prove his defense to a preponderance of the evidence and by several federal circuit courts. (Seven of the twelve federal circuits likewise require full-blown ADs for at least some of the defenses enumerated earlier.[6]) Minor variants on the full-blown defense can be found in Delaware, Georgia, and North Carolina, where the defendant

---

[5] Alabama, Alaska, Delaware, Illinois, Louisiana, Maryland, Ohio, Pennsylvania, Rhode Island, South Carolina, Texas, and West Virginia. In many more states, an insanity defense requires proof by the defendant to at least a preponderance of the evidence.

[6] The First, Fourth, Seventh, and Ninth Federal Circuits do not require full-blown ADs.

must prove certain defenses "to the satisfaction of the jury." In Kentucky, the evidence for an affirmative defense must be "convincing," whatever that means. Delaware, Georgia, and Oregon have required defendants to prove an insanity defense BARD, a requirement astonishingly upheld repeatedly by the U.S. Supreme Court.[7] However these unconventional standards are glossed, it is clear that all require of the defendant substantially more than that he raise a reasonable doubt about his guilt.

Second, many jurisdictions utilize what I call *modestly demanding affirmative defenses*. These defenses occur in about half the state criminal courts and in a slim majority of the federal circuit courts. Here are their principal features:

- The state is obliged to prove all the elements of the crime BARD.
- If the defendant asserts an AD, he is assigned a burden of producing sufficient evidence to raise a reasonable doubt that the AD falsely applies to him.
- If the judge is persuaded that the defendant has met these two conditions, he will instruct the jury about the existence of the defense in question, allow the jury to hear evidence relevant to the exculpatory defense, and inform the jury that the prosecutor must prove BARD that defendant's excuse is false.
- If the judge is not persuaded that a sufficient modicum of evidence exists for the AD, the judge will not allow the defense to raise it with the jury, the judge will not instruct the jury that the law considers the AD in question a form of exoneration, and the case will proceed before the jury as if the only germane issue is whether the prosecutor has proved the statutory elements of the crime BARD.

The sequence of events here is important to grasp correctly. The defendant claiming a modestly demanding affirmative defense must first persuade the trial judge – out of hearing of the jury – that he has a plausible case to make. That approach requires the defendant *both* to produce exculpatory evidence supporting his affirmative defense *and* to convince the judge that this evidence reaches the stipulated bar of creating a reasonable doubt about the hypothesis that his AD is false. Like the full-blown AD, this "light" version of an affirmative defense imposes a burden on the defendant of satisfying a standard of proof, although the latter standard – raising

---

[7] See *Rivera v. Delaware*, 429 U.S. 877 (1976), and *Leland v. Oregon*, 343 U.S. 790 (1952). Arizona requires clear and convincing evidence for an insanity defense.

a reasonable doubt – is clearly less onerous than the preponderance-of-evidence standard.[8]

It may seem strange to speak of a burden of proof at all when the obligation on the defendant is as weak as merely raising a reasonable possibility that he is innocent. But I am persuaded that there is no better way of describing it. Note that the defendant in these cases is not merely obliged to present relevant, exculpatory evidence. (This is the so-called burden of producing evidence.) That requirement alone would not be tantamount to imposing a burden of proof (although it would certainly be sufficient to give lie to the claim that the defendant is never obliged to present any evidence). What makes this into a genuine burden of proof is that a clear specification exists of the quantum of proof necessary for the defendant to get his case before the jury: He must present enough evidence to raise a reasonable doubt about his guilt.

Both these ADs are puzzling in different ways. What they share is an implicit renunciation of the presumption of innocence, and of the related thesis that the prosecutor alone bears the burden of proof in a criminal trial. Those points are too obvious to require elaboration here. Instead, I want to focus on some other epistemic puzzles posed by these two forms of AD. I begin with the full-blown version, where a defendant asserting a certain defense must persuade the jury that his AD is more likely than not, and I turn later to explore some of the puzzles associated with the more modest version of the AD.

3 The Problem with Full-Blown ADs: Misunderstanding the Function of a Standard of Proof

This practice of obliging the defendant to prove his exculpatory defense to be more likely than not poses conceptual problems aplenty. For instance, one might profitably explore how, if at all, such defenses can be squared with the presumption of innocence (because they appear to presume guilt rather than innocence) or how they can be reconciled with the thesis that the burden of proof in a criminal trial never shifts to the defendant. For today's purposes, I prefer to leave those intriguing questions to

---

[8] In some jurisdictions, it is enough to satisfy the requirement of raising a reasonable doubt if the defendant merely asserts that his action was in conformity with one of the available defenses. In most, however, the bare assertion of a defense is insufficient to put it on the table. In that case, there must be exculpatory evidence apart from defendant's plea of an AD. Hereafter, when I refer to modestly demanding ADs, I am alluding to the latter sort of case. The former one raises none of the problems that I discuss here.

one side.[9] Instead, I want to focus specifically on an issue that I think is more salient than either of those. It has to do with the kind of reasoning that lies behind the widespread belief that it is appropriate under certain circumstances to require that the defendant prove it to be more likely than not that his proposed defense or excuse is true.

For purposes of illustration, I focus on one familiar affirmative defense – that of self-defense – although my remarks apply alike to all those affirmative defenses that I earlier labeled as "justificatory."

In a case of self-defense, the law recognizes that if A – having done nothing to provoke B – is threatened or menaced by B, and if A has reasonable grounds to believe that B is about to do him grievous harm, then A is entitled to take whatever steps a rational person might deem necessary to protect himself, including, if necessary, immobilizing B by killing him. (I don't intend to explore the morality of the doctrine of self-defense, although it strikes me as plausible enough.)

What matters for our purposes is that relevant legislative bodies have specified that one is legally blameless if one genuinely acted in self-defense. There exists, in other words, no criminal liability associated with such actions. A person who harms another in self-defense is as innocent of a crime as if he had done no harm at all. That is the law.

The relevant epistemic question obviously is this: if action in self-defense is innocent behavior, then why must the defendant prove it more likely than not that he so acted?[10] If we believe, as a matter of general policy, that the state must prove one's guilt in the case of an alleged crime, why does the defendant have to prove his innocence in the case of self-defense? More specifically, in a usual criminal case (where an AD is not involved), the state must prove the defendant's guilt beyond a reasonable doubt, whereas in a case of self-defense, the defendant must prove his innocence by a preponderance of the evidence (POE). What we have in play here are two different standards of proof, BARD and POE, and two different parties carrying the burden: the state and the defendant, respectively.

It is widely accepted that proof BARD is an exacting standard. It requires a very impressive proof and it does so for a compelling reason: we regard

---

[9] I have explored such questions briefly in "The Presumption of Innocence: Material or Probatory?" *Legal Theory*, Fall 2005.

[10] In general, a defendant asserting self-defense in a murder case must show (to a preponderance) *each* of the following: (1) that he had reason to believe that he was under an imminent threat of death or serious bodily injury, (2) that he had not negligently put himself in this situation, (3) that he had no legal alternative to avoid the threatened harm, and (4) that it was reasonable to believe that his action would avoid the threatened harm.

the mistakes associated with a criminal trial verdict – a false acquittal and a false conviction – as exacting very different costs. As a society, we have reached a social consensus that false convictions are much more egregious than false acquittals. We have adopted an exacting standard of proof because we believe that a demanding standard is, other things being equal, more likely to reduce the probability of a false conviction than a less demanding standard would. Put differently: with BARD we expect a low rate of false positives, and we are willing to absorb a relatively high rate of false negatives, if necessary, in order to keep false convictions to an acceptable level. By contrast, the preponderance of the evidence standard implicitly but unequivocally denies that one sort of error is more egregious than the other. The Supreme Court has held on numerous occasions that the costs of a false conviction (defendant's loss of liberty and reputation, failure to convict the true culprit, imperfect deterrence of future crimes, and so on) are so serious that no one can be legitimately convicted in a criminal trial unless the standard in play is BARD.

No prosecutor could successfully argue that a jury was bound to convict a defendant provided that they thought his guilt was more likely than not. Indeed, no prosecutor could successfully argue that the defendant had to prove anything. But, when we turn to trials that involve full-blown affirmative defenses, all these familiar rules change. Where the affirmative defense is concerned, the prosecutor doesn't have anything to prove, except that the defendant committed an act whose commission the defendant himself concedes. The defendant, by contrast, must prove his innocence to a preponderance of the evidence. Failure to do so means a verdict of guilty.

The thesis I want to argue is simply this: if the rationale for the selection of a standard of proof like BARD is that we believe that standard incorporates a considered social consensus about the respective costs of errors, then the utilization of any *other* standard of proof for determining guilt and innocence violates the social contract that alone undergirds, and makes rational, the selection of the accepted standard. That is to say, with certain possible exceptions to be discussed, *the coexistence of rival standards of proof of guilt and innocence in a system of criminal justice speaks to a profound confusion (on the part of both legislators and judges) about what a standard of proof is and about whence its rationale derives.*

The argument is a straightforward one. It begins with the uncontroversial observation that the principal function of a standard of proof is to capture our shared social perceptions of the relative costs of the two sorts of mistakes to which criminal trials are subject: to wit, false convictions and false acquittals. Above all else, *a standard of proof is a mechanism for distributing the errors that are likely to occur.* A demanding standard of proof carried by the prosecution, such as proof beyond reasonable doubt, is much more apt to produce false acquittals than false convictions (assuming that defendants are as likely to be innocent as guilty).[11]

We accept the idea that such a high standard be imposed on the prosecution because we believe, in the classic and graphic metaphor of William Blackstone, that it is better that ten guilty men go free than that one innocent man is condemned to the gallows. Unless we think that the social costs of a false conviction are roughly ten times greater than the costs of a false acquittal, then we have no business setting the standard of proof as high as we do (supposing proof beyond a reasonable doubt to be in the neighborhood of 90–95% confidence of guilt). If, for instance, we regarded the two sorts of mistakes as roughly equally costly, a preponderance standard would obviously be the appropriate one because it shows no bias toward one sort of error over the other.

It is a given in the burgeoning literature on the logic of the standard of proof that such a standard must be set at a level that reflects the social *costs* associated respectively with false acquittals and false convictions. Essentially, the standard of proof is set sufficiently high (or low) to capture our shared social judgments about the respective costs of these errors. In still more rigorous treatments of this topic, the social *benefits* of true convictions and true acquittals are likewise factored into the utility calculation.

Whether one grounds the standard of proof simply on the respective costs of potential errors or on a more complex expected utility calculation, one conclusion is salient and inevitable: any modification in the criminal standard of proof that moves it significantly away from that point of confluence between the costs and probabilities of respective errors implies a drastic revision of assumptions about the costliness, and thereby the acceptability, of false convictions and false acquittals. Lowering the standard of proof imposed on the prosecution *or* imposing a significant standard of proof on the defendant would entail precisely such a revision of the social contract about the relative costs of the two sorts of error.[12]

It is not only legal scholars with a penchant for quantification who think of the standard of proof as a mechanism for distributing errors. When the Supreme Court in *Winship* settled that reasonable doubt was to be the

---

[11] I added the parenthetical expression because the actual ratio of errors in any long run of trials depends not only on the standard of proof but also on the distribution of truly innocent and truly guilty among those who come to trial.

[12] For a detailed articulation of such arguments, see Bell (1987), DeKay (1996), and Lillquist (2002).

constitutional standard of proof, the justices had explicitly in mind the idea that acquittals of the guilty were to be strongly preferred over convictions of the innocent.[13] Even before reasonable doubt acquired constitutional status, its near-universal occurrence in the common law likewise reflected an unequivocal judgment that, in criminal trials, false acquittals were preferable to false convictions.

The question before us involves asking what this "social contract" implies for affirmative defenses that require the defendant to establish more than a reasonable doubt about his innocence. The answer is clear: If a state requires that the defendant establish a certain defense (say self-defense or consent) to a preponderance of the evidence or even higher level, that state is saying that erroneously convicting someone who genuinely acted in self-defense or with the consent of the victim is no more egregious an injustice than acquitting someone who falsely alleges self-defense or consent. If a state says that a defendant claiming insanity must prove his insanity by clear and convincing evidence (let alone beyond a reasonable doubt), that state is saying that erroneously convicting someone who was truly insane is a vastly *lesser* injustice than erroneously acquitting someone who was sane when he committed the crime.

These, I argue, are curious and ultimately incoherent judgments of value. It is bad enough that they fly in the face of the Blackstonian thesis that false acquittals are less costly than false convictions. They add insult to that injury by undermining the presumption of innocence and the prosecutorial burden of proof when they insist that a defendant can win an acquittal only if he can prove his innocence to a relatively high standard.

It may be helpful to consider this hypothetical pair of examples: Jones and Smith are both on trial for first-degree murder. Jones offers an alibi (which is not an affirmative defense), presenting witnesses who claim to have been with him elsewhere at the time of the crime. The judge instructs the jury that, to win an acquittal, Jones has no burden of proof and the prosecutor must prove beyond a reasonable doubt (among other things) that Jones was at the crime scene. Smith, by contrast, claims to have been acting in self-defense. In many jurisdictions (significantly, not in federal

---

[13] As Justice Harlan (concurring) insisted in *Winship*, "In a criminal case, on the other hand, we do not view the social disutility of convicting an innocent man as equivalent to the disutility of acquitting someone who is guilty." He went on to quote Justice Brennan's earlier insistence that "the requirement of proof beyond a reasonable doubt in a criminal case [is] bottomed on a fundamental value determination of our society that it is far worse to convict an innocent man than to let a guilty man go free" (*In re Winship*, 397 US 358, at 369–70 (1970)). Similar sentiments have been voiced by the Supreme Court numerous times since *Winship*.

courts), Jones must present enough evidence to make it more likely than not that he was indeed acting in self-defense. Unless he does so, his evidence may be excluded or, even if admitted, the judge may elect not to give the jury a self-defense instruction. Jones and Smith, charged with the same crime, obviously enjoy quite different prospects. Jones will be acquitted if he can raise a reasonable doubt about his guilt. Smith, by contrast, is bound to be convicted if all he can do is to raise a reasonable doubt; to win an acquittal, he must prove it more likely than not that he acted in self-defense.

The crucial issue for us involves the message being sent by the justice system with these two cases. In Jones's trial, the message – implicit in the standard of proof beyond a reasonable doubt – is that it is far better to acquit the guilty than to convict the innocent. In Smith's trial, by contrast, the inescapable message is that convicting the innocent and acquitting the guilty are equally undesirable.

The salient question is a simple one: What is the *principled* difference between these cases that would justify such discrepant assessments of the relative costs of errors? Jones, it is true, is denying that he committed an act that the state is obliged to prove, whereas Smith is conceding that he committed the act but insisting that his behavior was justified and can point to a state statute on self-defense that stipulates self-defense to be a full justification for acting as he did. If Smith's action was genuinely one of self-defense, then he is every bit as innocent of the crime as Jones is, if his alibi is true. But when the state insists that Smith must prove it to be more likely than not that he acted in self-defense, it is saying that convicting an innocent Smith would be a much less egregious error than convicting an innocent Jones. This is nonsense. We must hew to the line that convicting a person innocent of a given crime brings the *same* costs, independently of the specific attributes that render innocent those who are wrongly convicted. Likewise, acquitting a guilty person arguably generally brings the *same* costs. To hold that convicting the innocent is sometimes much worse than acquitting the guilty while other times saying that convicting the innocent is no worse (or perhaps even better) than acquitting the guilty is to fall into babbling incoherence, especially when we are making reference to the same generic crime in the two cases. Before we can assert that conclusion, however, we need to consider briefly the plethora of arguments that have been proffered for putting in place a system that permits very different standards to apply in different criminal trials. (I might add in passing that virtually no legal system exists in the world that does not use such a combination of discrepant standards.)

4 Spurious Justifications of the Double Standard

Different legal systems offer different reasons for imposing a probatory burden on the defendant who adopts an affirmative defense, even while those systems invariably and explicitly subscribe to the doctrine that, in a criminal trial, the burden of proof falls officially on the prosecution. Here are the more common arguments for this policy:

1. *The "can" implies "ought" thesis.* In the United States, it is frequently argued that states are free to set whatever standard of proof they like for such defenses on the grounds that nothing obliges a state to make available the mechanism of an affirmative defense. It is (just) conceivable that a state might include *no* defenses at all in its penal code, not even self-defense. Because the defining and offering of defenses is purely at the option of the state legislature and because such defenses are not governed by *Winship*'s insistence on proof BARD for the key elements of the crime, some jurists, legislators, and academics – not to mention the U.S. Supreme Court itself – have argued that each state may impose, on either the prosecution or the defendant (though it is typically the latter), whatever sort of burden of proof or production it deems appropriate for such defenses. Because, the argument goes, an affirmative defense is a creature of the state legislature – in effect a gratuitous sop to the defendant – that body is at liberty both to provide it (or not) and, if provided, to specify the quantum of proof required to satisfy it. I do not dispute this assertion insofar as it concerns the thesis that states have *constitutional* authority to proceed in this fashion. But constitutional authority is one thing; good reasons are quite another. Our question must be not whether courts and legislators have the legal authority to proceed in this fashion but rather whether any right-minded person would choose to do so. The fact that states have the constitutional authority to impose a demanding burden of proof on the defendant in cases involving an affirmative defense patently does not in itself justify the imposition of such a burden of proof. The fact that doing so is arguably constitutional does not mean that it makes for coherent policy. That a state can do this is no reason to believe that it should.

2. *The initiator thesis.* If the defendant makes a positive assertion about the events surrounding the crime (as opposed to merely denying the prosecutor's allegations), it is only natural that he is obliged to establish that hypothesis. To put the doctrine in its traditional guise: "He

who alleges a fact must prove it."[14] This doctrine is, of course, traditional in civil trials, where we expect the party affirming some fact to
carry the burden of presenting enough evidence to make that allegation plausible. The question is whether its transposition into criminal
proceedings makes any sense.

The salient issue, of course, is *who* is doing the alleging and why. Imagine
the following situation: Sally accuses John of rape. Rape is clearly a criminal
offense. On the other hand, consensual relations between adults do not
count as rape because both morality and the law hold that consent annuls
culpability. So, a crucial issue in John's trial turns on the question of consent.
Suppose that the jurisdiction in question passes a statute saying that the
crime of rape occurs when one party has sexual intercourse with another
without the latter's consent. If the law is so constructed, then it falls to the
prosecutor to prove the two elements of the crime – sexual relations and lack
of consent. John does not have to prove anything or present any evidence.
Suppose, on the other hand, that the legislature in question defines rape as
sexual union and then specifies consent as a legitimate affirmative defense. If
that jurisdiction is likewise one in which the defendant is required to prove
his AD by a preponderance of the evidence, then John has some serious
work to do if he hopes to win an acquittal.

My point is that our decision as to whether the defendant is alleging a
fact (as opposed to simply denying an allegation of wrongdoing) is nothing
more than an artifact of how the relevant law in question is framed. If rape
is defined by statute as nonconsensual sex, then a defendant accused of rape
who believes that he had the consent of his alleged victim is not asserting a
fact but denying one element of the prosecution's case. On the other hand, if
rape is defined as sexual intercourse by a man with a woman who is not his
wife – with a side statute stipulating that consent is an affirmative defense –
then the defendant in a rape case is suddenly alleging an affirmative fact
and carries a burden of proof. In other words, it is in principle open to the
drafters of criminal legislation to stuff all the justificatory and exculpatory
clauses into an affirmative defense statute, thereby burdening the defendant
with having to prove what it would otherwise fall to the state to have to
to disprove: his innocence. Accordingly, the distinction between what the
defendant denies and what he alleges becomes hostage to the vicissitudes of
how the legislation was drafted. If society's considered view is that certain
circumstances are genuinely exculpatory or justificatory (as consent most

---

[14] *Reus excipiendo fit actor.*

surely is in the case of sexual relations), then it should fall to the state to have to prove BARD that those circumstances did not obtain in the case in question. The argument that the defendant should have to prove that those circumstances applied because he is alleging that they did is nothing but the result of a cynical legislative shell game obviously designed to undermine the presumption of innocence and to circumvent the state's obligation to prove the defendant's wrongdoing. Dressing up this sham by invoking the principle that he who alleges X must prove X is a disgrace.

3. *The thesis of the impossibility of "proving a negative."* For several centuries, jurists have been enamored of the silly idea that it is impossible to prove a negative. Indeed, for many of them, that is why the burden of proof in a criminal trial usually falls on the prosecution; because if the defendant had to prove that he did not commit the crime, the state would be asking him to do something which is difficult or impossible to do. This is, of course, sheer nonsense, which should never have been allowed to escape from Logic 101. Having to prove a negative, as legal vernacular conceives it, is no different epistemically from having to prove a positive and sometimes it is much easier. For instance, depending on the circumstances of the case, it may be much easier for the defendant to prove that he was not at the scene of the crime than for the prosecutor to prove that he was there. It may be as easy for the defendant to prove that the gun did not bear his fingerprints as for the prosecution to prove that it did.

This argument comes up in the context of affirmative defenses because, it is said, an affirmative defense involves, as its very name implies, a positive assertion. If – it is said – the prosecutor had to disprove an affirmative defense, this would oblige him to prove a negative. And that – it is alleged – is somewhere between very difficult and impossible. So, the argument continues, any positive assertion made by the defendant must be his to prove. Otherwise, the reasoning goes, he could simply allege his defense and the prosecutor could never disprove it. A familiar precept learned by every student of the law puts the point this way: "The proof lies upon him who affirms, not upon him who denies; since, by the nature of things, he who denies a fact cannot produce any proof."[15] This is patently absurd. If the prosecution alleges that John was at the scene of the crime, it is not in principle impossible for John to prove he was not. If the prosecution alleges

---

[15] This is an almost literal rendering of the classic Latin dictum: *Ei incumbit probatio, qui dicit, non qui negat; cum per rerum naturam factum negantis probatio nulla sit.*

that the television found in John's garage was stolen from Jack, it is not beyond the limits of conceivability that John may be able to prove that he bought it from Jack. Putting the shoe on the other foot, if the defendant alleges that he killed Jack in self-defense, it takes little or no imagination to conceive of circumstances in which the prosecution can produce witnesses who can disprove the defendant's excuse. Because we all understand that we prove negatives all the time – indeed that the proof of a positive is usually a proof of indefinitely many negatives (in this jargon), this point of view is too absurd to merit further discussion.

4. *The thesis of defendant's privileged epistemic access.* By the nature of the case, many affirmative defenses rest on allegations of fact about which the defendant is likely to have better evidence than the prosecution. Many of the affirmative defenses, including self-defense, sanity, mistake, and so on, depend crucially on the beliefs and mental state of the defendant at the moment of the crime. In most circumstances, he is obviously more likely to have evidence pertinent to his mental state than does the state. Thus, it is argued that it appropriately falls to the defendant to produce the evidence necessary to establish the assertion. As Jeffreys and Stephan have put it, a common "explicit justification for shifting the burden of persuasion to the defendant is often to ease the prosecutor's difficulty in disproving facts peculiarly perceived [to be] within the defendant's knowledge."[16] Although this may often be true, it is irrelevant to questions about the standard and the burden of proof. Consider briefly a normal trial that involves no AD. The definition of virtually every crime includes a *mens rea* element. The crime of murder, for instance, requires an intention to do grievous bodily harm. The crime of robbery requires a belief that the purloined item belonged to someone else. The crime of fraud requires a willful intent to deceive a third party. In all of these cases, the defendant enjoys privileged epistemic access but no court in the land takes that to mean that defendant must disprove these claims about his mental state. On the contrary, in a normal trial, we impose a probatory burden on the prosecution, even with respect to elements of the crime about which the defendant is more likely to have evidence than the prosecution. In *Tot v. US*, the Supreme Court put paid to this argument that the burden of proof should fall on the party with greater epistemic access: "It might, therefore, be argued that to place upon all defendants in

---

[16] Jeffries and Stephan (1979, p. 1335).

criminal cases the burden of going forward with the evidence would be proper. But the argument proves too much."[17] As Justice Black put it in a parallel case: "This would amount to a clear violation of the accused's Fifth Amendment privilege against self-incrimination."[18] Why, then, should we suppose that an AD requires the defendant to prove something about his state of mind rather than requiring the prosecution to disprove something about the defendant's state of mind?

Clearly, none of these arguments for imposing a burden of proof on the defendant in a case involving an affirmative defense is even plausible, let alone compelling. Simply, no reasoned basis exists for treating an affirmative defense that involves *justified* behavior any differently than we treat any other sort of defense. If BARD is the prevailing standard in a normal criminal trial, it should remain the standard in a trial in which the defendant claims that his action was justified.

5 The Problem with Modest ADs: The Burden of Production

I turn now to make a couple of observations about the seemingly less egregious form of AD, where the defendant has only to raise a reasonable doubt about his guilt. Recall that here, the defendant must present sufficient evidence to raise a reasonable doubt in the judge's mind about the prosecution's case. Failure to do so means that defendant cannot invoke the AD in question and that, if he does not dispute the prosecution's story about the elements of the crime, he will probably be convicted.

At first glance, one might suppose that this situation is scarcely different from the defendant's situation in a normal trial, where there is no AD. There, as here, it is appropriate to remark that the defendant, if he wishes to win an acquittal, must raise a reasonable doubt in the mind of the trier of fact about his guilt. Because the standard of proof imposed on the defendant appears to be the same in both cases, one might have supposed that the defendant's burden is not more onerous when he proposes an AD than it is in a usual trial. That would, I think, be an egregious mistake. It ignores the fact that, when an AD is in play, the defendant not only has a (modest) burden of proof; he likewise has a burden of the production of evidence, which is lacking in a non-AD trial. It should be clear that imposing a burden of production, along with a burden of proof, makes it less likely that an innocent defendant would be able to win an acquittal than if he faced

---

[17] *Tot v. US*, 319 US 463, at 469 (1943).
[18] *Turner v. US*, 396 US 398, at 432 (1979). Black dissenting.

only the burden of proof. This is because, in a non-AD trial, several routes are open to the defendant for satisfying his burden of raising a reasonable doubt. One of those routes, of course, involves the production of exculpatory evidence by the defendant. But others do not. For instance: one can raise a reasonable doubt by impeaching the credibility of prosecution witnesses, by arguing the implausibility of the prosecution's theory of the case, or by challenging the reliability of the prosecution's evidence. Given the sequence of events spelled out in a modestly demanding AD, however, none of these alternatives is open to the defendant until and unless he has already satisfied the burden of producing enough evidence to raise a reasonable doubt. This initial burden of production effectively closes off many avenues to satisfying the burden of proof that would otherwise be open to the defendant if he were not propounding an AD.

A little reflection readily shows that modestly demanding ADs lead to more false convictions than does BARD alone. Imagine the following situation: an innocent defendant alleges self-defense. The prosecution, let us suppose, lacks evidence sufficient to prove BARD that the defendant did not act in self-defense. In short, if the defendant can get self-defense on the table, he will be acquitted, given that the prosecution cannot prove BARD that he did not act in self-defense. However, there will be a certain proportion of defendants who genuinely acted in self-defense, but who will lack sufficient evidence of that fact (beyond their bare assertion of self-defense) to raise a reasonable doubt. Some subset of those defendants would be able to discredit a prosecutor's case against self-defense (thereby winning an acquittal) but of course that case will never be laid out if the judge believes that insufficient positive evidence exists for the AD to allow the case to proceed along that track.[19] The burden of producing evidence in an AD trial means that, even though the official standard (raising a reasonable doubt) is being observed, the defendant faces a much higher probatory hurdle than he would in a trial that did not involve an AD. Once again, we have to ask rhetorically why someone who is innocent because of his AD should face a tougher probatory task than that faced by someone who is innocent without invoking an AD.

The epistemic moral here is both important and novel: we see that it is not the height of the standard of proof alone that determines the frequency of

---

[19] We might note, in passing, that if a judge decides that the defendant has offered insufficient evidence to raise a reasonable doubt about his defense, this is tantamount to a *directed verdict of guilt* with respect to the alleged self-defense. This should be troubling in its own right, because judges are supposed to be able to issue only directed acquittals, not directed convictions.

false convictions. The burden of the production of evidence can likewise influence the ratio of false convictions to true acquittals. Accordingly, if we want to draw some plausible inferences about how frequently any given system of criminal justice would commit type I errors, we need to know not only the error frequencies associated with the standard of proof in play but we also need to know something about the system's allocation of burdens of proof and production of evidence.

6 When a Significant Burden of Proof on the Defendant May Be Appropriate

Having argued that nothing less than proof BARD should be the standard governing trials that involve what I called defenses of justification, I now consider whether there might not be a good argument for treating the other class of defenses – policy defenses – very much as they *are* treated by many states, that is, as requiring that the defendant prove his AD to a preponderance of the evidence. This seemingly abrupt change of orientation on my part reflects the fact that the costs of the mistakes involved in nonjustificatory defenses show a very different profile and character than we have seen thus far. Consider two obvious examples: the expiry of the statute of limitations and diplomatic immunity.

Associated with many crimes is a legislatively mandated point at which the state surrenders its right to prosecute. Technically, this period of grace does not belong to the defendant as a matter of right but is a gratuitous favor or indulgence awarded by the state to those who have committed crimes whose use-by date has expired. If a statute of limitations is associated with a specific crime, that is because the jurisdiction in question has decided as a matter of policy that pursuing suspected felons beyond a certain point in time is an unwise use of limited state resources.

If a defendant brought to trial maintains that the statute of limitations has expired with respect to the crimes with which he is charged, states routinely require him to prove that grace period has been reached. Would it be better to insist that the state prove BARD that the grace period has not arrived? As always, in considering what the appropriate standard of proof should be, we need to ask ourselves about the respective costs of the mistakes which may be committed. In the case of the statute of limitations, the mistakes are these: (1) the state might convict someone who committed the crime after the grace period had ensued and (2) the state might acquit someone who committed the crime but whose grace period had not yet started. The evaluation of the costs of these mistakes is surely very different from that which we used earlier, when we were considering such defenses as self-defense or insanity. Then, one of the mistakes into which we might fall was that of sending to jail someone whose actions were morally blameless. Here, however, we are dealing with a defendant whose moral culpability for the crime is beyond dispute. If we make a mistake in the state's favor, we are not sending a blameless person to jail but a guilty one. Although doubtless costs are associated with convicting a guilty person for a crime whose grace period has ensued, they are nothing as egregious as the costs associated with sending a morally blameless person to jail. Under such circumstances, where the costs of the two errors are roughly comparable, a standard such as preponderance seems appropriate. (One may argue, and I would not quarrel, that it would be better to require the prosecutor to prove to a preponderance of the evidence that the statute of limitations does *not* apply. But it seems preposterous to argue that the cost of a mistake in the state's favor here is so high that this should have to be proved beyond a reasonable doubt.)

I suggest that the same kind of arguments apply, *mutatis mutandi*, to most of the other ADs that I have called policy defenses. If someone charged with a crime alleges immunity by virtue of his diplomatic status, the state pays a cost if it convicts him when he is truly a diplomat. But requiring him to prove his diplomatic credentials seems not unreasonable because, even if a mistaken conviction is reached, it does not involve sending someone to prison who is blameless. Arguably, cases involving entrapment, prosecutorial immunity, and double jeopardy fall into the same category. Because the consequences of an error in favor of the state in such cases are much less costly than the consequences of convicting someone who is genuinely innocent of wrongdoing, it seems appropriate to entertain a less exacting standard for guilt than BARD. One way of doing that is to require the defendant to prove his defense to at least a preponderance of the evidence. (Alternatively, one could require that the prosecution disprove the defense not to the BARD standard but to a preponderance of the evidence.)

7 Conclusion

I have argued here that affirmative defenses provide unusual grist for the epistemologist's mill. One can not only use some rudimentary epistemic insights to criticize existing practices in the law, but we can also learn something important from those practices. A case in point: the structure of modestly demanding ADs vividly shows that the location of the burden of the production of evidence can be as important as the standard of proof is in determining how the ensuing errors are likely to be skewed. Epistemologists and decision theorists have talked at length about the role of the standard of proof (and to a lesser extent about the role of the rules of evidence) in shaping the distribution of errors, whether in the law or in statistical testing and clinical trials; but we have barely noticed how much the location of either the burden of proof or the burden of the production of evidence on one party rather than another can do to shape the likely profile of errors that may result. What is likewise clear from the discussion in the preceding section is that a one-size-fits-all standard of proof is appropriate in neither criminal law nor anywhere else *if* the costs of the associated errors vary significantly from one case or situation to another. In clinical trials, for instance, it is common to use the same standard for determining that a new drug is safe and that it is effective. It is quite unclear whether errors about safety and errors about efficacy have family resemblances sufficiently close to warrant the use of the same decision rule in both cases.

A more general query emerges from epistemic investigations in the law that it might be useful to raise, even if I haven't the space to give my resolution of the conundrum. It boils down to this: the law has, or at least aims at, clear standards of proof for the hypotheses at stake in a trial. The law likewise has a clear sense of the necessity of grounding standards on an evaluation of the costs of the respective errors that may be committed. Such concerns are notably lacking in much of the literature of the philosophy of science. Notoriously, neither the Popperians nor the Bayesians have acceptance rules or standards of proof. Classical confirmation theory (a la Carnap) paid no attention whatsoever to the question of whether accepting a false theory was better or worse than rejecting a true one. We philosophers (like the scientists themselves) blithely talk about "accepted scientific theories" and heap abuse on those who refuse to accept the theories accepted by the scientists. But if there are generally no canonical rules for the acceptance of scientific theories, if standards of acceptance vary from one scientist to another within a given specialty and from one science to another (as they patently do), then how do we go about describing and defending the conviction that science is a quintessentially rational, rule-governed activity? If theoretical scientists in particular are indifferent to differences in the costs of the mistakes they may be making, or if they adopt an idiosyncratic view about the respective costs of such mistakes (as Popper notoriously did, with his insistence that accepting a false theory was invariably more costly than rejecting a true theory), then why should we expect the rest of the intellectual community to accept the edicts of science as the embodiment of rationality?

References

Bell, R. (1987), "Decision Theory and Due Process," *Journal of Criminal Law & Criminology*, 78: 557.

DeKay, M. (1996), "The Difference between Blackstone-Like Error Ratios and Probabilistic Standards of Proof," *Law and Social Inquiry*, 21: 95.

Jeffries, J. and Stephan, P. (1979), "Defenses, Presumptions and Burden of Proof in the Criminal Law," *Yale Law Journal*, 88: 1325.

Kalven, H. and Zeisel, H. (1966), *The American Jury*, Little, Brown & Co., Boston, MA.

Laudan, L. (2005), "The Presumption of Innocence: Material or Probatory?" *Legal Theory*, 11: 333–61.

Lillquist, E. (2002), "Recasting Reasonable Doubt: Decision Theory and the Virtues of Variability," *UC Davis Law Review*, 36: 85.

Error and the Law
Exchanges with Larry Laudan

Deborah G. Mayo

As with each of the contributions to this volume, my remarks on Larry Laudan reflect numerous exchanges over a long period, in this case, since he was a colleague in the 1980s. Here, we put to one side the discussions we have had on matters of theory testing (although they arise in previous chapters) and shift to extending our exchanges to a new project Laudan has thrown himself into over the past decade: legal epistemology. Laudan's idea of developing a legal epistemology takes us to the issue of how philosophers of science may develop accounts of evidence relevant to practice. I heartily concur with Laudan's invitation to philosophers of science to develop a "legal epistemology" and agree further that such an "applied epistemology" calls for a recognition of the role of burdens of proof and the need to "have a self-correction device for identifying and *revising our erroneous beliefs*" (Laudan, this volume, p. 376). My goal will be to point out some gaps that seem to need filling if one is to clear the way for this interesting program to succeed. Some queries include:

1. *Experimental Reasoning and Reliability*: Does/should probability enter in distinct ways in standards of evidence in law? (e.g., do the standards BARD and POE differ by degree or by kind?)
2. *Objectivity and Rationality*: Do "standards of evidence" vary across sciences? Do differing assessments of "costs" of mistaken inferences preclude a single standard for determining what conclusions are warranted from evidence? Does the latitude in setting standards of evidence preclude viewing scientific inference as a unified, objective, rule-governed activity?
3. *Metaphilosophical concerns*: What are the responsibilities of the "two-way street" between philosophy and practice?

Laudan's discussion offers an excellent example of how "applied philosophy of science" involves us in a two-way street: (a) philosophical analysis can point out conceptual and logical puzzles that often go unattended in practice; at the same time, (b) studying evidence in practice reveals problems (and also solutions) often overlooked in philosophy. Under (a), the most basic issue that cries out for clarity concerns the very meaning of the terms used in expressing these standards of evidence: evidence beyond a reasonable doubt (BARD) and that of the preponderance of evidence (POE). This aspect of the work is of a traditional philosophical kind – alleviating the conceptual puzzles of others – but in a new arena, which has the added appeal of requiring astuteness regarding probabilistic concepts in evidential standards. Going in the other direction, (b), Laudan's discussion leads us to consider whether variability of standards in legal practice challenges "the conviction that science is a quintessentially rational, rule-governed activity" (p. 395). His provocative argument suggests the answer is no.

1 Explicating Standards of Proof in the Law: SoPs

Laudan plausibly links the two familiar standards of evidence – beyond a reasonable doubt (BARD) and the preponderance of evidence (POE) – to that of controlling and balancing the rates of error, much as in conducting statistical tests.

With BARD we expect a low rate of false positives, and we are willing to absorb a relatively high rate of false negatives, if necessary, in order to keep false convictions to an acceptable level. By contrast, the preponderance of the evidence standard implicitly but unequivocally denies that one sort of error is more egregious than the other. (Laudan, this volume, p. 383)

1.1 Interpreting BARD

Let us begin with the BARD requirement. The rationale in law for ensuring a small probability of erroneous convictions (false positives) is akin to the intended use of controlling the type I error in statistics. In statistics, one is to choose the null hypothesis, H_0, so that erroneously rejecting H_0 is the error that is "first" in importance (even though this is not always adhered to); hence, it is the "type I error" (for a discussion, see Mayo and Spanos, 2006). Failing to reject the null hypothesis with data $\mathbf{x}$ is akin to failing to convict a defendant on the evidence brought to trial. By fixing at a low value the probability of rejecting the null when it is true (i.e., committing a type I error), the null hypothesis, like the defendant's innocence, is being protected. Within that stipulation, as in statistical tests, one seeks the test with the greatest power to detect that the null hypothesis is false (i.e., the one with the smallest probability of failing to reject the null when it is false, the type II error).

In Laudan's recent, excellent book (Laudan, 2008), the analogy with statistical tests (using approximately normal distributions) is more explicit than in this chapter. Just as standard statistical tests may be seen to consist of accept/reject rules, a trial may be seen as a rule that takes (the sum total of) evidence into either one of two verdicts: acquit or convict a defendant. That is, a trial uses evidence to infer either

H: innocence (i.e., defendant comes from the innocent population)
or
J: guilt (i.e., the defendant comes from the guilty population).

The rule is determined by specifying the extent of evidence required to reach one or the other verdicts, which in turn is to reflect chosen rates for erroneous convictions (type I error) and erroneous acquittals (type II error).

Something like a ten-to-one ratio of type II to type I error probabilities, Laudan explains, is commonly regarded as the approximate upshot of BARD: "Unless we think that the social costs of a false conviction are roughly ten times greater than the costs of a false acquittal, then we have no business setting the standard of proof as high as we do." By contrast, Laudan goes on to say, "If, for instance, we regarded the two sorts of mistakes as roughly equally costly, a preponderance standard would obviously be the appropriate one because it shows no bias toward one sort of error over the other." (Laudan, this volume, p. 384)

Type I error probability: P(test *T* leads to conviction; *H*-defendant is innocent) = α (fixed at a small value, say .01).[1]

Type II error probability: P(test *T* leads to acquittal; *J*-defendant is guilty) = β (set at, say, 10 times α (i.e, around .1); the power of the test is then .9).

Note that the error probability attaches to the generic event that the trial outputs evidence that reaches the SoP that is set – not to a specific set of evidence. Setting the SoP to be BARD is akin to requiring the evidence *e* to be statistically significant at the .01 level – which may be abbreviated as SS(.01). We get:

P(*e* is SS(.01); *H*-defendant is innocent) = .01.

---

[1] This may be read: The probability that the test (trial) convicts the defendant, under the assumption that she is innocent, equals α.

We may understand "e is SS(.01)" in this context as the event that the evidence is sufficiently "far" from what would be expected under the null hypothesis (innocence) that such evidence would be expected no more than 1% of the time were the evidence to have truly come from an innocent person. The type II error rate, accordingly, is

P(e is not SS(.01); J-defendant is guilty) $= .1$.

1.2 Interpreting POE

Laudan regards a "preponderance standard," POE, by contrast to BARD, as holding "the two sorts of mistakes as roughly equally costly" and so, presumably, $\alpha = \beta$, although he does not say how small each should be. Should they both be .1? .01? .05? Setting $\alpha = \beta$ does not ensure both are small, nor does it preclude their being set even smaller than .01. So already we see a need for clarification to make out his position.

What Laudan is clear about is that he regards BARD and POE as merely differing by degree, so that whatever levels of evidence they require, they are not measuring different things. They refer to the same "bar," just raised to different heights. Some of his remarks, however, suggest slippage between other potential construals of POE. For example, Laudan also construes finding a POE for some hypothesis as evidence that the hypothesis is "more likely than not." In particular, in his "full-blown version" of an affirmative defense (AD), the defendant asserting a given defense must persuade the jury that his AD is more likely than not (pp. 379, 381).

Let H_{AD} be the hypothesis that the affirmative defense holds true. Then this "more likely than not" assertion, employing the statistical notion of likelihood, asserts the following:

$\mathrm{Lik}(H_{AD}; e) > \mathrm{Lik}(\text{not-}H_{AD}; e)$

where $\mathrm{Lik}(H_{AD}; e)$ abbreviates the likelihood of H_{AD} given e, which is defined as

P(e; H_{AD})

(see Chapter 7). It is important to see that in calculating likelihoods the evidence e is fixed – it refers to a specific evidence set; what varies are the possible hypotheses under consideration. So to say that H_{AD} is more likely than not is to say

(i) P(e; H_{AD}) $>$ P(e; not-H_{AD}).

The likelihood ratio is best seen as a "fit" measure, so that statement (i) says that H_{AD} fits e better than not-H_{AD} does. However, this does not entail anything about error probabilities α and β! So, perhaps the initial impression that Laudan regards both BARD and POE as error-probabilistic standards, differing only by degree, was mistaken. The POE appears now to refer to likelihoods. But likelihoods do not obey the laws of probability, for example, Lik(H_{AD}; e) + Lik(not-H_{AD}; e) need not add to 1 or any number in particular.

Finally, and most emphatically, that the likelihood ratio exceeds 1 does not say a Bayesian posterior probability of H_{AD} exceeds that of not-H_{AD}. That is, statement (i) does not assert

(ii) $P(H_{AD}$; $e) > P($not-H_{AD}; $e)$.

Yet statement (ii) may sound as if it formally captures the phrase "preponderance of evidence." So there is a good deal of ambiguity here, further showing that legal scholars must clarify, as they have not, the meanings of such evidential standards before we can disentangle arguments about them. For POE to be cashed out as statement (ii) requires a way to assign prior probability assignments (supplemented with an interpretation – frequentist, subjective, conventional). So, the meaning of POE needs settling before we proceed to tackle the question of whether POE and BARD reflect inconsistent SoPs, as Laudan argues. To take some first steps, I continue to draw out conceptual confusions that otherwise hound not just legal discussions, but trials themselves.

1.3 Prosecutor's Fallacies

The intended error-probability construal of BARD, (iii), is not immune to an analogous confusion – most notably, it may be confused with asserting that the probability of the trial outputting an erroneous conviction is .01 (iv). However, the error-probabilistic claim

(iii) $P($trial T convicts defendant; H-defendant is innocent$) = .01$
does not entail a posterior probability statement:
(iv) $P(H$-defendant is innocent; trial T convicts defendant$) = .01$.

Statement (iv) requires a prior probability assignment to H-defendant is innocent; statement (iii) does not. Confusing statement (i) with (ii), and (iii) with (iv), appropriately enough, are sometimes known as forms of "prosecutor fallacies"! If philosophers are going to help bring clarity into legal epistemology, they must be careful to avoid classic fallacies, and yet

work in probalistic confirmation theory is often mired in just such confusions (as we saw in Chapter 5). So, we must clean out some of our own
closets before we offer ourselves as legal epistemologists.

Whether a given rationale for an SoP holds up depends on how that
standard is interpreted. For example, Laudan remarks: "A demanding standard of proof carried by the prosecution, such as proof beyond reasonable
doubt, is much more apt to produce false acquittals than false convictions
(assuming that defendants are as likely to be innocent as guilty)"(p. 384).
Although it is true that if trial T requires BARD (defined as $\alpha \ll \beta$) then
it is true that the rate of false acquittals (type II errors) will exceed the
rate of false convictions (type I errors) – Laudan's use of "likely" in the
parenthetical remark alludes to Bayesian priors (the distribution of guilt
among defendants), according to his note 11. The claim, with the Bayesian
construal, is that BARD ensures that the proportion of guilty among those
acquitted exceeds the proportion of innocents among those convicted:

(v) $P(J$-defendant is guilty; test T acquits) $>$ $P(H$-defendant is innocent;
test T convicts),

where P is frequentist probability. To grasp statement (v), think of randomly
sampling among populations of acquitted and convicted: statement (v)
asserts that the probability of the property "guilt" among those acquitted
exceeds the probability of the property "innocence" among those convicted.
But (v) does not follow from BARD construed as $\alpha \ll \beta$, even assuming the
probability of .5 to "innocence" as Laudan does.

It is worth asking how one could justify such a prior probability of .5
to innocence – it may sound like a "fair" assessment, but such is not the
case. If it is to be a degree of belief, then this would not do, because we
are to presume innocence! Even if one knew the proportion of defendants
actually innocent of a crime, this would not give the probability that *this*
defendant is guilty (see the exchange with Achinstein, Chapter 5). One
might suggest instead using a subjective prior (degree of belief) of 1 to
innocence – but then we can get no evidence of guilt. I think any attempt
to stipulate a Bayesian interpretation would be ill-suited to capturing the
legal standards of proof now used. I assume Laudan agrees, but he seems
prepared to relegate such technical matters to one side. What I wish to
convince epistemologists of law is that attempting to scrutinize potential
conflicts of standards of evidence without also clarifying these probabilistic
notions is apt to do more harm than good. Nor is it just the probabilistic
notions that are open to equivocal interpretations – the very notion of SoP

is ambiguous, and one of Laudan's most serious allegations turns on this ambiguity, as will be seen in Section 3.

Conversely, there is much to be learned by trying to coherently fix the meanings of these standards of evidence. By striving to do justice to an intended interpretation in the case at hand (affirmative defenses), a position that at first blush seems problematic may turn out, once reinterpreted, to hold water after all. Let us see if this is true in Laudan's case.

2 Interpreting Preponderance of Evidence (POE) in Affirmative Defenses (ADs)

I continue to work on the assumption, following Laudan, that BARD is akin to the error-statistical requirement of controlling the type I error probability to a low value, and with that stipulation, minimizing the type II error probability. What we did not settle is which of the possible construals of POE to use. I will now propose one that I think readily lends itself to the kind of criticisms Laudan wishes to raise regarding SoPs in ADs. Whether his criticism holds is what we need to determine.

Claiming no expertise in the least regarding the legal issues, I will just follow Laudan (acknowledging, as he does, that different states in the United States follow different stipulations). Given that it has been established BARD that the given action was committed by the defendant, the question is whether a given excuse applies, such as self-defense. That is the circumstance under which ADs arise. Questions about whether an AD is warranted may be usefully construed in a manner analogous to questions of whether a test's *intended* low error rates are at least approximately equal to the actual ones. That is, we begin, following Laudan, with the supposition that the error rates of erroneous convictions and erroneous acquittals (α and β) are fixed on societal policy grounds – they are, he says, part of a social contract. We may call these the *primary error rates*; they are the ones chosen for determining the primary question: whether the (presumably illegal) action was committed. By contrast, Laudan's questions about burdens of proof in ADs ask which standards of evidence should apply (for deciding if the act is "excused"). In particular, he asks which AD standards promote or violate the primary error-probabilistic stipulations.

I cannot vouch that this is the thinking behind the legal statutes Laudan criticizes, but I will continue a bit further with this analogy from statistical tests to see where it may lead. In this analogy, substantiating an AD is akin to saving a null hypothesis H_0 from rejection by explaining away the observed anomaly. In the legal setting, H_0 plays the role of an assertion of "innocence"; analogously, in science, H_0 might be seen as asserting: the observed deviation of the data from some theory T does not discredit T (i.e., "T is innocent of anomalies"). In other words, or so I propose, the AD defense is akin to a *secondary* stage in statistical testing. The secondary stage, recall, concerns testing the underlying assumptions. In particular, it is imagined that a null hypothesis has been rejected with a small p-value – as with the legal case, that much is not in question. A violation of an experimental assumption plays the role of a claim of self-defense because if that excuse is valid, then that invalidates the inference to guilt. (Similarly, if the legal excuse is valid, the defendant is legally "innocent" – therefore not guilty – even though the action is not contested.)

Laudan wants us to focus on the rationale for requiring the defendant to prove a POE that his proposed defense or excuse is true. I focus on self-defense. Laudan's worry is that requiring a defendant to provide a POE for the excuse of self-defense seems to vitiate the requirement that the prosecutor provide evidence BARD of guilt. Now Laudan focuses on the small type I error rate that accompanies BARD – rate of erroneous conviction – but to tackle our present concern we must consider the rate of type II errors. Although the rate of a type I error (erroneous conviction) is set by the "social contract" to be smaller than that of the type II error (erroneous acquittals), surely we would not abide by large type II error rates, or even a type II error rate as large as .51. That would mean failing to convict a guilty person more than 50% of the time. Once it is remembered that the type II error rate must also be sensibly low, even if several times greater than the type I error rate, the arguments about AD can be more meaningfully raised (e.g., a type II error rate ten times that of the type I error rate associated with BARD would be approximately .1 or .2). If there is too much latitude in permitting such excuses, then the originally intended low type II error rate does not hold – the actual type II error rate can become too large.

Continuing to focus on the AD of self-defense, an overly promiscuous standard for excusing would clearly lead to a high (primary) type II error rate – the question is what is overly promiscuous. The answer would seem pretty plainly to be any standard that invalidates the primary requirement to avoid a high type II error rate. We can well imagine that, were there blatant grounds for suspecting self-defense, the case would scarcely have been brought to court to begin with (e.g., shooting a student who was gunning down people in a classroom).

Because we may assume that in realistic ADs the explanation for the "anomalous" data or action is not so obvious, it stands to reason that some evidence is needed. Were it sufficient for the defendant to give an excuse that is always available to a defendant whether guilty or innocent (e.g., a variation on "I believed I had to defend myself") – much as Velikovsky can always save himself from anomaly by appealing to amnesia (see Chapter 4) – the type II error rate would be considerably raised. I am not saying whether a given handling of AD would permit the type II error rate to increase, but rather I suggest that this would be a productive way to address the concern Laudan raises.

A good analogy is the reasoning about the cause of the observed deflection effect – anomalous for Newton's law. "The many Newtonian defenders adduced any number of factors to explain the eclipse effect so as to save Newton's law of gravity" (Mayo, 1996, p. 287). Suppose all they had to do was assert that the observed deflection is due to some Newton-saving factor N (e.g., shadow effect, corona effect, etc.), without giving positive evidence to corroborate it. The result would be a very high probability of erroneously saving Newton. How severely the proposed excuse would need to pass to avoid overly high type II error rates, even approximately and qualitatively determined, is what would matter.

3 Is the Interpretation of Evidence Relative to Costs of Errors?

In analyzing standards of evidence in law, Laudan elicits a number of novel insights for epistemology and philosophy of science. The variable standards of proof in the law, he observes, stand in contrast to the classical image of evidential appraisal in science. Laudan draws some dire lessons for the practice of science from variable standards of proof in the law that I want now to consider:

If there are generally no canonical rules for the acceptance of scientific theories, if standards of acceptance vary from one scientist to another within a given specialty and from one science to another (as they patently do), then how do we go about describing and defending the conviction that science is a quintessentially rational, rule-governed activity? (Laudan, this volume, p. 395)

I would concur in denying that scientists apply uniform rules for the acceptance of scientific theories, but I find it no more plausible to suppose that scientists possess individual rules for accepting theories on the basis of their chosen cost-benefit analysis. I do not see scientists going around applying rules for theory acceptance in the first place – whether theory acceptance is given a realist or an antirealist interpretation. Given that Laudan's discussion here is about standards of evidence for reaching an inference (e.g., grounds to acquit, grounds to convict), I take him to be saying roughly the following:

Laudan's variable standards thesis: The variability of standards of evidence, growing out of the differing assessments of costs of mistaken inferences, poses a serious threat to the conviction that there is a single standard for determining what conclusions are warranted from evidence.

Considering Laudan's variable standards thesis presents us with the opportunity to consider the possibility of objective criteria for warranted evidence and inference more generally. Laudan's provocative charge seems to be that there are no uniform, overriding standards for scrutinizing the evidential warrant of hypotheses and claims of interest; any such standards are determined by choices of errors and the trade-offs between costs of errors. Note that his charge, if it is to carry a genuine provocation, must not be, that individual scientists prefer different methods or worry more or less about different errors, much less that there invariably are different personal benefits or harms that might accrue from a hypothesis being warranted by evidence. These facts do not compel the conclusion that I take Laudan to end up with: that evidential warrant is relative to cost-benefit assignments varying across scientists and fields. We may call this view that of the "cost relativism" of standards of evidence.

But do standards of evidence patently vary across sciences? The only way Laudan's provocative thesis would follow is by trading on an equivocation in the meaning of "standard of evidence" – one that is perhaps encouraged by the use of this term in the legal context. If a "rational rule" or a "standard of evidence" is understood to encompass costs and utilities of various sorts, then, given utilities vary, rational rules vary. For example, we know that in the U.S., OSHA operates with different standards than does the EPA: for the former workplace settings, a statute may set the risk increase before a suspected toxin is open to a given regulation as 1 additional cancer per 10,000 – anything worse is declared an "unacceptable risk." The EPA operates with a more stringent standard, say 1 additional cancer in 1,000. The *agency standards* operate with different cut-offs for "unacceptable risks." (Such discretionary judgments are sometimes referred to as risk assessment policy options; see Mayo and Hollander, 1991.) But this does not mean the two agencies operate with different "standards of evidence" understood as criteria for determining whether the data warrant inferring a .0001 risk or a .001 risk or some other risk. If they did, it would be impossible to discern with any objectivity whether their tests are sensitive enough to inform us whether the agency's standards (for acceptable risk) are being met!

That is why Rachelle Hollander and I (1991) explicitly distinguish "acceptable risk," which involves a policy or value-laden judgment about costs, from "acceptable *evidence* of risk," which does not. Given the chosen agency standard, whether the data do or do not indicate that it is met is a matter of the extent of the evidence. The equivocation is exacerbated in the legal setting because there the term "standard of proof" is used to refer to a social policy judgment. Debating whether to require BARD or POE for a given legal context is to debate a policy question about how high to set the bar. To emphasize this legal usage of "standard of proof," we write it as SoP. A distinct evidential standard would refer to the standards of evidence for determining if a given policy standard is met. Once the SoP is chosen, whether or not given evidence reaches the standard chosen is not itself a legal policy question. As difficult as it may be to answer it, the question is a matter of whether the evidence meets the standard of stringency given by the error probability corresponding to the SoP chosen on policy grounds.

The analogy with risk assessment policy options is instructive. In legal epistemology an SoP is precisely analogous to a standard of acceptable risk – it is understood as a risk management concept. Altering a level of acceptable risk, we may grant, alters what counts as a "rational" policy or decision (e.g., reduce exposure to the toxin in the case of an EPA statute; perhaps "do nothing" in the case of an OSHA statute). But then Laudan's claim about the existence of different risk management settings becomes trivially true and would scarcely rise to the height of threatening "the edicts of science as the embodiment of rationality," as Laudan declares. Setting the SoP in cases of evidence-based policy is a matter of social, economic, pragmatic, and ethical values, but, given those specifications (e.g., given by fixing error rates), whether an inference is warranted is not itself relative to those values.

Without deciding whether Laudan really means to be espousing "cost relativism," we can pursue my argument by considering an analogous position that frequently arises in the area of science-based policy. It is often couched in terms of claims that there are different types of rationality, in particular "scientific rationality" and something more like "ethical rationality." The argument comes in the form of statistical significance tests, and they link up immediately with previous discussions.

In science, the null hypothesis is often associated with the status quo or currently well-supported hypothesis, whereas a challenger theory is the rival alternative.

(Scientific context) H_0: No anomalies exist for an established theory T.

The small type I error rate serves to make it difficult for the tried-and-true accepted hypothesis to be too readily rejected. In the realm of risk assessment, some argue that such standards, while fine for "scientific rationality," may be at odds with "ethical rationality." In particular, suppose our null hypothesis is the following.

(Risk policy context) H_0: No increased risks are associated with drug X.

Because the consequences of erroneous acceptance of H_0 would lead to serious harms, it may be recommended that, in risk policy contexts, it is the type II error probability that needs to be controlled to a small value, since this approach would be more protective.[2]

The concern is that, by securing so small a probability of erroneously declaring an innocent drug guilty of increasing a risk (type I error), studies may have too high a probability of retaining H_0 (the drug is innocent) even if a risk increase of δ is present:

The probability of a type II error: $\beta(\delta) = $ P(test T accepts H_0; increased risk δ is present).

The concern is with cases where the probability of a type II error is high. We should use this information to argue via the weak severity principle:

If P(test T accepts H_0; increased risk δ is present) is very high, then accepting H_0 with test T is poor evidence that an increased risk δ is absent.

Although H_0 "passed" test T, the test it passed was not severe – it is very probable that H_0 would pass this test even if the increased risk is actually as large as δ (see Chapter 7).

The ability to critique which risks are or are not warranted is the basis for avoiding "cost relativism." Laudan's own critique of the consequences of varying legal standards would seem to require this.

4 Recommendations for Legal Epistemology

Laudan's contribution provides valuable grist for three issues that would arise in an epistemology of law. First is the need to clarify the standards of evidence. In a formally specified problem, there are relationships between equally probable hypotheses, equally likely hypotheses, and equal error probabilities (in the sense of frequentist statistics), but they say very different

---

[2] In fact this should be the nonnull hypothesis, because erroneously inferring the drug is safe would be deemed more "serious," but in practice it is typically the null hypothesis.

things and, unless these notions are pinned down, it will be difficult to evaluate arguments about what consequences for error rates would accrue from adopting one or another standard of proof. Second, there is a need to address Laudan's questions about the consistency/inconsistency of legal SoPs, in particular, standards for ADs and the intended BARD standard for guilt and innocence. I have proposed we do so by considering whether they ensure or violate the stipulated error rates for guilt/innocence. Third, there is a need to distinguish the specification of acceptable risks of error – an issue of policy or management – from that of appraising the acceptability of the evidence, given those standards.

References

Laudan, L. (2006), *Truth, Error and Criminal Law: An Essay in Legal Epistemology*, Cambridge University Press, Cambridge.

Mayo, D.G. (1996), *Error and the Growth of Experimental Knowledge*, University of Chicago Press, Chicago.

Mayo, D.G., and Hollander, R., (eds.) (1991), *Acceptable Evidence: Science and Values in Risk Management*, Oxford University Press, New York.

Index

Abduction, 91–92
Achinstein, P., 2, 9–10, 27, 55, 87, 170ff,
  189ff, 200–201, 327, 354, 402
Achinstein exchange
  college readiness example, 186–188,
    195–196
  error-statistical critique, 192
  evidence in, 183, 189–190
  objective epistemic probability rule,
    179–180, 194–200
  probability in, 192, 194
Active inquiry framework in error statistics,
  18–19
Affirmative defenses
  burden of production in, 391–392,
    393
  burden of proof in, 376–378, 381,
    393–394
  can implies ought thesis, 387
  defendant's privileged epistemic access
    thesis, 390–391
  double standards, justifications for,
    387–391
  impossibility of proving a negative thesis,
    389–390
  initiator thesis, 387–389
  insanity pleas, 379, 380
  overview, 378–386, 394–395
  probabilities in, 401–403
  risk assessment, 405–408
  self-defense, 382, 385–386, 392
  standard of proof function, 381–386,
    405–408
Amnesia, collective, 128–129, 149–150,
  156–158, 166

Arguments from coincidence, 67–68, 70,
  73–74, 76–77
  from conspiracy, 74–77
  from error, 24–26, 33
  Hacking and, 68

Bayesian statistics
  and frequentist induction, 280, 318,
    324–327
  criticisms of, 192–194, 200, 298–302
  Duhem problem in, 37–38, 360–362
  inference, 280
  nuisance parameters in, 300
  objectivity in, 276–277, 304
  probability in, 9–10, 280, 334–335
  reference priors, 298–302
  severity testing vs., 37–38
  treatment of adjustable constants,
    49–51
Bayesian philosophy of science, 37
  epistemology, 199, 334–335, 360–362
  Glymour on, 334–335
  Howson on, 131, 194, 200
  future directions, 200
  Myrvold and, 334–335
Berkeley, B., 104
Bernoulli trials, 279, 284–286
  binomial vs negative binomial
Birnbaum, A., 325
Birnbaum's argument (SLP), 307–308
  Mayo's challenge to, 310–313
  see also weak conditionality principle
Birnbaum's evidence function, 287
BOLD signals, 342–348
Brans-Dicke theory, 49–51, 52, 61–62

Carnapian confirmation logic, 318
  Neyman's criticism of, 318
Catchall hypothesis, 37–38
  Earman and, 29–30
Cathode ray experiments, 64–67
Causal modeling
  Cartwright on, 368
  directed acyclic graphs (DAGs), 342–348,
    368–369, 370
  faithfulness condition, 366–368
  Markov condition, 337–348, 366–368
Chalmers, A., 2, 9, 25, 26–27, 28, 38–39, 55,
    58ff, 71, 73ff, 106–108, 111, 124, 168,
    353, 360
Chalmers exchange
  on arguments from coincidence, 68, 73
  on comparativist-holism, 38–39
  induction in, 119
  on severe testing, 73–77, 111
  theory testing/explanation in, 9, 73–74,
    79–80, 106–107, 167
Chance regularities, 213–216, 366–367
Chatfield, G., xvi
College-readiness example, 186–188,
    195–196
Comparativism, severe testing vs., 29,
    39–43
Comparativist-holism
  Chalmers on, 59–60
  criticism of, 38–39
  Musgrave on, 107–108
Conditionality principle, weak (WCP),
    197–198, 200, 295–298, 305, 324–325
Conditioning, in statistics (See also WCP)
  by model formulation, 282
  to achieve UMP tests, 292
  to achieve frequentist goals, 298
  to induce relevance, 294
  for separation from nuisance parameters,
    291
Confidence intervals, 265–266, 288–291
Conservative PC (CPC), 342–348
Copernican theory, 127–128, 140–141, 143
Corroboration and severity, 25
Counterexamples, philosophical role of, 162
Cox-Mayo exchange, Spanos on
  linking FEV and SEV, 320
  model adequacy and error statistics, 321
  new philosophy of frequentist statisics,
    315
  revisiting Bayesian criticisms, 324

Cox, D. R., 10, 37, 55, 86, 124, 207, 211, 242,
    247ff, 276ff, 283, 295, 300, 304–306,
    313–314, 315ff, 328
Critical rationalism
  overview, 88–90, 109–110
  predictive success in, 95–105
  progressive, 117–119
  realism and, 90–91, 94–95
  referential (entity) realism, 100
  severity testing in, 105–111,
    113–115
  skepticism vs. irrationalism, 119–122

Data mining, 272
Deduction
  inference as, 91, 102–103
  from the phenomena, 130–144
Demonstrative induction, 130–135
Descriptive statistics, 226–227
Double-counting (double-use of data),
    See also use-novelty
  Musgrave on, 29
  and novelty, 11, 13, 24, 129ff, 155, 168,
    234, 240, 322–324
  and severity, 155, 168
  and selection effects, 11, 240, 266ff
Duhem problem
  and Bayesianism, 37–38, 360–362 ADD
  and confirmation, 135–144
  and hypothetical-deductive inference in,
    250–251
  and IBE, 94–95
  in new experimentalism, 58
  error statistics and, 19–20
  severe testing and, 105–111
  Worrall on, 125–130; 145–153

Econometric modeling, third way
  Box, G.E.P. and G.M., Jenkins, 236–238,
    242
  defined, 7, 204–205, 236–238
  Granger, C.W.J, 204–205, 236, 243
  Hendry, D.F., 204–206, 226, 237–238,
    240, 242–243, 304, 323, 329
  Sargan, J.D., 204–205, 237–238, 245
  Sims, C., 204–205, 237, 245
Econometrics
  Cowles Commission, 230–232
  data-driven modeling, 6–7, 236–238
  empirical modeling, 6–7, 212, 214, 215,
    216, 235–236

error statistical modeling and, 205–210, 211–218, 240–241
Frisch, R., 227–229
Haavelmo, T., 229–230, 231
induction, model-based, 210–211
Koopmans, T., 230, 232, 243–244
local errors, discovery of, 218–219
textbook, issues in, 232–235
theory/data-driven modeling, 202–205, 240–241
theory testing, philosophy of, 207
underdetermination in, 42–43, 67–70, 371–374
Econometrics, textbook approach
anemic confirmation and, 233–235
Gauss-Markov theorem and its problems, 233
Leamer, E.E., 204, 234–235, 244
Econometric Society, 227–229
Economic methodology, literature on
Backhouse, R.E., 225, 242
Blaug, M., 223, 242
Hands, W.D., 6, 27
Hoover, K.D., 219, 240, 243, 349, 365, 367, 374–375
Maki, U., 244
Rosenberg, A., 8, 27
Epistemic probabilists,
futuristic suggestions for, 199
Equivalence principles
Einstein, 46–47, 64–67
strong, 51–52
ERROR 06, xiv, 2
Error and the Growth of Experimental Knowledge (EGEK), 15
Errors, canonical, 19
Error probabilities
arguing from error, 24–26
in BARD, 398–400, 401–403
confidence intervals, 288–291
controlling, 10, 85–87, 168, 326, 353–354
in econometrics, 234–235, 236–238, 257–258
evaluation of, 19–20, 79–80, 163–164, 167, 364–365
in frequentist induction, 210–211, 249, 273, 276–277, 316–317, 319
in inductive conclusions, 175–184, 191–192, 237–238, 247–249, 253–254
and model adequacy, 321–322
overview, 16–19, 21–23, 32, 209–210

an reference posteriors, 300–302
and selection effects, 266–272
in test rules, 119–122
type I and II, 34, 317–318, 319–321, 408
Error statistics
defined, 16
econometrics modeling, 205–210, 211–218, 240–241, 364–365
future directions, 168
local errors, discovery of, 218–219
meta-methodology in, 85–87 ADD
and Mill, 190–191
overview, 31–35, 73
philosophy of, 16–19, 23–24, 83–85, 247–248
probability in, 191–192, 249
role of in philosophy of science, 19–20
severity principle (*See* Severe testing)
stability in, 41–42
statistical knowledge in, 206, 211–218
underdetermination and, 42–43, 67–70, 371–374
E.R.R.O.R. research, xii
Evidence, minimal scientific principle for, 3, 24
Experimental economics, 219–221
Guala, F., 219–221, 243, 374–375
Smith, V., 205, 220, 246
Sugden, R., 205, 220–221, 246
Experimental gravitation testing. *See* General relativity theory
Experimental reasoning (empirical inference) and reliability, 8–12
Explanation
abduction, 91–92
critical rationalism (*See* Critical rationalism)
faithfulness condition, 337–348
Glymour on, 359–360
Hempel on, 332–333
issues in, 331–332
and testing (*see* Theory testing/explanation)
Musgrave on, 88–90, 93–94, 102–111, 116–122, 333–334, 336
predictive power and, 353–354
probability in, 334–335
and severe testing, 337–338, 351–353, 359–360
symmetry thesis, 332–333, 355–357
and unification, 335–337, 357–358

Fallacies
  in statistical tests, 32–33, 318
  of acceptance, 187, 209, 255, 317–319,
    370
  of probabilistic instantiation, 179–180,
    196–197
  of rejection, 34, 234, 256, 317–319, 370
  prosecutor's, 193
  statistical vs. substantive significance,
    259–260, 264–265
Falsificationism
  methodological, 119–122
  Popper on, 318, 327
  and minimal principle of evidence, 3
FEV principle, 254, 256, 319–321
Fisher-Neyman-Pearson philosophy
  See behavioristic vs. evidential construal
  See also error-probabilities
  conditioning and, 291–294
  criticisms of, 227–229
  FEV and, 319–321
  foundations of, 248–250, 277–278
  history of, 248, 317–319
  model-based, 210–211
  misspecification testing/respecification,
    211–218
  null hypothesis in, 252–253
  post-data extension of, 229–230
  similar tests, 219
  Spanos on, 319–321
  test statistics, 288–291
Frequentist induction
  conditioning for relevance, 294–298
  error probabilities in, 210–211, 249, 273,
    276–277, 316–317, 319
  error statistics and, 247ff, 321–322, 327
  FEV principle, 254, 256, 319–321
  formal analysis, 282–284
  history of, 317–319
  hypothetical-deductive inference and,
    250–251
  model adequacy, 302, 321–324
  null hypotheses (*See* Null hypotheses)
  objectivity in, 276ff
  overview, 280–303, 304
  philosophy of, 23, 247–249, 316–317,
    327–328
  probability in (inductive behavior),
    249–250, 273
  p-values in (*See* P-values)
  reference posteriors and, 300–302

  revisiting Bayesian criticisms of, 324–327
  roles of, 277–278
  selection effects, 266–272
  statistical significance testing, 251–257,
    269–270
  sufficiency (*See* Sufficiency)
  weak conditionality principle, 197–198,
    200, 295–298, 324–325

General relativity theory (GTR)
  Chalmers on, 65, 69–70, 76, 80–83
  constants, adjustable, 49–51
  Earman, J., 43, 50, 55, 133
  four periods in, 43
  gravitational radiation effects, 53–54
  history of, 43
  new domain exploration, 79, 80–83
  new experimentalism, 61–62, 69–70
  Nordvedt effect, 51–52
  PPN framework in, 43, 45–47, 115–116
  severity testing in, 44, 49–51, 61–62,
    107–108, 359–360
  solar system tests, 52
  use-constructed hypotheses in, 50–51,
    156–160
  viable theory criteria, 48–49
  Will, C. M., 36, 44–57, 65–70, 72, 79,
    81–83, 87, 133, 265, 275
Glymour, C., 9–10, 38, 55, 79, 84, 200, 331ff,
    350, 351ff, 364ff, 374–375
Glymour exchange, (*See also* Explanation)
  Bayesian epistemology, 360–362
  on explanation, 359–360
  theory testing/explanation in, 355–358
Graphical causal (GC) modeling, (*See also*
    causal modeling)
  empirical validity appraisal, 367, 371–374
  faithfulness condition, 338, 366
  faithfulness testing, 374
  Glymour on, 337ff
  Markov condition, 338, 366
  overview, 364–365
  severe testing and, 370–371
  Spanos on, 364ff
  statistical model in, 365–368
  structural model in, 368–370
  Zhang's uniform convergence result,
    338–342

Hypothesis testing (*See* Neyman-Pearson
    testing)

Identification restrictions, 230–232
Induction/ampliative inference
  criticisms of, 113–115
  justificationist interpretation of, 116–122
  reliability, 122
  warranting, 119
Induction
  behavior *vs.* inference, 17, 21, 120,
    253–254
  frequentist, 249–250, 273
  justificationism and, 88–90
  Mill-Newton views on, 172–175, 178,
    180, 190–191
  model-based, 210–211
  philosophy of, 247–249
  probability in, 175–184, 186–188,
    189–190
  p-values in (*See* P-values)
  severity principle and, 32
  uniformity of nature principle, 184–186
Inference to the best explanation (IBE)
  criticisms of, 26, 122–123
  in Duhem thesis, 94–95
  justificationism and, 93–94
  Musgrave on, 90–92, 333–334
  realism and, 94–95
  van Fraassen on, 90–91, 93–96
Information, statistical vs. substantive, 206,
  211–214, 219, 236, 239, 318

Justificationism
  in explanation, 88–90, 93–94, 105–111,
    116–122
  in IBE, 93–94
  induction and, 88–90, 116–122
  in new experimentalism, 88–90, 93–94,
    105–111
  Popper on, 88–90, 117–119
  probability and, 9–10

Kepler, J., 334–337
Keynes, J.M., 250, 267, 274
Kuhn, T., 1, 12, 15, 27, 84, 87, 96, 126–129,
  142, 144, 153, 157, 168, 203–205, 357
Kuru, 355–357, 359–360

Lakatos, I., 1, 11–12, 38, 120, 123, 141, 153,
  157, 168, 204–205, 220
Large-scale theory testing, 35–38, 39, 83–85,
  115–116, 207–210. *See also* General
  relativity theory

Laudan exchange
  analogy between AD defense and
    (secondary) statistical tests of
    assumptions, 404
  standards of proof (SoP) in law and type I
    and II errors in statistics, 398–399
Laudan, L., 9, 28–29, 39, 41, 52, 56, 68, 71,
  83, 87, 99–101, 106–108, 111, 354, 363,
  376ff, 396, 397ff, 409
Legal epistemology
  affirmative defenses in (*See* Affirmative
    defenses)
  beyond a reasonable doubt, 378–381, 385,
    386, 391–393, 398–400
  burden of proof in, 376–378, 381, 393–394
  overview, 378–386, 394–395
  probabilities in, 401–403
  recommendations for, 397, 408
  risk assessment and, 405–408
  self-defense, 382, 385–386
  standard of proof function, 381–386,
    405–408
Lehmann, E.L, 211, 244, 248–250, 274, 293,
  304, 307, 314
Likelihood, 279–280, 400
Law of Likelihood, 280
Likelihood Principle (strong), 287, 288, 298,
  300, 305, 316–317, 324–325

Markov condition, causal, 337–348,
  366–368
Maxwell equation testing, 64–67, 335–337
Mayo and Spanos, 5, 22, 34, 206, 208, 209,
  234, 272, 320, 370, 398
Metaphilosophical themes
  overview, 8, 14
  use of counterexamples, 155, 162
  philosophy-laden philosophy of science,
    73, 113, 189
  tasks for philosophers, 14, 351, 360, 397
  see also two-way street
Methodological falsificationism, 119–122
Methodology, philosophy of
  objectivity in, 10–12
  trends/impasses, 1–5
  volume's relevance to, 5–8
Mill, J. S., 1, 91, 112, 170, 172–180, 188–191,
  201, 223–225, 244–245, 267, 274
Mill philosophy
  criticisms of, 170–172, 189–190, 192–194
  on deduction, 91

Mill philosophy (*cont.*)
  error statistics, 190–191
  on inductivism, 172–175, 178, 180,
    190–191
  probability in, 175–184, 191–192
  propositions, deductively derived, 224
  uniformity of nature principle, 184–186
Minimal scientific principle for evidence, 3,
    11, 24, 42, 319
  and weak severity, 21, 114, 155, 194, 197
Miracle Argument, 95–105
Mis-Specification (M-S) Testing
  Fisher vs. Neyman-Pearson, 211–218
  in frequentist induction, 322
  GC modeling and, 364–365
Model adequacy,
  statistical, 214–216, 219, 233–234, 236,
    239, 318, 366
  substantive, 217–219, 22, 227, 234, 318,
    373
Model validation (statistical adequacy),
  ancillarity and, 323–324
  and error-reliability, 321
  M-S testing and, 217, 302, 322
  sufficiency and, 322
Models
  ARIMA, 236–238, 241
  autoregressive distributed lag (ADL), 237,
    241
  data-driven, 6–7, 236–238
  econometrics (*See* Econometrics)
  estimable (structural), 212
  statistical adequacy of, 302, 321–324
  graphical causal (GC) (*See* Graphical
    causal (GC) modeling)
  multivariate Normal, simple, 365–368
  hierarchy of, 34–35, 55, 157, 208–209
  simultaneous equations, 230–232
  statistical, 86, 116, 205, 210, 212, 365
  vector auto-regressive (VAR), 237, 241
Musgrave, A., 9–11, 25, 27, 83, 88ff,
    112–113, 113ff, 124, 155, 163, 167–169,
    190, 193, 201, 226, 245, 327, 333–336,
    353–354, 360
Musgrave exchange
  Glymour on, 333–334
  Mayo on, 113ff
  Musgrave on Cartwright, 103–104
  Musgrave on Laudan, 99–101
  Musgrave on severity, 105–111
  progressive critical rationalism, 116–122
  contrasts with David Miller, 89, 117, 135

Naturalistic philosophy of science, 3–5
New experimentalism
  Chalmers on, 70–71
  challenges to, 30–31, 64–67, 71
  confirmation/reliability/
    underdetermination in, 67–70, 363
  Duhem thesis in, 58
  future directions, 168
  goals of, 115–116, 208, 363
  Musgrave on, 106–108
  overview, 3–5, 28–29, 58
  severe testing (*See* Severe testing)
  theory, role of, 29, 62–64, 111
New experimentalists/new modelers
  Cartwright, N. 4, 27
  Morgan, M., 4, 27, 35
  Morrison, M., 4, 27, 35
  Galison, P.L., 3, 27
  Hacking, I., 3, 27
Newtonian philosophy
  on inductivism, 172–175, 178, 180,
    190–191
  mechanics testing, 61, 64–67
  probability in, 175–184
Neyman, J., 1, 210–211, 247–250, 255–256,
    265, 274, 316–318, 329
Neyman-Pearson philosophy. *See*
    FisherNeyman-Pearson philosophy
Neyman-Pearson (N-P) tests (*See also*
    behavioristic vs. evidential construal)
  Lakatos on, 119–120
  an error-statistical interpretation of,
    317–319
  and methodological falsificationism,
    119–122, 317–319
  testing, within vs. without, 324
Nordvedt effect, 51–52
Novel-evidence
  novelty-severity relationships, 145ff,
    155–156
  See also UN rule, use-constructed
    hypotheses
N-P tests, 119–122, 317–319
Null hypotheses
  absence of structure, 259, 263
  confidence intervals and, 265–266
  dividing, 258–259, 263
  embedded, 257–258, 262–263
  in Fisher-Neyman-Pearson philosophy,
    252–253
  model adequacy, 259, 264
  overview, 257, 260–261

p-values (*See* P-values)
test statistics, 288–291, 340

Objective (reference) Bayesians
   Berger, J., 6, 27, 50, 55, 277, 301, 304–305,
      308, 311, 314, 325–328
   Bernardo, J., 299–300, 304
   Cox and Mayo on, 298–302
   Jeffreys, H., 250
   Kass and Wasserman on, 37, 56
Objectivity
   and conditionality, 276 ff
   and controlling error probabilities,
      17
   and costs of mistakes, 397
   and double counting, 155
   in Bayesian accounts, 189, 278
   in frequentist induction, 276–277
   in philosophy of methodology, 10–12
   in testing assumptions, 364
   and theory change, 74, 113
   Popper on, 10–12
   rationality of science and, 8–13, 17–18
Optional stopping
   stopping rule, 304, 306ff
   Savage, L.J., 274, 306–307
Overidentifying restrictions, 217, 372–373

Parameterized Post Newtonian (PPN)
   framework
   development of, 115–116
   gravitational radiation effects, 53–54
   Nordvedt effect, 51–52
   overview, 43, 45–47
   viable metric theory criteria, 48–49
Pearson, E.S., 1, 210, 245, 249, 254–257,
      265, 274, 316–317, 329
Peirce, C. S., 1, 91–92, 112, 118, 170–172,
      175–176, 179, 186, 188, 274, 376
Philosophy of science
   current trends and impasses, 3
   naturalism in, 4
   assumed irrelevant for practice, 2–4
Philosophy of statistics,
   error statistical, 5–6, 247ff, 277–278,
      315ff
   Bayesian vs. error-statistical, 6, 200,
      301–302
Planetary orbits/retrogressions, 127–128,
      140–141, 143, 334–336, 337
Popper, K., 1–2, 10–12, 15, 26, 38, 40, 43,
      59–60, 84–85, 89, 113–114, 117–120,

      122–123, 144–145, 165, 168, 180,
      204–205, 225, 248–251, 267, 274, 327,
      358, 395
Popper philosophy. *See also* Critical
   rationalism
   comparativism in, 38–39
   hypothetical-deductive inference in,
      250–251
   Mayo on, 113–115, 168
   on justificationism, 88–90, 117–119
   objectivity in, 10–12
   probability in, 9–10, 249–250
   on reliable rules, 122
Pre-eminence of theory viewpoint, 6,
      202–203, 238–239
Preponderance of evidence standard,
      378–381, 382, 386, 400–401, 403–405
Principle of common cause (PCC), 338,
      366–367, 368
Prion theories, 357
Probability
   in Achinstein philosophy, 192, 194
   in affirmative defenses, 401–403
   in Bayesian statistics, 9–10, 334–335
   (*See* Error probabilities) in error statistics,
      16–19
   in explanation, 334–335
   in frequentist induction, 249–250, 273
   inductive, 175–184, 186–188
   in inductivism, 175–184, 186–188,
      189–190
   Mayo on, 178–179, 186–188, 195
   in Mill, 175–184, 191–192
   Musgrave on, 9–10, 336
   philosophical foundation of, 360–362
   in Popper, 9–10, 249–250
Ptolemaic theory, 127–128, 140–141, 143,
      334–336, 337
P-values
   in inductive reasoning, 319–321
   overview, 253, 261–266, 317–319
   and selection effects, 266–272
   and test statistics, 22, 253, 256, 288–291

Questions addressed in this volume, 12–14

Randomness testing, 210–211
Referential (entity) realism, 100
Reichenbach, H., 15, 366–368, 375
Rejection fallacy, 317–319 (*See also* fallacies)

Sampling theory, philosophy of, 16–19

SAT test, 131–132, 145–146, 153, 163–164,
  195–196
Scientific research methods
  goals of, 115–116
  philosophers vs. practitioners, 7–8
Selection effects, 11, 240, 249, 266ff, 316
  (*See also* double-counting)
Severe testing
  arguing from error, 24–26, 33
  arguments from conspiracy, 74–77
  Bayesian philosophy *vs.*, 37–38
  Chalmers on, 73–77, 111
  and comparativist-holism, 29, 39–43,
    107–108
  in critical rationalism, 105–111, 113–115
  defined, 22, 156
  and explanation, 337–338, 351–353,
    359–360
  GC modeling and, 370–371
  in general relativity theory, 44, 49–51,
    61–62, 107–108, 359–360
  induction as, 119
  informativeness and, 79–80
  large-scale theory testing, 35–38, 39,
    60–62, 115–116
  Mayo on, 58–59, 108–109, 145–153, 167,
    180–181, 337–338, 353–354
  overview, 17–18, 21–23, 28–29, 31–35, 40
  severity principle, 21, 32
  use-construction rule, 161
  weak rule, 197–198, 200, 295–298,
    351–353
Simultaneity problem, 230–232
Spurious (statistical) results, 203, 366–367
Statistical significance testing (*see also*
    P-value)
  (*See also* fallacies)
  overview, 252–253
  redefining, 271–272
  selection effects in, 269–270
  testing, within vs. without, 324
Statistical vs. substantive information, 203,
    212–213, 217–219, 232, 371–372
Statistics (*See also* error statistics)
  formal analysis, 281
  history of, 317–319
  philosophical foundations of, 247–249
  roles of, 18, 277–278
Stopping rule (*See* optional stopping)
Sufficiency
  confusions about, 286

  data reduction by, 284–285
  in model adequacy, 322–324
  nuisance parameter conditioning,
    291–294
  principle, 286–287, 308
  relevance induction conditioning,
    294–298
  role of, 286
  in sampling theory, 287
  strong likelihood principle (SLP), 288
  test statistics, 288–291
  weak conditionality principle, 197–198,
    200, 295–298, 324–325
Symmetry thesis, 355–357

Tacking problem/irrelevant conjunction, 14,
    38, 110, 123, 351, 359, 361
Theory, large scale
  arguments from coincidence, 67–68, 70,
    73–74, 76–77
  arguments from conspiracy, 74–77
  as component of knowledge, 64–67
  data in formation of, 130–135, 145–153
  deduction/instantiation, 161–162
  Mayo on role of, 62–64, 76–77, 83–85
  new domain exploration, 79
  unexplained coincidences, avoidance of,
    77–79
Theory-data confrontation in economics,
  Cairnes, J.E., 223–225, 242
  in classical economics, 222–225
  current views on, 240–241
  Friedman, M., 226, 243
  Keynes, J.N., 223–224, 243
  Kydland, F., and P. Prescott, 239, 244
  Leontief, W.W., 235–236
  Lucas, R.E., 238–239, 244
  Malthus, T., 222–223, 225, 244
  Marshall, A., 225, 244
  Mill, J.S., 224–225, 244–245
  in neo-classical economics, 225–226
  Ricardo, D., 222–224, 240, 245
  Robbins, L., 225–226, 229, 245
Theory testing/explanation
  assumptions in, 362–363
  Chalmers, 9, 73–74, 79–80, 106–107, 167
  Duhem's problem, 125–130
  Mayo on, 351ff
Transdiction, 172–175
Two-slit experiment. *See* Wave theory of
    light

Two-way-street
  linking philosophy and practice, 5, 12, 14,
    168, 307, 327, 398
  promoted by statistical science, 4

Underdetermination, (*See also* Duhem's
    problem)
  and arguments from coincidence, 67–70,
    75
  and error statistics, 8, 20, 37, 42–43, 207,
    371–374
  new experimentalism and, 58, 363
Uniformity of nature principle, 184–186
Use-Novel (UN) rule (no double-use rule)
  Charter, 125
  defined, 156
  flaws/equivocations in, 164–167
  Giere on, 165, 168
  Mayo on, 145–153
  overview, 125–130
  Worrall on, 125 ff
  predictive value of, 129–130
  refutations of, 130–135, 146, 158–161
Use-constructed hypotheses
  in blocking anomalies, 158

gellerization, 146, 159
overview, 158–161
rule, stringent, 161
test procedure, 165
theory deduction/instantiation, 161–162

Velikovsky controversy, 128–129, 135–136,
    149–150, 158–161, 166
Velikovsky's scotoma dodge, 158–159
Verisimilitude, 101
Very long baseline radio interferometry
    (VLBI), 31–32

Wave theory of light
  confirmations of, 135–144
  overview, 127
  refutations of, 130–135
Weak conditionality principle, 295–298,
    305, 308, 324–325
Worrall, J., 9, 11, 38, 49, 57, 68, 72, 87, 124,
    125ff, 154, 155ff, 169, 361
Worrall exchange
  on Duhem's problem, 125
  conditional confirmation, 146
  on severe tests, 145